INTRODUCTION TO INSECT PEST MANAGEMENT
Robert L. Metcalf and William H. Luckman, Editors

OUR ACOUSTIC ENVIRONMENT
Frederick A. White

ENVIRONMENTAL DATA HANDLING
George B. Heaslip

THE MEASUREMENT OF AIRBORNE PARTICLES
Richard D. Cadle

ANALYSIS OF AIR POLLUTANTS
Peter O. Warner

ENVIRONMENTAL INDICES
Herbert Inhaber

URBAN COSTS OF CLIMATE MODIFICATION
Terry A. Ferrar, Editor

CHEMICAL CONTROL OF INSECT BEHAVIOR: THEORY AND APPLICATION
H. H. Shorey and John J. McKelvey, Jr.

MERCURY CONTAMINATION: A HUMAN TRAGEDY
Patricia A. D'Itri and Frank M. D'Itri

POLLUTANTS AND HIGH RISK GROUPS
Edward J. Calabrese

SULFUR IN THE ENVIRONMENT, Parts I and II
Jerome O. Nriagu

FOOD, CLIMATE, AND MAN
Margaret R. Biswas and Asit K. Biswas, Editors

ENERGY UTILIZATION AND ENVIRONMENTAL HEALTH: METHODS FOR PREDICTION AND EVALUATION OF IMPACT ON HUMAN HEALTH
Richard A. Wadden, Editor

METHODOLOGICAL APPROACHES TO DERIVING ENVIRONMENTAL AND OCCUPATIONAL HEALTH STANDARDS
Edward J. Calabrese

Methodological Approaches to Deriving Environmental and Occupational Health Standards

Methodological Approaches to Deriving Environmental and Occupational Health Standards

Edward J. Calabrese, Ph.D.
Division of Public Health
University of Massachusetts
Amherst, Massachusetts

A Wiley-Interscience Publication
JOHN WILEY & SONS
New York Chichester Brisbane Toronto

Library of Congress Cataloging in Publication Data:

Calabrese, Edward J., 1946–
Methodological approaches to deriving environmental and occupational health standards.

(Environmental science and technology)
Includes index.
1. Environmental health—Standards. 2. Industrial hygiene—Standards. 3. Industrial toxicology—Standards. I. Title.

RA566.C27 363.6 78-17861
ISBN 0-471-04544-6

Printed in the United States of America

10 9 8 7 6 5 4 3 2 1

To my wife, Mary

Series Preface

Environmental Science and Technology

The Environmental Science and Technology Series of Monographs, Textbooks, and Advances is devoted to the study of the quality of the environment and to the technology of its conservation. Environmental science therefore relates to the chemical, physical, and biological changes in the environment through contamination or modification, to the physical nature and biological behavior of air, water, soil, food, and waste as they are affected by man's agricultural, industrial, and social activities, and to the application of science and technology to the control and improvement of environmental quality.

The deterioration of environmental quality, which began when man first collected into villages and utilized fire, has existed as a serious problem under the every-increasing impacts of exponentially increasing population and of industrializing society. Environmental contamination of air, water, soil, and food has become a threat to the continued existence of many plant and animal communities of the ecosystem and may ultimately threaten the very survival of the human race.

It seems clear that if we are to preserve for future generations some semblance of the biological order of the world of the past and hope to improve on the deteriorating standards of urban public health, environmental science and technology must quickly come to play a dominant role in designing our social and industrial structure for tomorrow. Scientifically rigorous criteria of environmental quality must be developed. Based in part on these criteria, realistic standards must be established and our technological progress must be tailored to meet them. It is obvious that civilization will continue to require increasing amounts of fuel, transportation, industrial chemicals, fertilizers, pesticides, and countless other products; and that it will continue to produce waste products of all descriptions. What is urgently needed is a total systems approach to modern civilization through which the pooled talents of scientists and engineers, in cooperation with social scientists and the medical profession, can be focused on the development of order and equilibrium in the presently disparate segments of the human environment. Most of the skills and tools that are needed are already in existence. We surely have a right to hope a technology that has created such manifold environmental problems is also capable of solving them. It is our hope that this Series in Environmental Sciences and Technology will not only serve to make this challenge more explicit to the established professionals, but that it also will help to stimulate the student toward the career opportunities in this vital area.

Robert L. Metcalf
James N. Pitts, Jr.
Werner Stumm

Preface

The development and implementation of standards for environmental and occupational pollutants has become a critical focal point at which science becomes operational within society. The medical, social, and economic implications that emanate from the promulgation of specific standards are quite extensive; every person's life is affected in some way. Despite the obvious importance of developing safe and economically realistic standards, there has remained a critical need to develop a reliable methodology that can be applied to a wide variety of substances. Numerous articles have been written about various components of the standard-setting process, such as toxicological and epidemiological aspects, high-risk groups, animal extrapolation, and others but a general synthesis of this material has not been put forth. Perhaps the best-known approaches are embodied in the Criteria Documents for the ambient air pollutants, the NIOSH Criteria Documents for industrial chemicals, and the documentation information for TLVs by the American Conference of Governmental Industrial Hygienists. Despite the excellent contributions of these publications in the area of standard setting, their approach has been to focus on the individual pollutant and not on the development of a methodological approach to standard setting.

To accomplish the task of developing a broadly applicable biomedical methodology for standard setting, the following scheme has been developed.

Chapter 1 presents a model for standard setting for both environmental and occupational pollutants. Chapter 1 is actually a summary of the subsequent chapters, since their intention is to explain each component of the model in detail. Chapter 2 critically reviews the basis on which extrapolations can be made from animal models to humans. Chapter 3 discusses the concept of high-risk groups and their role in standard derivation, with particular emphasis on several of the national ambient air standards. Chapter 4 develops an assessment of the extent to which chemical interactions such as synergism and additivity occur in the air environment and how they should be considered in the development of safe levels of exposure. Chapter 5 focuses on various toxicological issues such as threshold, nonthreshold, latency versus dosage, carcinogen versus noncarcinogen standards, and NIOSH's approach to standard setting for carcinogens. Chapter 6 deals with a comprehensive assessment of the drinking water standards from an historical perspective, provides a critical review of the present rationale of individual standards, and compares United States standards with those of other countries. In a similar fashion, Chapter 7 reviews occupational health standards from an historical perspective to the present. Particular emphasis is placed on the role of the ACGIH and the OSH Act of 1970 in this process. Chapter 8 is a further assessment of occupational health standards, emphasizing the role of biological indicators in assessing toxic exposures to pollutants and developing appropriate standards. Chapter 9 further develops a relatively new concept in occupational health standards; that is, how should novel work schedules be considered in the standard derivation process? Chapter 10 explains the historical reasons why ambient air standards were established. It also critically evaluates the basis on which the ozone, nitrogen dioxide, sulfur dioxide, and carbon monoxide standards were derived. Chapter 11 examines how foreign countries approach the problem of standard setting. Particular emphasis is directed toward the Soviet Union and the philosophical and experimental basis on which their standards are set. Chapter 12 uses the case study approach by which one substance, asbestos, is followed through the standard development and implementation process. Chapter 12 integrates the information presented in the previous 11 chapters to produce a rational and orderly process by which standards may be developed within our society of competing interests.

Finally, it should be emphasized that this book deals with the biomedical basis on which standards may be derived. It does not specifically treat the economic and social components in this process. However, Chapter 12 does provide a noneconomist's version of the role of economic factors in the development of the OSHA standard for asbestos.

I would like to acknowedge the contributions others have made in assisting in the completion of this study. Much of Chapter 1 was originally published in the *Journal of Biological Education* Volume 10, 1976, with Sally Jansen and Alfred Sorensen as my co-authors and is reproduced with permission of the publishers. I would like to thank Franz Reichsman for his numerous discussions on problems in animal extrapolation techniques. The assistance of Regina Brown in the researching of Chapter 8 on biological indicators is deeply appreciated. Also, the effort involved in writing Chapter 12 was considerably lightened by the outstanding job Mary Ann Clasby performed in researching this area. I would also like to thank Dr. Robert W. Tuthill of the University of Massachusetts for contributing his valuable critical assistance on the section in Chapter 6 concerning the sodium drinking water standard. Dr. Tuthill and I have previously published a less technical and shortened version of this presentation in the *Journal of Environmental Health* and with the publisher's permission much of that article is reprinted. In addition, many thanks to Dr. Tuthill for the many hours of discussions we have had on epidemiologic methodology and its role in standard setting. I would also like to thank Floyd Taylor, Chief of the Water Supply Branch, EPA, Region I, for opening his files to me.

I would also like to give a special note of thanks to my wife, Mary, for editing and proofreading this manuscript. Finally, while acknowledging the vital contributions of so many, I take full responsibility for the contents herein.

EDWARD J. CALABRESE

Amherst, Massachusetts
August, 1978

Contents

1

A Model For Deriving Standards For Environmental/Occupational Pollutants

THE ESTABLISHMENT OF APPROPRIATE environmental and occupational health standards is an important aspect of preventive medicine. Since most cancers are believed to be associated with environmental factors (NAS, 1975), the setting of standards is a reasonable approach for the control of cancer rates. Toxic effects other than cancer, which are caused by various pollutants (e.g., neurological disorders caused by mercury toxicity), may also be controlled in this manner. The development of standards is a logical outcome of the study of pollutants, since ultimately the levels of human exposure to such substances must be regulated. This chapter presents a conceptual model that summarizes the process of how health standards for a variety of environmental and industrial pollutants are developed. Figure 1-1 represents a model that outlines the process of standard derivation. Subsequent chapters expand on each component of the model.

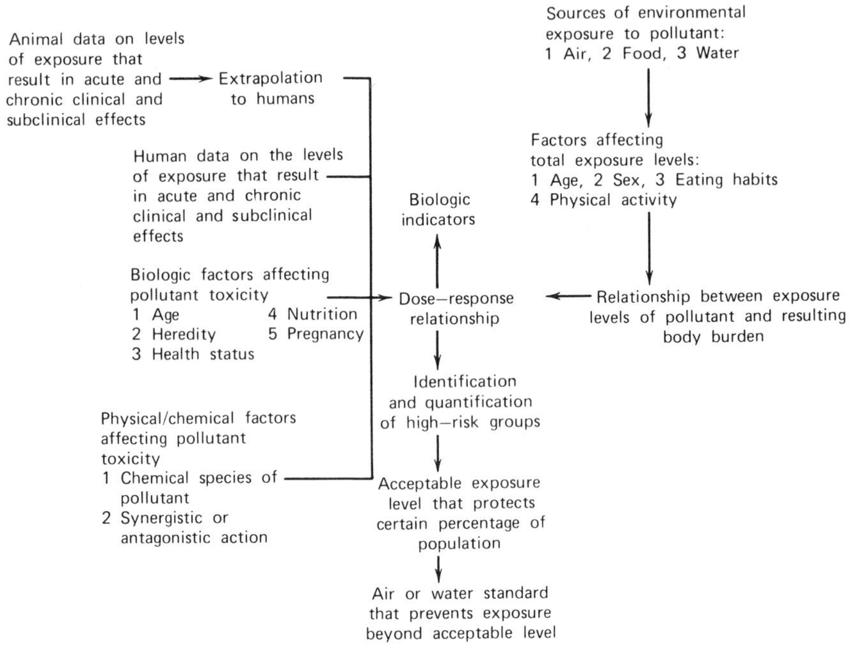

Figure 1-1. A model for deriving an air or water quality standard.

INTER- AND INTRASPECIES EXTRAPOLATION

Certain health data are required before environmental and occupational health standards can be derived. A knowledge of the effects of the particular pollutant on animals and, if possible, on humans is vital to the determination of the levels of the pollutant that may adversely affect health. Since toxicity depends on the route of administration, dose, and duration of exposure reported, research should be examined to determine whether exposures to high doses over a short period of time (acute exposure), low doses over a prolonged period (chronic exposure), and exposure to both high and low doses administered intermittently, have occurred. Effects resulting from such exposures are consequently either acute (occurring or developing rapidly after a single administration of a substance) or chronic (manifested after elapse of some time) (Casarett and Doull, 1975). The organ systems affected by acute exposure are not necessarily the same as those affected following repeated exposure over long periods of time.

Although several routes of administration can be utilized, inhalation and oral administration are the most important, since these are typical modes of entry of environmental pollutants for the general population. The observation of both clinical (observable) and subclinical (without clinical manifestations, said of the early stages or a slight degree of a disease) effects and the doses at which these occur is essential. The decision maker should examine closely the number and type of animal species used in reported studies; the number should be sufficient for statistical confidence. Data gathered from studies utilizing animals whose biochemical and physiologic responses are most similar to those of man are preferred for extrapolation (WHO, 1975).

When considering the effects of a chemical on animal systems, it is important to note whether healthy adult animals were utilized or whether animals with defects similar to those found in the human population considered at high risk to the pollutant were studied. For example, individuals with various genetic disorders, including glucose-6-phosphate dehydrogenase deficiency, acatalasemia, methemoglobin reductase deficiency, and cystic fibrosis, are thought to be at high risk to various environmental pollutants (NAS, 1975). Therefore, animals that show similar genetic disorders should be used to simulate the human condition as much as possible.

The appropriate selection of the animal model is also important when considering the toxicity of pollutants on the human fetus and neonate (Weisburger and Rall, 1972). For example, polychlorinated biphenyls (PCBs) are excreted principally by an enzymatic process called conjugate glucuronidation (Dutton, 1961). The conjugate glucuronidation process is not fully functional until the infant is about 2–3 months old. Thus it is thought that the fetus and neonate may accumulate PCBs to a greater degree than older children and adults. Cats are known to have a partial deficiency of the conjugate glucuronidation process (Oehme, 1969). Thus, to a certain extent, cats tend to simulate the condition of the human fetus and neonate and may offer a useful model for investigating neonatal response to PCBs. Experiments with specific animals, selected because of deficiencies that correspond to human diseases, should be more heavily weighed when only animal studies are available.

Another approach to improving the validity of data extrapolations from animals to man is to consider toxicity as a function of the general metabolic rate (WHO, 1975). Usually, the faster the metabolism, the lower the toxic effects of the pollutant on the animal. Since small animals often have faster metabolic rates than larger animals, similar quantities of a pollutant (milligrams of pollutant per kilogram of body weight of the animal) are more toxic in a large animal. Because body weight and body surface area are often directly related to general metabolic rate, investigators have found that toxicity is often a function of these two easily measurable variables. Thus extrapolations can be made from animal studies to man. At present, however, the most reliable extrapolation appears to be an integration of data derived from both specific animal models most closely simulating the actual human condition and the type of study that considers toxicity as a function of general metabolic rate.

In the evaluation of controlled human experiments, it is important to realize that most laboratory exposure studies employ young healthy volunteers. The response of these individuals to a pollutant is quite different in most cases from that of one who is hypersusceptible to the pollutant. Consequently, if the standard is to protect those individuals at high risk, other information specifically relating to such high-risk individuals must be evaluated, or another type of extrapolation, from healthy to high risk, must be utilized.

EPIDEMIOLOGIC STUDIES

One of the most valuable sources of information is the epidemiologic study in which the effects of the specific pollutant on an exposed population are studied. Its value lies in the fact that data are derived from human subjects as compared with animal models. However, in contrast to controlled animal or human experiments, the epidemiologic study incorporates numerous uncontrolled variables that tend to weaken the confidence of any disease–pollutant association. Despite their limitations, epidemiologic studies have provided much of our knowledge of toxicity to humans of arsenic, asbestos, lead, mercury, and numerous other pollutants. For toxic substances that have been in the environment for many years, primarily as a result of industrial processes, the epidemiologic studies are probably the most important source of information. For newly developed chemicals that have not been widely distributed in the environment, data from animal model studies may provide the best indication of potential toxicity to humans.

In the case of pollution episodes, people are exposed to very high levels of a pollutant for a short period of time. The results of such studies must be extrapolated downward to determine what, if any, measurable effect the pollutant will have at a lower level of exposure.

Well-known water pollution episodes include the tragic mercury poisoning incidents in Minimata and Niigata, Japan, resulting from the consumption of contaminated fish. The victims suffered neurological disorders and even death. Blood data from Japanese victims, in conjunction with Swedish studies, were used to arrive at a Food and Drug Administration (FDA) guideline for an allowable daily intake for mercury in fish (Expert Group, 1971).

During a 5-day period in December 1930, the Meuse Valley in Belgium, an industrialized area, was the scene of a disastrous air pollution episode in which several hundred people suffered serious respiratory symptoms and 63 died of acute heart failure. Although no precise measurements were made during the episode, it was estimated that a number of pollutants, including sulfur dioxide and sulfuric acid, rose to high levels (Firket, 1931).

An air pollution episode in Donora, Pennsylvania, in October 1948, affected 43% of the population and resulted in the death of 20 persons. Again, no measurements were available during the episode, but a later study revealed that no single agent was responsible, although sulfur dioxide and its oxidation products were significant contaminants (Schrenk et al., 1949).

The most disastrous air pollution episode occurred in London in December 1952, when about 4000 excess deaths occurred. The cause of the largest excess of deaths was related to bronchitis. Eight times more people died from bronchitis than would normally be expected at that time of year, and three times the predicted "normal" number of deaths from pneumonia occurred. There also appeared to be a significant increase in death from myocardial degeneration and coronary heart disease (Ministry of Health, 1954).

Epidemiologic studies conducted under nonepisodic conditions reveal the health effects due to exposure to typical urban air pollution levels. For example, the United States Environmental Protection Agency's Community Health and Environmental Surveillance System (CHESS) Program was designed to provide dose-response information relating short-term pollutant exposures to adverse health effects (USEPA, 1974). CHESS studies in New York and the Salt Lake Basin, as well as studies in Idaho-Montana, Chicago, and Cincinnati, considered the health effects of sulfur oxides and the relative contribution of sulfur dioxide, total suspended particulates, and suspended sulfates to observed disease frequencies.

The health indicators of long-term pollution effects employed in these studies included:

1. Increased prevalence of chronic bronchitis in adults
2. Increased acute lower respiratory infections in children
3. Increased acute respiratory illness in families
4. Subtle decreases in ventilatory function in children

The health indicators for short-term pollution effects were the aggravation of cardiopulmonary symptoms and of asthma. Threshold estimates were developed for the effects of the pollutants considered. The conclusions of these studies are not definitive, but contribute to more refined quantitative and scientific hypotheses concerning pollutant/health effect association in a real-life environment.

These threshold estimates supported the existing National Primary Air Quality Standards for long-term exposures, insofar as these standards could be assessed in terms of the health indicators employed. Short-term exposures revealed that adverse effects were experienced on days on which sulfur dioxide and total suspended particulate levels were below the 24-hr standards. However, it was found that adverse health effects seemed to be associated with suspended sulfate levels rather than with the observed concentrations of sulfur dioxide and total suspended particulates (USEPA, 1974).

Health data derived from nonepisodic occupational exposure have also played an important role in the identification of the toxic potential of various pollutants to humans. Data derived from occupational exposure usually relate to adult subjects between 18 and 65 years of age, exposed for 8 hours a day for 5 days a week. In contrast, when one establishes an environmental standard, the exposure is considered for 24 hours a day, and all people within reason are to be protected. Thus it is necessary to extrapolate from industrial data if used in environmental health standards. This is also why occupational standards are much higher than environmental standards. For example, much of our knowledge of the toxicity of radium 226 is derived from the radium dial painters during World War I. Based on the exposure of dial painters to radium 226, a drinking-water standard for radium 226 was derived in part (Miller et al., 1969; Riddiough et al., 1977).

HIGH-RISK SEGMENTS

Certain biologic factors affect toxicity and result in a greater susceptibility of some individuals to a pollutant. Age, genetic makeup, health, nutritional status, and conditions of stress to which an individual is subjected may influence the toxicity of a pollutant. Because of the vulnerability and immaturity of their developing biologic systems, fetuses, infants (especially premature infants), and young children are at greater risk than adults. For example, several Japanese children born to mothers exposed to methylmercury in fish in Minimata suffered congenital malformations even though symptoms of mercury poisoning in the mothers were minimal or absent (Expert Group, 1971). This suggests that the fetus may be more sensitive to methylmercury toxicity than the pregnant woman.

The lack of gastric acidity in newborn infants allows nitrate-reducing organisms to thrive in the upper gastrointestinal tract where nitrates are reduced to nitrites before the former can be completely absorbed (Cornblath and Hartmann, 1948). Nitrites cause methemoglobin formation in infants' blood. A temporary deficiency of NADPH methemoglobin reductase in infants slows down the reduction process to hemoglobin (Lee, 1970). Thus the presence of nitrates in water can result in methemoglobinemia in infants, whereas a healthy adult is less likely to suffer these consequences.

Not only the very young suffer; the elderly are also more susceptible. At all ages, the thymus is necessary for the normal differentiation and maturation of thymus-derived (T) cells. Cell-mediated immunity, in turn, requires T cells. After birth the size of the thymus decreases with increasing age. Cell-mediated immunity also decreases with age (Marx, 1975). Many investigators think that these events are causally related to the increased incidence of infections, autoimmune disease, and cancer that accompanies aging. Since certain environmental carcinogens, such as a variety of hydrocarbons, depress humoral and cell-mediated immune reactivity (Szakal and Hanna, 1972), the aged are subject to an even greater likelihood of environmental cancers.

Several inherited metabolic disorders may result in hypersusceptible reactions to pollutants. For example, individuals with a serum α_1 antitrypsin (SAT) deficiency are predisposed to pulmonary disease when exposed to respiratory irritants (Waldbott, 1973). Trypsins and other proteases are capable of digesting human and animal lung alveolar walls, a necessary condition for the development of emphysema. In normal individuals, antitrypsin in the serum inhibits this proteolysis.

Individuals with the deficiency of the red cell enzyme glucose-6-phosphate dehydrogenase (G-6-PD) are hypersusceptible to the action of many hemolytic chemicals (Stokinger and Scheel, 1973). G-6-PD is essential for the production of hydrogen to maintain sufficient amounts of reduced glutathione (GSH), which is necessary for the integrity of the red cell membrane. Even in this enzyme-deficient condition, there is sufficient G-6-PD to maintain membrane integrity under normal circumstances. However, under stress of hemolytic chemicals that utilize extra hydrogen already in short supply, the added demand for hydrogen leads to insufficient amounts of GSH which in turn leads to methemoglobin formation, red cell rupture, hemolysis, and hemolytic anemia.

The nutritional status of the exposed population must also be considered. A nutritional deficiency increases susceptibility to adverse health effects of some pollutants (Shakman, 1974). Inadequate protein in the diet of rats increases the toxicity of most pesticides, but decreases, or fails to alter, the toxicity of other agents. Vitamin E deficiency increases ozone toxicity, whereas vitamin E supplementation decreases ozone toxicity in rats (Shakman, 1974). Dietary deficiencies of calcium and iron, quite common in the United States, significantly potentiate the toxicity of lead (Mahaffey, 1974).

The health status of exposed individuals predetermines their ability to tolerate environmental stress. For example, those suffering from cardiovascular or respiratory disease are at greater risk from the effects of CO or SO_2, respectively, than are healthy individuals (Carnow, 1966). Also, individuals with severe kidney ailments have disrupted lung-clearance mechanisms (Goldstein and Green, 1966).

The additional stress of pregnancy may result in greater susceptibility. Detoxification mechanisms and pathways of protein, carbohydrate, and lipid metabolism are altered. Dietary habits and nutritional requirements are also modified (Hunt, 1975).

PHYSICOCHEMICAL FACTORS AND TOXICITY

Toxicity also depends on the chemical properties of the pollutant and its possible interaction with other pollutants or variables in the medium (air, water) in question. These factors must be considered when investigating the health effects of the pollutant. For example, organic mercury (methyl mercury compounds) is more toxic than inorganic mercury compounds (Expert Group, 1971). Among organic mercury compounds, alkyl mercury compounds are more toxic than aryls. An environmental or industrial chemical may react additively, synergistically, or antagonistically with other chemicals, possibly altering its toxicity.

An example of synergism has been reported in experiments with "normal" human volunteers exposed to both ozone (O_3) and SO_2. Using the maximal expiratory flow rate (MEFR) at 50% vital capacity as an index, it was found that a combination of 0.37 parts per million (ppm) of SO_2 and 0.37 ppm of O_3 had a much greater debilitating effect on respiratory function than either gas alone. With 0.37 ppm O_3, a 2-hour period is needed to show significant effect on the MEFR, but if the same concentration of SO_2 is present at the same time, the effect occurs within 30 minutes. It should be noted that it would take 0.75 ppm of SO_2 alone 30 minutes to produce similar changes in the MEFR (Bates and Hazucha, 1973). Such experiments make it quite clear that the establishment of a standard for O_3 must take into account the levels of SO_2 present in the atmosphere as well as other modifying factors.

Humidity and temperature may also affect the toxic potential of some environmental pollutants including lead, parathion, antimony, and others. It is important to consider these interactions, since the atmospheric and aqueous environments are composed of many interacting pollutants (Baetjer, 1968). The intergration of the impact of pollutants simultaneously affecting human health is necessary if truly holistic standards are to be derived.

DETERMINATION OF EXPOSURE LEVELS

To assess the degree of exposure of a population to a pollutant, the sources of the pollutant must be known and monitoring devices and analytic techniques must be available to measure the ambient concentrations. Ambient air, water, and food may all contribute to the daily intake of pollutants. Additional exposure may result from occupation.

Several factors may affect the individual's total exposure to the pollutant. For example, because of their smaller size and often greater physical activity, young children may take in greater amounts of a pollutant per kilogram of body weight. The elderly, on the other hand, are often less active and require less oxygen (therefore, individuals of this susceptible population subgroup are exposed to lesser amounts of a pollutant, at least via air). Males, who are more often employed in occupations involving physical labor, are more likely to inhale larger volumes of air thereby influencing pollutant exposure.

Certain individuals or large segments of the population may be subject to greater exposure to certain pollutants because of their dietary habits. For example, people following certain fish diets, such as "Weight Watchers" and various ethnic groups, are likely to ingest more methylmercury than meat eaters. Another example of this phenomenon is seen in the fact that PCBs may concentrate in human breast milk (Berglund, 1972). Consequently, breast-fed neonates are expected to have higher than average amounts of PCB exposure.

To derive a standard for a pollutant based on body burden, the total daily intake of a pollutant must be estimated. Dietary consumption for various food groups and the concentration of pollutants in several foods have been determined in market basket surveys (Waters, 1975). Data on concentrations of many pollutants in public water supplies are available, and the average daily intake of water by the average individual has been estimated (USEPA, 1973). The average volume of air inhaled by the "standard" man and woman and by more active individuals has also been estimated (Diem, 1962). If the air concentration of the pollutant has been measured, the daily intake of the pollutant by inhalation can be estimated. If the rates of absorption via the lung and digestive tract and the rate of excretion are known, the resulting body burden can be estimated for cumulative pollutants. This estimate allows determination of that amount of the pollutant which can be inhaled without exceeding an acceptable level. The allowable concentration in air can thus be established (Figure 1-2).

For noncumulative pollutants such as O_3 and other gaseous irritants, the total body burden concept gives way to an indicator of total exposure, which is equal to the concentration of the pollutant times the duration of exposure to the pollutant. For example, experiments have shown that O_3 is toxicologically accumulative in its effect. The effective dose (D) is contingent on the concentration (C) or intensity and the time of exposure (T) for short-term single exposures. Therefore

$$D = f(CT)$$

King (1973) showed that the effective dose of a single O_3 exposure is determined by the following equation:

$$D = 2.8 + \log_{10} C.$$

D is expressed on a scale of toxicologic severity where $D = 1$ is the limit of perception and $D = 5$ is rapidly fatal.

If the pollutant is a known carcinogen, such as long-fibered asbestos, the derivation of a standard is more complicated. The problem with standards for carcinogens is that it is unresolved whether there is a nonthreshold effect in operation. Until this point of contention is resolved, no level of carcinogen is really "acceptable." If the carcinogenic activity acts in a nonthreshold manner, any exposure, regardless of how low, may result in an eventual adverse effect. In contrast, if a threshold exists, there is a level below which no adverse effect occurs. In

TD
177
S37
1974

RA
566
1034
1978

ouch-ouch Disease
Bone deformation

Bone pain &
convulsions

Health
Env
pollutants
Walbott

Health & IND. Growth

1975 Ciba Foundation
Symposium 32
Associated Scientific Pub.
Amsterdam · Oxford · NY

Known Data

P = pollutant
188 μg = daily intake of P via food (DIF)
12 μg = daily intake of P via water (DIW)
25 μg = acceptable total daily absorbed intake of P for the high-risk portion of the population (ADI)
0.10 = absorption rate of P in the gastrointestinal tract (AGI)
0.50 = absorption rate of P in lung (AL)
20 m^3 = amount of air inhaled per day (AI)

Unknown Data

X = amount of pollutant P that can be inhaled without exceeding the acceptable level.

Calculations

Amount of P absorbed via consumption of food and water per day (AF + AW)

$$\text{AF} + \text{AW} = (\text{DIF})(\text{AGI}) + (\text{DIW})(\text{AGI})$$
$$\text{AF} + \text{AW} = (188)(0.10) + (12)(0.10)$$
$$\text{AF} + \text{AW} = 20.0\ \mu\text{g}$$

Amount of P absorbed through the lung without exceeding a total exposure for the three routes of exposure (air, food, and water) of 25 μg per day (AAL).

$$\text{AAL} = \text{ADI} - (\text{AF} + \text{AW})$$
$$\text{AAL} = 25 - 20$$
$$\text{AAL} = 5\ \mu\text{g}$$

Ambient air standard for P (AS)

$$\text{AS} = \text{AAL}/(\text{AI})(\text{AL})$$
$$\text{AS} = 5/(20)(0.50)$$
$$\text{AS} = 0.5\ \mu\text{g}\ P/\text{m}^3$$

Figure 1-2. A procedure by which an ambient air standard for numerous cumulative pollutants may be derived.

a nonthreshold case, the latent period may be extended such that the effects are not manifested within the lifespan of most individuals. A recent report (Jones and Grendon, 1975) reviewing the available literature on certain carcinogens such as radium 226 and a variety of well-known hydrocarbons has indicated that if the dose is lowered by a factor of 1000, the latent period is extended by a factor of 10. More research is necessary before this "dose-response" relationship is adopted in the standard derivation process.

WHAT IS AN ADVERSE RESPONSE?

The definition of adverse response is a problematic area. Is the first evidence of a biochemical or physiologic change to be considered adverse? Or is the appearance of clinical symptoms necessary for the definition of adverse response? One school of thought considers a biologic response, of whatever kind of intensity, as evidence of impending loss of health, or regards the response itself as a direct expression of injury (Hatch, 1973). Consistent with this is the idea that any measured concentration of the offending agent in body tissues, however low, is evidence of excessive exposure, since the body is stressed in its attempt to metabolize, store, or excrete the agent. In contrast, many assume that the homeostatic and compensatory mechanisms in the body make it possible to offset such insults so that the body is able to deal with toxic agents or exposure levels above zero without a threat to health (Hatch, 1973).

Although a compensatory mechanism may be adequate to protect the healthy individual, this mechanism may operate insufficiently in high-risk individuals. For example, recent investigations (Buckley et al., 1975) concerning the effects of O_3 on normal males have indicated that a 0.5 ppm exposure for 2¾ hours causes a 14% reduction in GSH levels. Concomitant with the reduction of glutathione levels is the initiation of a compensatory response by which G-6-PD activity is increased by 20%. Such a response helps stabilize GSH levels. However, for individuals already deficient in G-6-PD, this compensatory mechanism may not be sufficient to restore GSH to adequate levels. These individuals are thus subjected to the possibility of a hemolytic crisis because of the absence of an adequate compensatory adaptation to the chemical stress.

In contrast, individuals with deficiencies of the enzyme catalase are known to have a compensatory mechanism with respect to oxidant stress. For example, under nonstress conditions acatalasemic erythrocytes metabolize glucose through the hexose monophosphate shunt at three times the normal rate. Under hydrogen peroxide stress conditions, glucose is metabolized at 12 times the normal rate in the shunt, thus providing the underlying mechanism for oxidative protection (Jacob et al., 1965). Therefore, the concept of homeostatic compensatory mechanisms may not be equally applicable to all segments of the population.

Using available data from animal experiments, human exposure studies, and epidemiologic information, a dose-response relationship can be derived. It is recognized that this is often a crude relationship. Such a relationship, however, allows one to estimate that dose which evokes an adverse or toxic response, either acute or chronic.

Biologic indicators such as pollutant levels in the blood or urine reveal degrees of exposure and, in many cases, biologic response. For example, certain blood lead levels are associated with changes in enzymatic activity, undue absorption, or toxic effects. Indicators, when reliable, can be used to screen an exposed population and prevent a toxic level from being reached.

Once the toxic dose to the "normal healthy" population is derived, consideration must be given to high-risk groups such as fetuses, infants, young children, the elderly, pregnant woman, the nutritionally deprived, the psychologically or physically debilitated, individuals with genetic disorders, and those exposed to excessive amounts of the pollutant. These groups must be quantified to determine the proportion of the population at greater risk than the so-called "normal" population from a certain pollutant. The Federal Clean Air Act requires that primary ambient air quality standards be set to fully protect both specifically susceptible subgroups and healthy members of the population (Finklea et al., 1973). This is the goal for which we should strive in selecting an "acceptable" exposure level for standard setting. If sufficient data are available on the level at which effects occur in high-risk groups, an acceptable level may be that which does not result in subclinical physiologic changes of a potentially damaging nature; or, depending on the available data, it may involve the application of a safety factor from that dose which is toxic to the general population such that susceptible populations may be protected.

The percentage of the population that is chosen for protection is necessarily a subjective decision. Ideally, the entire population should be protected, but reality necessitates a cost-benefit analysis such that the ultimate choice is based on economic, social, and political considerations as well as on adverse health effects.

REFERENCES

Baetjer, A. M. (1968) Role of environmental temperature and humidity in susceptibility to disease. *Arch Environ Health* 16:565–570.

Bates, D. V. and Hazucha, M. (1973) In *Proceedings of the Conference on Health Effects of Air Pollutants,* pp. 507–540. Assembly of Life Sciences, National Academy of Sciences, National Research Council. Washington, D.C.: US Government Printing Office.

Berglund, F. (1972) Levels of polychlorinated biphenyls in foods in Sweden. *Environ Health Perspect* 1:67–71.

Buckley, R. D., Hackney, J. D., Clark, K., and Posin, C. (1975) Ozone and human blood. *Arch Environ Health* 30:40–43.

Carnow, B. (1966) Air pollution and respiratory diseases. *Scientist and Citizen* 8:1–5.

Casarett, L. J. and Doull, J. (1975) *Toxicology: The Basic Science of Poisons,* pp. 13–14. New York: Macmillan.

Cornblath, M. and Hartmann, A. F. (1948) Methemoglobinemia in young infants. *J Pediatr* 33:421–425.

Diem, K., ed. (1962) *Scientific Tables,* p. 625. Ardsley, N.Y.: Geigy Pharmaceuticals.

Dutton, G. J. (1961) The mechanism of glucuronide formation. *Biochem Pharmacol* 6:65–71.

Expert Group, Report from an (1971) *Methyl Mercury in Fish: A Toxicologic-Epidemiologic Evaluation of Risks.* Stockholm: Nordisk Hygienisk Tidskrift.

Finklea, J. F., Shy, C. M., Moran, S. B., Nelson, W. C., Larsen, R. I., and Akland, G. G. (1973) *The Role of Environmental Health Assessment in the Control of Air Pollution,* p. 9. Triangle Park, North Carolina: EPA Research.

Firket, J. (1931) The cause of the symptoms found in the Meuse valley during the fog of December 1930. *Bull R Acad Med* (Belgium) 11:683–739.

Goldstein, E. and Green, G. M. (1966) The effect of acute renal failure on the bacterial clearance mechanisms of the lung. *J Lab Clin Med* 68:531–542.

Hatch, T. F. (1973) Criteria for hazardous exposure limits. *Arch Environ Health* 27:231–235.

Hunt, V. R. (1975) *Occupational Health Problems of Pregnant Women. A Report and Recommendations of the Office of the Secretary.* Washington, D.C.: Department of Health, Education and Welfare.

Jacob, H. S., Ingbar, S. H., and Jandl, J. G. (1965) Oxidative hemolysis and erythrocyte metabolism in hereditary acatalesemia. *J Clin Invest* 44:1187–1199.

Jones, H. B. and Grendon, A. (1975) Environmental factors in the origin of cancer and estimation of the possible hazard to man. *Food Cosmet Toxicol* 13:251–268.

King, G. S. (1973) Ozone and air conditioning. *Soc Health J* 93:84–86.

Lee, D. H. K. (1970) Nitrates, nitrites and methemoglobinemia. *Environ Res* 3:484–511.

Mahaffey, K. R. (1974) Nutritional factors and susceptibility to lead toxicity. *Environ Health Perspect* Exp. Issue No. 7, p. 107.

Marx, J. L. (1975) Thymic hormones: inducers of T-cell maturation. *Science* 187:1183–1185, 1217.

Miller, C. E., Hasterlik, R. J., and Finkel, A. J. (1969) *The Argonne Radium Studies—Summary of Fundamental Data.* Springfield, Va: Clearing house for Federal, Scientific and Technical Information.

Ministry of Health (1954) *Mortality and Morbidity During the London Fog of December, 1952.* Reports on public health and related subjects, No. 95. London: Her Majesty's Stationary Office.

NAS (National Academy of Sciences) (1975) *Principles for Evaluating Chemicals in the Environment.* A report of the Committee for the Working Conference on Principles of Protocols for Evaluating Chemicals in the Environment, ch. IX, pp. 134–155. Environmental Studies Board, National Academy of Sciences, National Academy of Engineering, and Committee on Toxicology. Washington, D.C.: National Research Council.

Oehme, F. M. (1969) *A Comparative Study of the Biotransformation and Excretion of Phenol.* PhD dissertation, University of Missouri.

Riddiough, C. R., Musselman, R., and Calabrese, E. J. (1977) Is EPA's ^{226}RA drinking water standard justified? *Med Hypoth* 3(5):171–174.

Schrenk, H. H., Heimann, H., Clayton, G. O., Gafafer, W. M., and Wexler, H. (1949) Air pollution in Donora, Pennsylvania. Epidemiology of the unusual smog episode of October 1948. *Public Health Bull* No. 306. Washington, D.C.: Federal Security Agency.

Shakman, R. A. (1974) Nutritional influences on the toxicity of environmental pollutants—a review. *Arch Environ Health* 28:105–113.

Stokinger, H. E. and Scheel, L. D. (1973) Hypersusceptibility and genetic problems in occupational medicine—a consensus report. *J Occup Med* 15:564–573.

Szakal, A. K. and Hanna, M. G., Jr. (1972) In *Conference on Immunology of Carcinogenesis.* National Cancer Institute Monograph 35, pp. 173–182. US Department of Health, Education and Welfare, Public Health Service, National Institute of Health, Bethesda, Md.

US Environmental Protection Agency (1973) *1973 Drinking Water Standards.* Advisory Committee on the Revision and Application of the Drinking Water Standards. Washington, D.C.: Environmental Protection Agency.

US Environmental Protection Agency (1974) *Health Consequences of Sulfur Oxides: a Report from CHESS, 1970–1971.* Triangle Park, North Carolina: EPA Research Publication, No. EPA—650/1-74-008.

Waldbott, G. L. (1973) *Health Effects of Environmental Pollutants.* St. Louis, Mo.: C. V. Mosby.

Waters, E. P. (1975) Checking the retail shelves. *FDA Consumer* 9:4–7.

Weisburger, J. H. and Rall, D. P. (1972) In *Environment and Cancer,* pp. 437–452. Baltimore: Williams and Wilkins.

WHO (World Health Organization) (1975) In *Methods used in the USSR for Establishing Biologically Safe Levels of Toxic Substances.* Geneva: World Health Organization.

Animal Extrapolation in Standard Setting

2

THE THEORETICAL BASIS OF ANImal extrapolation as a predictor of human responses is derived from the well-founded assumption (i.e., evolutionary theory) that all animals are related phylogenetically to a greater or lesser degree. According to Beyer (1960), the similarities among the different classes of higher vertebrates have evolved to satisfy similar requirements of growth, maintenance and reproduction. The end result of this process of evolutionary adaptation has been the occurrence of a striking orderliness of function across the broad range of different species. Consequently, the process of speciation from a common ancestor is not only the cause of the amazing similarities among species, but also the source of their differences. The application of the principles of phylogenetic relationships in the selection of animals for predictive toxicologic testing has been suggested by Sterling (1971), who criticized the excessive use of rodents as predictive models because they are phylogenetically far removed from humans compared to other species such as the dog or monkey (see Wood, 1972 for a critique of Sterling, 1971).

The animal species that best predicts the response of man is a major issue in predictive pharmacology and toxicology. This is a difficult question to answer because the situation is much more complicated than would be surmised. For example, would the same species be equally predictive for all chemicals and drugs to be tested? Would species differ in the degree to which they can predict toxicity based on the specific organ system; that is, could the rabbit's kidney be most like that of man whereas the monkey's lungs be the best to simulate respiratory challenge? Also, one is faced with the question which humans are being considered? For example, which animal model best simulates the neonate, the pregnant female, the person with an inadequate diet, or individuals with genetic deficiencies? Is it truly reasonable to expect that one species can offer the broad range of predictive potentials needed to assess the responses of the highly diverse human population to any foreign substance? This is the central issue on which this chapter focuses. To accomplish the task of evaluating the process of animal extrapolation, we consider in detail (1) the metabolic basis governing why species differ in their sensitivity to drugs, (2) how these species differences affect the extrapolation process, (3) the value of the quantitative and qualitative predictions of human drug and pollutant exposure based on clinical anticancer drug trials, (4) the use of specific animal models to study various human conditions (neonates, blood disorders, etc.), and (5) the Soviet approach to animal extrapolation techniques.

INTERSPECIES DIFFERENCES

In theory, there are five possible factors that may be responsible for the variations in the sensitivity of different species to the toxic effects of drugs or environmental pollutants. These factors include absorption, distribution, storage, metabolism, and excretion. According to Rall (1969, 1974), the major factor affecting the occurrence of variations in species sensitivity is primarily differences in metabolism. The remaining factors (i.e., absorption, distribution, storage, and excretion) are thought to be generally similar among species and therefore do not contribute significantly to species variations with respect to drug or pollutant response. Factors such as absorption, tissue localization, and penetration across barriers depend for the most part on physiochemical properties that are common to both animals and humans (Schein et al., 1970; Brodie, 1962, 1964a, 1964b; Brodie et al., 1965). For example, the rates of absorption and penetration across the blood-brain barrier that are accurately predicted based on the physiochemical characteristics of the substances in question are most often quite similar in animals and humans. The subsequent section considers the biological chemical differences between various animal species with respect to the five parameters (metabolism, absorption, etc.) mentioned by Rall. An appreciation of the extent of the physiologic differences with respect to drug and pollutant toxicity assists the reader in developing an understanding of the usefulness and limitations of animal extrapolation.

Variations in Metabolism

There are both qualitative and quantitative differences among animal species concerning responses to various types of drugs. Brodie (1962) pointed out the highly unpredictable responses of various species to morphine. For instance, morphine is a depressant in rats, dogs, and man, but, quite to the contrary, it stimulates cats, goats, and horses. In addition to this example of a qualitative difference between the responses of various animal species, there are a number of examples that illustrate quantitative differences in drug metabolism between species.

Quantitative

1. Merperidine, a narcotic analgesic, is metabolized at a rate of 17% per hour in man, with a typical dose lasting 3–4 hours. In contrast to humans, the dog rapidly metabolizes (70–90% per hour) this drug. Brodie (1962) suggested that the rapid degradation of meperidine by dogs is probably the reason why there is such a difficulty affecting tolerance or addiction to meperidine in this species.

2. Sulfamethazine is absorbed following oral ingestion with a similar efficiency in the dog, cow, and man, yet man excretes it quickly, whereas the dog's rate is intermediate and the cow is very slow (Beyer, 1960).

3. Phenylbutazone, an antirheumatic drug, has been extensively studied in a variety of animals and man. Its biologic half-life has been reported by Beyer (1960) to be 3 hours in rabbits, 6 hours in dogs, and 72 hours (3 days) in man. According to Brodie (1962), the antirheumatic effect of phenylbutazone was initially discovered in man. This is explained by the fact that exceptionally large doses of the drug are needed to affect an antiinflammatory response in rats because of their rapid metabolism of the drug.

4. Carcinomide and probenecid are both active inhibitors of penicillin; they are conjugated and excreted at markedly different rates in man and dogs. The end result is a 10-fold difference in effective dose between these two species (Beyer, 1960).

5. Ethylbiscoumacetate (EBC), an anticoagulant, presents an interesting situation of how researchers must consider not only the similarity between man and the model in terms of rate of metabolism, but also with regard to the metabolic pathway. The rabbit was thought to offer an excellent predictive model in the case of EBC because it metabolizes EBC at the same rate as man. However, Brodie (1962) reported that the similarity ended with the rate of metabolism because man and rabbit metabolize the drug by different pathways. Humans and rabbits metabolize EBC via hydroxylation of the aromatic ring and deesterification, respectively; curiously, the dog metabolizes EBC by the identical pathway as man, but considerably slower (Figure 2-1).

6. Quinn et al (1958) reported species differences (mouse, rat, guinea pig, rabbit, dog, and man) for the metabolism of hexobarbital,

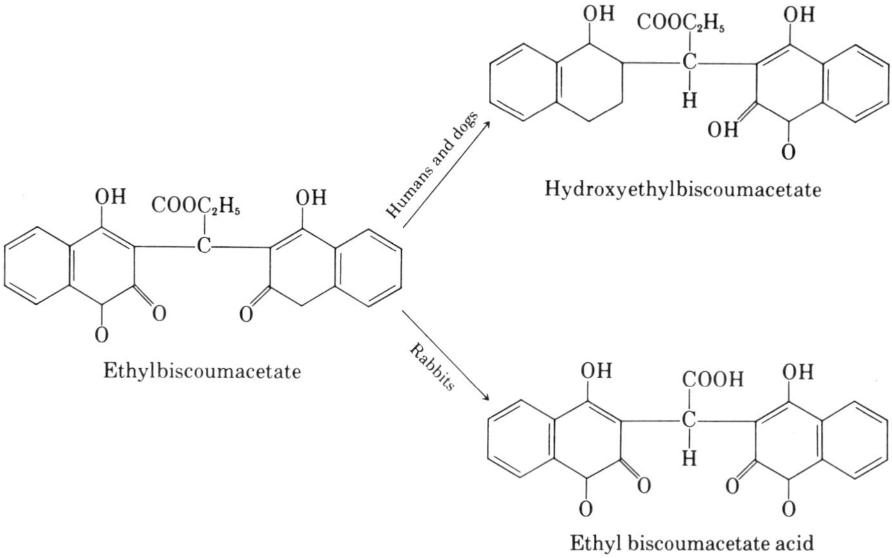

Figure 2-1. Metabolic pathways for the transformation of ethylbiscoumacetate in man, dogs, and rabbits. Source: Brodie, B. B. (1962). Difficulties in extrapolating data on metabolism of drugs from animals to man. ***Clin Pharmacol Ther*** 3(3):374–380.

antipyrine, and aniline (Table 2-1). The data indicated that the mouse biotransformed hexobarbital and antipyrine 20 and 60 times, respectively, more quickly than man. According to Brodie (1962), there was no obvious relationship between the size of the animal and the rate of biotransformation. Brodie noted that the activity of the liver detoxification enzymes with respect to hexobarbital in the mouse is approximately 17 times greater than in dogs. He concluded that species differences in the duration of hexobarbital activity can be accurately viewed as a function of the microsomal enzyme activity. This perspective is in striking agreement with Rall (1969), who commented on the study of Freireich et al. (1966) concerning the dose-effect relationship of 18 anticancer agents (see discussion later in this chapter).

Metabolic Pathways—Relative Use

Brodie (1962) noted that it is not uncommon for a substance to be metabolized via several pathways. This may become important in the extrapolation process if the species in question use the different pathways according to different rates with different resulting effects. Axelrod (1955a, 1955b) reported that amphetamines can undergo N-demethylation, hydroxylation, and deamination in vivo. Deamination is the principal route of metabolism in rabbits, whereas demethylation is the most important pathway in dogs; however, rats metabolize amphetamines principally via hydroxylation (Brodie, 1962). Brodie

Table 2-1. Species Differences in Metabolism of Hexobarbital, Antipyrine, and Aniline

	Biologic half-life (min)		
Species	Hexobarbital	Antipyrine	Aniline
Mouse	19 ± 7 (12)[a]	11 ± 0.25 (6)	35 ± 4 (6)
Rat	140 ± 54 (10)	141 ± 44 (6)	71 ± 1 (3)
Guinea pig		110 ± 27 (5)	45 ± 8 (7)
Rabbit	60 ± 11 (9)	63 ± 10 (7)	35 ± 22 (6)
Dog	260 ± 20 (8)	107 ± 20 (8)	167 ± 66 (6)
Man	360		

[a] Figures in parentheses refer to number of animals tested.

Source: Brodie, B. B. (1962) Difficulties in extrapolating data on metabolism of drugs from animals to man. *Clin Pharmacol Ther* 3(3):374–380.

(1962), commenting on Parke and Williams (1956), reported that ortho- and para-hydroxylation of aniline use different enzymes. Consequently, rabbits produce appreciable quantities of the para-derivative, but only minute amounts of the ortho-derivative. Cats are just the opposite, producing large quantities of the ortho- and small quantities of the para-derivative. Finally, Oehme (1969) reported that cats are much less capable of transforming phenols to glucuronides compared to dogs, goats, sheep, and all other mammalian species. This is thought to explain the relative sensitivity of cats to phenol.

Variation in Absorption (Digestive Tract and Topical)

Pyrvinium chloride is known to be selectively toxic to mice via oral ingestion. This results from its high rate of absorption through the gastrointestinal tract (Oehme, 1969).

In contrast to humans, who absorb lead via the digestive tract at a rate of about 10%, cattle and sheep are known to absorb lead at the rate of only 1–2% (Blaxter, 1950; White and Cotchin, 1948; Oehme, 1970). Thus cattle and sheep are recognized to have a greater tolerance to ingestion of large quantities of lead than do humans (Allcroft, 1950; Allcroft and Blaxter, 1950; Scharding and Oehme, 1973).

Oehme (1970), commenting on Kruckenberg (1969, personal observation), reported that an organophosphate insecticide (specific name not mentioned) that was relatively nontoxic to a variety of animal species including horses was found to be highly toxic to pigs when they were exposed via dermal application.

Variations in Distribution of Drugs/Pollutants

Placental Sensitivity

Koppanyi and Avery (1965) noted that the placental barrier is not identical among different species. In fact, studies have revealed differences in microscopic structure and in the capability to screen the fetus from a variety of chemical substances. Chemical substances may pass across the placenta by a number of mechanisms, including diffusion, active transport, and gradient mechanisms. Koppanyi and Avery (1965) concluded by stating that "the effect of a given teratogen varies from species to species which must be in some part due to selective distribution of this membrane.

Blood-Brain Barrier Selectivity

Another possible source of variation in drug/pollutant distribution among different species is the occurrence of differences in the functioning of the blood-brain barrier. Way (1967) noted that the blood-brain barrier is known to develop at different rates in various species. Such a difference is quite critical when one considers the effects of substances acting on the central nervous system of young animals. The extent of these variations between different species must be evaluated more fully.

Plasma Protein Binding

The extent to which species vary in their capability to bind various drugs/pollutants is of critical importance for both pharmacology and environmental toxicology, because this affects, in part, the degree of penetration to the site of action and the subsequent metabolism and excretion of the substance in question. Consequently, a bound substance is considered to be involved in an inactive role in the body. On release, a formerly bound substance is usually metabolized and excreted in a normal fashion. Albert (1973) compared the differences between various species (mouse, rat, rabbit, horse, and human) in the capability of binding various drugs (Table 2-2). Note, for example, the difference between the mouse and human with respect to the binding of sulfisoxazole and sulfadiazine.

Biliary Excretion and Enterohepatic Circulation

Smith (1970) noted that drugs of molecular weight from 300 to 500 exhibit the greatest interspecies variation in biliary excretion. Smaller

Table 2-2. Binding of Drugs by Serum Albumin

	Percentage of Drug Unbound in Plasma			
Species	Benzypenicillin	Cloxacillin	Sulfadiazine	Sulfisoxazole
Human	49	17	67	16
Horse	59	30	—	—
Rabbit	69	22	45	18
Rat	—	—	55	16
Mouse	—	—	93	69

Source: Albert, A. (1973) *Selective Toxicity*, 5th ed, p. 41. London, Chapman and Hall.

Table 2-3. Biliary Excretion in Various Species of Substances of Molecular Weight 300–500 (Percentage of Dose Excreted in Bile)

Species	Methylene disalcylic acid (mol. wt. 288; 10 mg/kg IV, 6 hours)	Succinyl-sulfathiazole (mol. wt. 355; 20 mg/kg IV, 6 hours)	Stilbestrol glucuronide (mol. wt. 445; 10 mg/kg IV, 3 hours)
Rat	54	29	95
Fruit bat	—	25	—
Hen	—	25	93
Dog	65	20	65
Cat	—	7	77
Sheep	—	7	—
Rabbit	5	1	32
Guinea pig	4	1	20
Pig	—	0.2	—
Rhesus monkey	—	0.2	—

Source: Abou-EL-Makarem, M.M., Millburn, P., Smith, R.L., and Williams, R.T. (1967) Biliary excretion of foreign compounds. Species differences in biliary excretion. *Biochem J* 105:1289.

substances are usually thought to be excreted via the urine, whereas those substances larger than 500 molecular weight tend to be readily excreted in the bile in all species. Table 2-3 shows the interspecies differences in the percentage of three substances excreted in the bile in several species. The differences may be almost as large as a factor of 150 in certain cases. For example, compare the pig and rhesus monkey with the rat with respect to the biliary excretion of succinyl sulfathiazole.

Okita (1967) demonstrated that the occurrence of species variation with regard to the duration of action of cardiac glycosides (digitalis and digoxin) are best understood by differences in enterohepatic circulation as compared to differences in metabolism or protein binding of the compounds. It should be noted that this explanation holds for only nonpolar glycosides because charged molecules are not usually reabsorbed in the intestine.

Sensitivity of the Target Organ

Brodie (1962) reported that following exposure to hexobarbital, mice, rats, and rabbits recover the righting reflex at about 60 μg of the drug per milliliter of plasma (μg/ml), whereas the effect in dogs and man lasts until the levels decline to about 20 μg of drug per milliliter of plasma.

Thus, to the extent that animal models differ from humans with respect to the five parameters (metabolism, absorption, excretion, etc.) listed by Rall, there are difficulties in accurately extrapolating effects in animals to man.

The examples of species differences discussed here illustrate the hazards inherent in predicting the response in one species based on the response in another. This is especially true when the data are not sufficient to illustrate the underlying mechanisms of biotransformations among the comparison species. Because species differ in their rates of metabolism of a specific substance and therefore have different toxic responses to the compound in question, it does not necessarily imply that these species lack usefulness in the extrapolation process. Once the comparative biochemistries of the species are known, correction factors can be derived that interpret the response. However, at present, there is considerable uncertainty in the area of comparative biochemistry, and therein lies the problem with predictive pharmacology. Thus, although we know the general reasons why species differ in their responses, the all-important specifics are often vague. The problem is also one of relative expectations. Most people would like predictions that are exceptionally accurate with little uncertainty. However, if a researcher says that he or she feels that their predictions are accurate plus or minus a factor of 3, is this accurate enough? The general public may say no, whereas the researchers who may have worked 10 years to get this far may say yes. Thus Rall (1969) noted that "the problem of predictability lies primarily in the fact that the stakes are so high and not that the [predictive] systems are so bad." In the next section, attempts to derive accurate quantitative extrapolations are evaluated in light of current research methodology in environmental toxicology.

PREDICTIONS OF SAFE EXPOSURE LEVELS BASED ON BODY WEIGHT OR SURFACE AREA DIFFERENCES BETWEEN ANIMAL MODELS AND HUMANS

The relationship of body size to physiologic function relates to numerous physiologic parameters that have been used, in large part, by pharmacologists and toxicologists for many years in predicting the effects of a wide variety of drugs and toxic substances on animals and humans. The subsequent portion of this section explains the relationship of body size (i.e., body weight and/or surface area) to different physiologic parameters and how this knowledge may be used in the process of extrapolation of animal data to human responses.

Small animals have been found to consume relatively more oxygen and produce relatively more heat compared to larger animals. The smaller animals are known to have a larger surface area to volume ratio than larger animals. However, when standardized for body surface area, no significant differences in oxygen utilization and caloric production for a variety of mammalian species and differently sized individuals of the same species exist. As a result of these and numerous similar observations, it has been generally accepted that the human basal metabolic rate should be characterized in terms of body surface area (Pinkel, 1958).

Other parameters that demonstrate a mathematic relationship to body surface area have been reported. Early studies by Dreyer and Ray (1910, 1912) indicated that the ratio of blood volume to body surface area was constant in rabbits, guinea pigs, and mice, but it decreased with increasing body weight. Griffin et al. (1945) reported good correlations between body surface area with plasma volume, available thiocyanate space, and total circulating plasma proteins in normal adults. Furthermore, Baker et al. (1957) reported that body surface area is the best criterion for predicting the total blood volume of the individual based on studies with 150 normal adults. Pinkel (1958), commenting on earlier studies by Grollman (1929), noted that body surface area could be reliably correlated with cardiac outputs of normal 20–30-year-old individuals in both basal and resting states.

Pinkel (1958), commenting on studies by Smith (1951), noted that body surface area is directly related to the total number of glomeruli and to the kidney weight of different mammalian species (rat, dog, and human). He pointed out that the Addis area excretion ratio is directly related to kidney weight in the rat, rabbit, and dog and to the projected kidney weight of man. Pinkel (1958) noted that this permitted the development of a direct relationship of renal function and body surface area for a variety of mammalian species, including potential animal models for man. Additionally, he indicated that urea clearance and glomeruli filtration roles are also related to body surface area in both children and adults.

Talbot et al. (1953) reported that clinical medicine has made practical use of the body surface area/bodily function ratio relationship. They noted that physicians use body surface area to estimate the requirement of patients for parenteral fluids and electrolytes. Consequently, the usual daily water needs of infants (100 ml/kg) and adults

Pinkel (1958) reported the investigation of a series of studies in which he tried to use body surface area as the experimental criterion for providing the rationale for predicting the proper human dose of anticancer drugs (five drugs used) based on animal studies. As a result of his striking results and their consistency, only one example is necessary to demonstrate the predictive potential his work suggests. Table 2-4 compares the therapeutic doses of mechlorethamine in the mouse, hamster, rat, and man. The data indicate that when the dose is considered on a per unit surface area basis, there is very little difference between the four species. Thus, in this case, the three animal species would have been excellent predictors of the human response.

In an effort to further test the quantitative predictability of animal drug studies to human usage, Freireich et al. (1966) attempted to standardize various toxicologic studies for 18 anticancer drugs on the mouse, rat, hamster, dog, monkey, and human so that the reliability of animal extrapolation models could be evaluated. The report by Freireich et al. is the most comprehensive study concerning quantitative drug prediction/extrapolation and should be read by individuals concerned with the techniques of animal extrapolation. According to Rall (1969), these 18 anticancer agents are usually not affected by the liver microsomal enzyme detoxification system. Consequently, there was considerably reduced variability in the metabolism of these substances between the different test species. Therefore, this study provided the opportunity to consider interspecies variation as a function of size, absorption, distribution, excretion, and mechanism of action (and not differences in drug-metabolizing systems). The results lead to the

Table 2-4. The Use of Body Surface As A Predictive Devise For Safe Dose of Therapeutic Doses

			Mechlorethamine Dosage		
Subject	Weight (kg)	Surface area (sq m)	Total dose (mg)	Total dose/kg (mg)	Total dose/ sq m (mg)
Mouse	0.018	0.0075	0.072	4.0	9.6
Hamster	0.050	0.0137	0.15	3.0	10.9
Rat	0.25	0.045	0.5	2.0	11.1
Man	70.0	1.85	21–28.0	0.3–0.4	11.3–15.1

Source: Pinkel, D. (1958) The use of body surface as a criterion of drug dosage in cancer chemotherapy. *Cancer Res* 18:854.

conclusion that the toxic effects of an agent are similar among species when the dose is measured on the basis of surface area. Furthermore, the ratio of animal/human toxicity for the mouse, hamster, dog, and monkey was extraordinarily close to 1. Consequently, in these specific cases, each species is a good predictor of toxicity in man. Rall (1969), commenting on the work of Freireich et al. (1966), noted that the correlation coefficient between one strain of mice and man and between the rhesus monkey and man exceeded 0.9.

Rall (1969) indicated that the relationship on a milligram of drug per kilogram of body weight basis (mg/kg) is actually comparable to the milligram of drugs per square meter basis that had been emphasized by Pinkel. The fundamental reason for this is the interconvertibility of body weight to surface area following the general formula of $A = KW^{2/3}$ where A equals the surface area (sq cm), K equals a constant for each animal species, and W equals the weight in grams. Thus on a milligram per square meter basis, the minimal toxic dose (MTD) is almost identical in each of the experimental animal species. However, on a milligram per kilogram basis, the MTD in humans is equal to $\frac{1}{12}$ the LD_{10} in mice, $\frac{1}{9}$ the LD_{10} in hamsters, $\frac{1}{7}$ the LD_{10} in rats, $\frac{1}{3}$ the MTD in rhesus monkeys, and $\frac{1}{2}$ the MTD in dogs.

Reports by Owens (1962), the Cancer Chemotherapy National Service Center (1959), Litchfield (1961, 1962), Golberg (1963), Rall (1969), Schien et al. (1970), and Weisburger and Rall (1972) have discussed the nature of qualitative predictions in animal extrapolation. Based on studies with mice, rats, and dogs, Owens noticed that the predictive value was good for toxic effects on the bone marrow, gastrointestinal tract, liver, and kidney. In contrast, these animal models were less reliable for the nervous system, including peripheral neuropathy, extraocular palsies, and CNS toxicity. Dermatitis and alopecia could not be predicted from studies with rodents, dogs, or monkeys.

The data that have been presented were designed to evaluate the effectiveness of animal models for predicting potential toxic effects of anticancer agents on humans. Of what application are these studies to the field of occupational and environmental toxicology? Studies on the development of effective anticancer drugs have a different scope and purpose than animal studies concerned with environmental/industrial pollutants. For example, the doses employed are much higher and the length of exposure considerably less than would be encountered in environmental/industrial exposures. Furthermore, whereas the animal models are usually healthy during the study of anticancer agents and environmental toxicology studies, the human counterparts are considerably different; that is, in cancer studies, the experimental subjects are usually terminally ill individuals who have been previously subjected to a variety of x-ray and chemotherapy treatments and whose general health status is often poor. However, in human exposure studies with environmental pollutants in controlled settings, one often finds healthy young adults as volunteers, thereby creating further difficulty in directly transposing the anticancer-type experiments into the environmental field with much precision. Anticancer studies are usually concerned with the efficacy of the drug as a therapeutic agent and try to balance the beneficial aspect with toxic side effects. Furthermore, the precise classification of a MTD for a patient is dependent on the subjective judgments of physicians who are treating the patient, but they usually include some objective reports such as white blood cell counts.

From these examples, it can be seen that there are definite obstacles impeding the direct application of pharmacological anticancer studies to environmental/industrial toxicology. An even greater obstacle in directly applying these pharmacologic-anticancer studies is that most of the chemical agents used are alkylating agents or antimetabolites, which are not affected by the detoxification enzymes of the liver. In contrast, most of the environmental/industrial pollutants are chemically transformed in the liver and thereby have a modified toxicity and residence time following the transformation in the liver. Thus the extent to which studies such as those of Freireich et al. (1966) are applicable to environmental toxicology is very debatable. Despite its questionable application to environmental toxicologic extrapolation, the Freireich et al. paper is the classic in the field and is constantly cited (see Rall, 1969; Dixon, 1976) as a major basis for quantitative animal extrapolations.

Dixon (1976) made an attempt to bridge the gap between clinical pharmacologists and environmental toxicologists. In his article, he spoke of the "need to establish reasonable and safe maximally permissible concentrations for various environmental chemicals. . . ." along with the need to develop predictive animal models on which extrapolations can be based. Dixon based his analysis on the well-known works of Freireich et al. (1966) and Schien et al. (1970). With the assistance of the Environmental Biometry Branch at the National Institute of Environmental Health Sciences (NIEHS), Dixon has tried to relate the data concerning clinical/experimental dose ratios for anticancer drugs to the general population and environmental pollutants as well. Table 2-5 indicates the probability of environmental chemicals exceeding the human toxic threshold based on extrapolation from experimental animal toxicology. Consider the MTD for the dog (mg/kg) under the 5% column. Dixon (1976) interpreted these data to indicate that at 1/10 (or 9.46/100 to be precise) of a known toxic dose, 5% of these drugs or environmental chemicals would exceed the human MTD. The identical procedure can be used interpret the remaining dose-response probability relationships suggested by the Table 2-5.

This type of extrapolation technique represents an interesting approach to the development of various models that may be used in the standard setting process. It is also important because it tries to incorporate the many years of experience in drug development and apply them to the new but related field of environmental toxicology. However, as Dixon himself pointed out, this extrapolative model must be validated. The previously discussed difficulties of transposing the clinical anticancer experience to the field of environmental toxicology with a high degree of precision were not specifically addressed during the development of the Dixon model. Before a high degree of confidence can be conferred on this model, such objections must be dealt with—possibly through the creation of "correction factors."

Table 2-5. Probability of Environmental Chemicals Exceeding Human Toxic Threshold Based on Extrapolation from Experimental Animal Toxicology

Experimental Toxicologic Values	Percentage of Experiment Dose for Varying Degrees of Safety					
	10[a]	5[a]	1[a]	0.1[a]	0.01[a]	0.001[a]
MTD						
Dog (mg/kg)	17.13	9.46	3.05	0.84	0.29	0.114
Dog (mg/sq m)	34.59	19.14	6.19	1.71	0.59	0.233
Monkey (mg/kg)	14.00	8.36	3.13	1.02	0.40	0.179
Monkey (mg/sq m)	32.06	17.97	5.94	1.68	0.59	0.237
LD						
Dog (mg/kg)	1.51	0.65	0.13	0.020	0.004	0.001
Dog (mg/sq m)	2.76	1.16	0.22	0.034	0.007	0.002
Monkey (mg/kg)	0.78	0.35	0.07	0.013	0.003	0.001
Monkey (mg/sq m)	5.48	2.76	0.75	0.168	0.049	0.017

[a] Percentage of total chemicals estimated to produce toxicity on human exposure.

Source: Dixon, R.L. (1976) Problems in extrapolating toxicity data for laboratory animals to man. *Environ Health Perspect* 13:43–50.

SPECIFIC ANIMAL MODELS

Rats in Bladder Cancer Studies

Since rats are known to concentrate their urine to an extremely high specific gravity, chemical substances are likely to stay in the urinary bladder for relatively long periods of time prior to excretion. Consequently, rats often form bladder stones that theoretically may initiate tumorigenesis via the action of physical irritation of the bladder wall. This unique condition inherent in the rat physiology and anatomy has been implicated as a possible factor in the saccharin-cancer controversy. More specifically, since it is thought that saccharin is not metabolized in rats prior to excretion, it has been suggested that when rats are fed extraordinarily large amounts of saccharin (e.g., up to 5% of the total diet), it is possible that the accumulated quantities of unmetabolized saccharin could become a physical irritant. Thus the question naturally arises, is saccharin actually a chemical carcinogen in the traditional sense, or does it effect carcinogenesis by only physical irritation? Although these questions must be resolved, it calls into question the validity of the rat model in bladder cancer studies when massive doses are employed (Culliton, 1977).

Rat versus the Mouse for Hydrocarbon-Induced Skin Cancer

Oncologists have long known that it is futile to study the development of skin cancer in rat models. Years of painting "known" hydrocarbon carcinogens on skin proved the rat to be a highly insensitive model for the study and prediction of skin tumors. According to Eckardt (1976), it was not until rabbits were selected as animal models that painted coal tar was first found to cause skin cancer in animals. Subsequently, the mouse was also found to effectively predict coal tar-induced skin cancers and thus became the model of choice for studying skin cancer. Why do mice and rabbits develop skin cancer on exposure to coal tar while rats fail to do so? Why can one substance cause cancer in one type of organism and not in another? Could the incorrect choice of an animal model lead to tragic situations in humans? Certainly the teratogenic effects of thalidomide were not prediced from rodent models (Litchfield, 1958). How should the scientist select the proper animal model? Are there any special clues that assist scientists in knowing which models best simulate the human condition?

Eckhardt (1976) provided an interesting framework on which one may proceed to answer, within the limits of scientific knowledge, the questions posed. He indicated that polynuclear aromatic hydrocarbons (e.g., those found in coal tar) are thought to be noncarcinogenic by themselves; however, they become carcinogenic when, in theory, they become activated by the enzyme aryl hydrocarbon hydroxylase, which is known to be present in the tissue of different animal species. Because the rat does not develop skin cancer after repeated applications of coal tar, Eckhardt (1976) logically concluded that the skin of the rat may not contain sufficient amounts of the aryl hydrocarbon hydroxylase enzyme. In contrast, the rabbit and the mouse would be expected to contain sufficient aryl hydrocarbon hydroxylase to effect the conversion of the "precarcinogen to the ultimate carcinogen."

It has been demonstrated recently that human lung tissue also contains aryl hydrocarbon hydroxylase. Kellermann et al. (1973) have shown that the relative risk of smokers with respect to developing bronchogenic carcinoma is related to the capacity to induce the activity of aryl hydrocarbon hydroxylase. More specifically, if one has a high capacity to induce this enzyme, he/she has a 36 times greater risk than someone with a low capacity for the development of bronchogenic carcinoma.

Thus mice offer an excellent model for the study of hydrocarbon-induced skin cancer. However, the real value emerges with the understanding of how the carcinogens effect tumorigenesis. For example, that mice can predict coal tar-induced cancers in humans is of little value if the cellular mechanisms are not identical. If they are not the same, the occurrence is not a truly predictive model. However, if the biochemistry is similar, predictive models can be derived that may be of use in future situations. Thus the discovery of the role of aryl hydrocarbon hydroxylase in carcinogenesis in mice has led to its application to a variety of tissues in which polynuclear aromatic hydrocarbons are known to cause cancer (e.g., scrotum, lung, etc.)

The Dog as a Model for Human Carcinogenesis

The FDA panel (1971) on carcinogenesis did not recommend the general use of the dog in the testing of chemical carcinogenesis because of its large size and relatively long lifespan (7–12 years). However, it did recommend that all chemicals that are related to aromatic amine carcinogens be tested in the dog. According to Deichmann and Radomski

(1969), evidence emerged in the 1930s that the dog was the best animal model for screening aromatic amines for possible bladder tumorigenesis in humans. Historically, of five animal models (mice, rats, guinea pigs, rabbits, and female mongrel dogs) originally tested by Hueper et al. (1938), only the dog developed bladder cancers similar to the "aniline tumors" that occurred in some dye workers following exposure. As a result of such data, Hueper et al. (1938) were quoted by Deichmann and Radomski (1969) in the following passage as strongly supporting the use of the dog model for bladder carcinogenicity.

> The known species specificity of chemical carcinogens and the distinct influence of species factors upon the target organ in which they may display their action determine to an important degree the outcome of bioassays performed on chemicals to be introduced into the human economy. Adequate experience with this experimental screening method has shown that the dog represents the species of choice in testing aromatic amino and nitro compounds for carcinogenic properties to the bladder of man because dogs seem to metabolize these chemicals into products identical with or similar to those produced by man. This fact accounts also for the far reaching identity between the histopathologic character of the preneoplastic, precancerous and cancerous lesions observed in the bladder of man and dog exposed to these urinary tract carcinogens.

In agreement with Hueper et al. (1938), Deichmann and Radomski (1969) noted that the dog simulates the carcinogenic response of humans only to aromatic amines. Thus, although man and the dog probably have a close similarity in the metabolism of aromatic amines, there is no evidence supporting the hypothesis that, in general, the dog metabolizes foreign substances in a manner more comparable to humans than do other available animal models.

In the same issue of the *Journal of the National Cancer Institute,* Bonser (1969) strongly challenged the qualified endorsement of the dog model by Deichmann and Radomski (1969). Bonser stated that the value of using the dog model must be weighed along with its deficiencies. Is the dog truly a unique model that alone predicts the response of humans to certain bladder carcinogens? Bonser noted that, with but one exception (aniline),* a rodent model has ultimately been able to predict

* According to Bonser, the evidence supporting the carcinogenic effect of aniline in dogs is equivocal.

Table 2-6. Substances Carcinogenic in Dog and Rodent Models (Bonser, 1969)

Benzidine
1-phenylazo-z-naphthal
2-acetylaminofluorene
0-aminoazotoluene
p-dimethylaminoazobenzene
Aromite
2-napthylamine
4-aminodiphenyl

the tumorigenic activity of the substance in question. Table 2-6 lists eight compounds to which both dog and rodent models exhibited carcinogenic responses. Bonser questioned whether much knowledge would have been lost if rodents were used exclusively; in the cases of these carcinogens, they exhibit relatively long latency periods (2–10 years) in dogs. In comparison, the knowledge of carcinogenic effects could be known considerably sooner and cheaper if rodent models were used instead of dogs. Bonser concluded that the routine testing of dogs with respect to environmental/occupational carcinogens rests primarily on historical grounds. She suggested that the expenditures used to conduct the dog studies be redirected to rodent studies. This suggestion is in fundamental agreement with a National Academy of Sciences (NAS) report (1975) that concluded that rodents are the only feasible model for large-scale screening. The basis for the criteria were not similar physiology to man, but economic feasibility, short latency, and susceptibility to the tumorigenic characteristics of many environmental agents.

Low Levels of G-6-PD in Studying High Risk to Ozone*

Human Disease

A recent report by Calabrese et al. (1977) identified individuals with the genetically inherited condition of glucose-6-phosphate dehydrogenase (G-6-PD) deficiency as a potential high-risk population to elevated ambient ozone exposure.

* This section was previously authored by E. J. Calabrese, *Am J Pathol* 91:409–411, 1978 and is reprinted with permission of the publisher.

Animal Model

Inbred strains (C57L/J and C57BR/cdJ) of mice (*Mus muculus*) have been bred to have low levels of the enzyme G-6-PD in their red blood cells (Hutton, 1971).

Comparison with the Human Disease

Hutton reported that the genetic regulation of G-6-PD activity in 16 inbred mouse strains falls into three distinct classes: high (i.e., "normal"), intermediate, and low. The ratios of G-6-PD activity in circulating red blood cells of the three activity classes are 3:2:1 (Hutton, 1971). Consequently, mice with low G-6-PD (C57L/J or C57BR/cdJ) have approximately 33% the G-6-PD activity in their red blood cells as mice with normal G-6-PD. The level of G-6-PD activity in human individuals characterized as "deficient" (i.e., Negroes with the A-variant strain, the most common G-6-PD deficient strain) may range from 8 to 20% that of the normal (Beutler, 1972). Thus the mouse strain described here is comparable to that of the human A-strain variant with regard to relative deficiency levels. It is also important to note that the mouse G-6-PD enzyme is identical to the A-variant G-6-PD deficiency with regard to optimal pH activity, thermal stability, and Michaelis constants. The G-6-PD activity units in normal humans as measured by millimoles of NADP are 14.2–22.0, whereas the G-6-PD activity units in the normal mouse class are 15–17 (Beutler, 1972; Chan and Kai, 1971). Additionally, the level of G-6-PD in mouse red blood cells decreases with cellular age in a fashion similar to that of man (Beutler, 1972; Hutton, 1971). Therefore, in both relative and absolute comparisons for a number of important physicochemical parameters, the enzyme G-6-PD in the mouse strain and human A-strain are very similar.

Despite the numerous similarities between the animal model and human condition, there are important parameters where the similarities diverge. For example, the amount of reduced glutathione (GSH) is actually higher in the low G-6-PD strain of mice than in the high G-6-PD strain (13.2–13.7 μM/g Hb vs. 12.5–12.9 μM/g Hb) (Hutton, 1971). In contrast, humans with a G-6-PD deficiency have been found to have significantly lower amounts of reduced GSH than do normal individuals (38–51 mg/100 ml vs. 53–84 mg/100 ml) (Zinkham et al., 1958; Beutler et al., 1955). Since reduced GSH is known to protect red cells from oxidative damage, this lack of comparable GSH levels is of concern. Also, in contrast to the sex-linked genetic condition of human G-6-PD, the quantitative differences in red cell G-6-PD activity of the mouse strains are regulated by alleles on at least two autosomal loci (Hutton, 1971).

Usefulness of the Model

Strains of mice with low levels of G-6-PD remain to be tested by environmental pathologists with respect to their efficacy as a model simulating exposure of a human high-risk group to oxidant (i.e., ozone) stress. To date, animal exposure studies with ozone have tended to ignore the use of animal models that simulate human high-risk groups. However, it is becoming more evident that environmental health standards will focus on protecting high-risk groups (Calabrese et al., 1976). Consequently, animal models simulating hypersusceptible groups such as humans with G-6-PD deficiency offer a significant potential growth to future environmental pathological research.

Availability of Low G-6-PD Mice

The two strains of low G-6-PD mice are available from the Jackson Laboratory, Bar Harbor, Maine.

Gunn Rats to Simulate Exposure to Polychlorinated Biphenyls*

Several years ago Cornelius and Arias (1972) reported on an animal model, the Gunn rat, that simulated the human disease, the Crigler-Najjar Syndrome. This human disease is characterized by a genetically induced unconjugated hyperbilirubinemia in infants which results in severe nonhemolytic acholuric jaundice and often kernicterus. A deficiency in hepatic uridine diphosphate (UDP) glucuronyl transferase activity is responsible for these clinical features.

Animal Model

The intention of this section is to encourage the utilization of the Gunn rat model in the toxicologic testing of the effects of polychlorinated biphenyls (PCBs), a widespread environmental contaminant.

* This section was previously authored by E. J. Calabrese, *Am J Pathol* 91:405–407, 1978 and is reprinted with permission of the publisher.

Rationale for Model

In attempting to establish health standards for environmental pollutants (e.g., PCBs), it is important to consider the health effects of such substances on human high-risk groups. A crucial component in this process is the development of an animal model that will allow predictions and extrapolation to these hypersusceptible segments of the population. Considerable evidence indicates that the phenolic and biphenolic compounds such as PCBs are detoxified and excreted primarily via the glucuronidation process in mammals including man (Goldberg, 1974; Block and Cornish, 1959; Hutzinger et al., 1972; Clarke and Clarke, 1967; Jones, 1965; Stecher, 1968; Wilkenson, 1968; Oehme, 1970). Since fetuses, neonates, and individuals with either Gilbert's Syndrome or the Crigler-Najjar Syndrome have recognized functional deficiencies in their capacities to excrete toxic compounds via conjugate glucuronidation (Gillette, 1967; Nyhan, 1961; Smith and Williams, 1966; Berglund, 1967; Lester and Schmidt, 1964; Billing, 1970; Arias et al., 1969; Lathe and Walker, 1958; and Arias, 1969), it has been predicted recently that these groups of individuals may be biochemically predisposed to accumulate PCBs (Calabrese, 1977).

Fetuses and neonates have been reported to have a developmental (i.e., temporary) deficiency of glucuronyl transferase (Wilkenson, 1968; Nyhan, 1961; Smith and Williams, 1966), whereas individuals with the Crigler-Najjar Syndrome have a permanent genetic defect of this trait (Cornelius and Arias, 1972). Individuals with Gilbert's Syndrome are characterized by a partial defect in bilirubin conjugation, the mechanism of which is not well defined (Schmid, 1972).

Evaluation of the Gunn Rat Model

The Gunn rat model offers a unique opportunity to simulate the manner by which PCBs are excreted in humans with decreased functional capacities to conjugate biphenolic compounds. In fact, the homozygous Gunn rat, with its total inhibition of the bilirubin conjugation process, simulates those with the Crigler-Najjar Syndrome, whereas the heterozygote carrier may at least functionally, if not structurally, represent the milder and relatively common forms of hyperbilirubinemia associated with Gilbert's Syndrome.

Fetuses and neonates, with their immature, but developing enzyme detoxification scheme (e.g., glucuronyl transferase), may be best represented by either the homo- or heterozygous condition, depending on

(50 ml/kg) are satisfied by approximately 2 liters of water per square meter of body surface. Pinkel (1958), commenting on Stickler and Pinkel (unpublished), further confirmed the body surface area to body function ratio relationship by reporting that the caloric needs of infants and children are nearly the same for all ages and weights when compared on a per body surface area basis.

A very interesting experiment that supports the body surface area concept was performed by Crawford et al. (1950), who selected patients into various treatment groups on the basis of their different body surface areas. The subsequent treatment of these patients with identical quantities of the experimental drug (sulfadiazine or acetylsalicylic acid) resulted in a linear relationship of blood levels of the drug and surface area, thus providing the basis for further drug therapy based on this surface area criterion.

Several other researchers have also reported a relationship between dose and body size (either body weight or surface area). For example, Broom et al., (1932), Bliss (1936), and Durham et al. (1929) noted mathematical relationships of dose to body weight varying from direct and linear through direct and exponential to inverse, respectively. Clark (1937) and Done (1964) have likewise demonstrated that the effects of drugs on different animal models can be mathematically related to a function of both body weight and surface area.

It should be noted that not all researchers have been able to verify the mathematical association of dose to body weight. For example, Rall and North (1953) reported that in rats treated with alpha naphthylthiourea (ANTU), there was an independence of dose from body weight; similar reports for the independence of dose and body weight have been summarized by Lamanna and Hart (1968), including the melanophore expanding activity of frog pituitary extracts (Deutsch et al., 1956), histamine lethality in mice (Angelakos, 1960), acetoxycycloheximide in rats (Pallotta et al., 1962), dysentery bacillus toxin in mice (Zahl et al., 1943), botulinal toxin administered intraperitoneally to mice (Lamanna et al., 1955).

Lamanna and Hart (1968) suggested that there is an experimental difficulty in testing the dose/weight relationship. This results from the occurrence of a large degree of individual variation. Very large numbers of animals within various weight limits are necessary to demonstrate the existence of statistically significant variations between the different weight groups. Lamanna and Hart mentioned the difficulty that emerges from the dependence of weight on age in small rodents, most markedly in in-bred strains. Thus it is important to distinguish the dose/weight relationship from the dose/age relationship.

the degree of developmental immaturity. Although the Gunn rat may not permit a completely accurate representation of the proposed human high-risk groups, its use would offer a researcher in environmental pathology a more selective model for appraising the health effects of PCBs on a high-risk segment of the population.

EXTRAPOLATION TO HUMAN POPULATIONS—THE SOVIET APPROACH

General Principles

1. No one species offers an excellent preditive model for human responses for the broad variety of pollutants on the different systems of the body.
2. The use of multiple species in toxicologic testing provides the greatest chance for accurate qualitative and quantitative predictions.
3. There is a linear relationship between toxicity responses and body weight.
4. The social nature of man limits the accuracy of the extrapolation process.
5. Extrapolation techniques usually are not designed to represent the total population; the animals used in most studies are usually homogeneous, whereas the human population is highly heterogeneous.
6. Special consideration should be directed to those who are at high risk to the toxic effects of pollutants; consequently, animal models that simulate high-risk segments of the population should be given a high priority.
7. The Soviets have developed extrapolation or "transition" coefficients to make the comparisons between animals and humans more realistic.

The application of animal extrapolation techniques in the development of maximum permissible limits (MPLs) has been an important avenue of biomedical research in the Soviet Union (Brochkov, 1974; Sanotskiy, 1974; Van Nordwijk, 1964; Ulanova et al., 1969). The most striking feature of the Soviet approach toward animal extrapolation techniques is their use of the relationship of biologic parameters to body weight to predict toxicity responses in humans. The Soviets have found what they call "a general biological regularity" whereby more than 100 highly diverse biologic parameters of mammals are linearly related to body weight. It is possible that the reason that this phenomenon is a "general biological regularity" rather than a "natural law" (e.g., the law of gravity) is the occurrence of a few obvious deviations from the regularity, such as the human life span; the relative weight of the brain, and the amount of oxygen consumed by the brain. Nevertheless, despite these exceptions to the "regularity," the Soviets appear to have developed a powerful technique useful in extrapolation problems.

Krasovskii (1976) reported that the relative weight of the internal organs (liver, kidneys, and adrenals) for a broad range of mammals is inversely related to body weight. Similar comparisons have also been made for a variety of physiologic parameters including pulse rate, breathing rate, consumption of oxygen, food, water, and air, and the activity of the liver microsomal enzymes. (Phosphatase activity was an exception, since it was not found to be a function of body weight.) A direct linear relationship exists between body weight and life span, cholinesterase activity, length of pregnancy, number of simultaneously born offspring, latent period of tumor manifestation, nerve and muscle cell dimensions, erythrocyte life span, and other physiologic parameters. Krasovskii (1976) took 86 biologic parameters and calculated the average human values based on regression equations derived from animal models. The computed and actual values differed in one direction or the other by no greater than 1.2–1.5 times (with the exceptions of life span, brain size, and oxygen consumption of the brain).

Based on such an impressive series of "regularities of biological parameters" Krasovskii (1976) said that is would seem logical to suggest that responses to toxicants should also vary according to body weight. In testing his hypothesis, Krasovskii reported that the lethal dose of 34 substances for dogs was calculated based on regression equations for a number of biologic parameters (including toxicity indices) for four species of small laboratory animals. The results were quite striking in that the calculated toxicity values differed, on the average, from the actual laboratory dose by a range of only 0.2–1.2 times.

In a further attempt to test the toxicity/body weight relationship theory, Krasovskii compared the acute toxicity indices of 107 toxicants for humans and for four to six laboratory animal species. The results indicated that, when the data from white rats were applied to man, there was exaggeration of effect by factors of 5.5–40; however, when the most sensitive species was used, predictions were somewhat improved (4.3–31.7). With the regression equation approach, the predictions were exceptionally good, with a difference of calculated response to actual toxic response ranging from 1.5 to 3.4 times. These data led Krasovskii to conclude that, when regression equations are used for multiple species to predict human toxicity, "the transfer error of experimental data for the average human will be not greater than 3–4 times."

REFERENCES

Abou-EL-Makarem, M. M., Millburn, P., Smith, R. L., and Williams, R. T. (1967) Biliary excretion of foreign compounds. Species differences in biliary excretion. *Biochem J* 105:1289.

Albert, A. (1973) *Selective Toxicity,* 5th ed, p. 41. London, Chapman and Hall.

Allcroft, R. (1950) Lead as a nutritional hazard to farm livestock. IV. Distribution of lead in the tissue of bovines after ingestion of various lead compounds. *J Comp Pathol* 60:190–208.

Allcroft, R. and Blaxter, K. L. (1950) Lead as a nutritional hazard to farm livestock. V. The toxicity of lead to cattle and sheep and an evaluation of the lead hazard under farm conditions. *J Comp Pathol* 60:209–218.

Angelakos, E. T. (1960) Lack of relationship between body weight and pharmacological effect exemplified by histamine toxicity in mice. *Proc Soc Exp Biol Med* 103:296–298.

Arias, I. M. (1969) A defect in microsomal function in non-hemolytic acholuric jaundice. *J Histochem* 7:250.

Arias, I. M., Gartner, I. M., Cohen, M., Ezzer, J. B., and Levi, A. J. (1969) Chronic non-hemolytic unconjugated hyperbilirubinemia with glucuronyl transferase deficiency. *Am J Med* 47:395.

Axelrod, J. (1955a) The enzymatic demethylation of ephedrine. *J Pharmacol Exp Ther* 114:430–438.

Axelrod, J. (1955b) The enzymatic deamination of amphetamine (benzedrine). *J Biol Chem* 214:753–763.

Baker, R. J., Kozoll, D., and Meyer, K. A. (1957) The use of surface area as a basis for establishing normal blood volume. *Surg Gyn Obstet* 104:183–189.

Berglund, F. (1967) Levels of polychlorinated biphenyl in food in Sweden. *Environ Health Perspect* 1:67.

Beutler, E. (1972) Glucose-6-phosphate dehydrogenase deficiency. In Stanbury, J. B., Wyngaarden, J. B. and Fredrickson, D. S., eds. *The Metabolic Basis of Inherited Disease,* p. 1358. New York: McGraw-Hill

Beutler, E., Dern, R. J., Flanagan, C. L., and Alving, A. S. (1955) The hemolytic effect of primaquine. VII. Biochemical studies of drug-sensitive erythrocytes. *J Lab Clin Med* 45:286–295.

Beyer, K. H., Jr. (1960) Transposition of drug studies from laboratory to clinic. *Clin Pharmacol Ther* 1(3):274–279.

Billing, B. H. (1970) Bilirubin metabolism and jaundice with special reference to unconjugated hyperbilirubinemia. *Ann Clin Biochem* 7:69–74.

Blaxter, K. L. (1950) Lead as a nutritional hazard to farm livestock. II. The absorption and excretion of lead by sheep and rabbits. *J Comp Pathol* 60:140–159.

Bliss, C. I. (1936) The size factor in the action of arsenic upon silkworm larvae. *J Exp Biol* 13:95–110.

Block, W. D. and Cornish, H. H. (1959) Metabolism of biphenyl and 4-chlorobiphenol in rabbit. *J Biol Chem* 234(12):3301.

Bonser, G. M. (1969) How valuable the dog in the routine testing of suspected carcinogens. *J Natl Cancer Instit* 43(1):271–274.

Brochkov, N. P. (1974) Questions of extrapolation data in evaluating the mutagenic action of environmental factors on man. *Proceedings First US/USSR Symposium on Comprehensive Analysis of the Environment.* March 25–29, 1974. pp. 55–58.

Brodie, B. B. (1962) Difficulties in extrapolating data on metabolism of drugs from animal to man. *Clin Pharmacol Ther* 3(3):374–380.

Brodie, B. B. (1964a) Distribution and fate of drugs; therapeutic implications. In Binns, T. B., ed., *Absorption and Distribution of Drugs,* pp. 199–251. Baltimore: Williams and Wilkins.

Brodie, B. B. (1964b) Kinetics of absorption, distribution, excretion, and metabolism of drugs. In Nodine, J. H., and Siegler, P. E., eds., *Animal and Clinical Pharmacologic Techniques In Drug Evaluation,* pp. 69–88. Chicago: Year Book Medical Publishers.

Brodie, B. B., Cosmides, G. J., and Rall, D. P. (1965) Toxicology and the biomedical sciences. *Science* 148:1547–1554.

Broom, W. A., Burn, J. H., Gaddum, J. H., Trevan, J. W., and Underhill, S. W. F. (1932) The variation in the susceptibility of different colonies of mice towards the toxic action of aconite. *Q J Pharm Pharmacol* 5:33–36.

Calabrese, E. J. (1977) Inefficient conjugate glucuronidation as a possible factor in PCB toxicity. *Med Hypoth* 3(4):162–165.

Calabrese, E. J., Jansen, S. J., and Sorensen, A. J. (1976) A model useful in deriving health standards for environmental pollutants. *J Biol Educ* 10(6):249–257.

Calabrese, E. J., Kojola, W. H., and Carnow, B. W. (1977) Ozone: a possible cause of hemolytic anemia in G-6-PD deficient individuals. *J Toxicol Environ Health* 2:709–712.

Cancer Chemotherapy National Service Center (1959) Specifications for preliminary toxicological evaluation of experimental cancer chemotherapeutic agents. Cancer Chemotherapy Rep. 1:89.

Chan, T. K., and Kai, M. C. S. (1971) Double heterozygosity for G-6-PD deficiency. *J Med Genet* 8:149.

Clark, A. J. (1937) General pharmacology. In Heffter, A. ed., *Handbuchder Experimentellen Pharmakologie.* Ergansungswerk, Bd. 4, pp. 165–176. Berlin: Springer.

Clarke, E. G. C. and Clarke, M. L. (1967) *Garner's Veterinary Toxicology,* 3rd ed. pp. 156–158. Baltimore: Williams and Wilkins.

Cornelius, C. E. and Arias, I. M. (1972) Animal model: hereditary nonhemolytic unconjugated hyperbilirubinemia in Gunn rats. *Am J Pathol* 69(2):369–371.

Crawford, J., Terry, M., and Rourke, G. (1950) Simplification of drug dosage calculation by application of the surface area principle. *Pediatrics* 5:783–796.

Culliton, B. J. (1977) Saccharin: a chemical in search of an identity. *Science* 196:1179–1183.

Deichmann, W. B. and Radomski, J. L. (1969) Carcinogenicity and metabolism of aromatic amines in the dog. *J Natl Cancer Instit* 43(1):263–269.

Deutsch, S., Angelakos, E. T., and Loew, E. R. (1956) A quantitative method for measuring melanophore expanding activity. *Endocrinology* 58:33–39.

Dixon, R. L. (1976) Problems in extrapolating toxicity data for laboratory animals to man. *Environ Health Perspect* 13:43–50.

Done, A. K. (1964) Developmental pharmacology. *Clin Pharmacol Ther* 5:432–479.

Dreyer, G. and Ray, W. (1910) The blood volume of mammals as determined by experiments upon rabbits, guinea pigs and mice and its relationship to the body weight and to the surface area expressed as a formula. *Philos Trans R Soc Lond (Biol Sci)* 201:133–160.

Dreyer, G. and Ray, W. (1912) Further experiments upon the blood volume of mammals and its relation to the surface area of the body. *Philos Trans R Soc London (Biol Sci)* 202:191–212.

Durham, F. M., Gaddum, J. H., and Marchal, J. E. (1929) Toxicity test for Novarsenobenzene (Neosalvarsan). *Med Res Council Rep Ser* 128.

Eckardt, R. E. (1976) Extrapolating from animals to man: carcinogenesis. *J Occup Med* 18(7):492–494.

FDA (1971) Panel on carcinogenesis report on cancer testing in the safety evaluation of food additives and pesticides. *Toxicol Appl Pharmacol* 20:419–438.

Freireich, E. J., Gehan, E. A., Rall, D. P., Schmidt, L. H., and Skipper, H. E. (1966) Quantitative comparison of toxicity of anticancer agents in mouse, rat, hamster, dog, monkey and man. *Cancer Chemother Rep* 50(4):219–244.

Gillette, J. R. (1967) Individually different responses to drugs according to age, sex and functional or pathological states. In Wolstenholme, G. and Proter, R., eds., *Drug Responses in Man,* p. 24. London: Churchill.

Golberg, L. (1963) The predictive value of animal toxicity studies carried out on new drugs. *J New Drugs* January–February, 7–11.

Goldberg, L., ed. (1974) The toxicity of polychlorinated polycyclic compounds and related chemicals. *Crit Rev Toxicol* 2(4):445.

Griffin, G., Abbott, W., Pride, M., Muntwyler, R., Mautz, F., and Griffith, L. (1945) Available (thiocyanate) volume and total circulating plasma proteins in normal adults. *Ann Surg* 121:352–360.

Grollman, A. (1929) Physiological variations in the cardiac output of man. VI. The value of the cardiac output of the normal individual in the basal, resting condition. *Am J Physiol* 90:210–217.

Hueper, W. C., Wiley, F. H., and Wolfe, D. H. (1938) Experimental production of bladder tumors in dogs by administration of beta naphthylamine. *J Indust Hyg Toxicol* 20:46–84.

Hutton, J. J. (1971) Genetic regulation of G-6-PD activity in the inbred mouse. *Biochem Genet* 5:315–331.

Hutzinger, O., Nash, D. M., Safe, S., Defreitas, A. S. W., Nortstrom, R. J., Wildish, D. J., and Zitko, V. (1972) Polychlorinated biphenyls: metabolic behavior of pure isomers in pigeons, rats and brook trout. *Science* 178:312.

Jones, L. M. (1965) *Veterinary Pharmacology and Therapeutics,* 3rd ed., pp. 442–447. Ames, Iowa: Iowa State University Press.

Kehoe, R. A. (1960) The Harben Lectures. The metabolism of lead in man in health and disease. II. The metabolism of lead under abnormal conditions. *J R Instit Public Health Hyg* 19–53 (1961).

Kellermann, G., Shaw, C. R., and Luyten-Kellermann, M. (1973) Aryl hydrocarbon hydroxylase inducibility and bronchogenic carcinoma *N Engl J Med* 289(18):934.

Koppanyi, T. and Avery, M. A. (1965) Species differences and the clinical trial of new drugs: a review. *Clin Pharmacol Ther* 7(2):250–270.

Krasovskii, G. N. (1976) Extrapolation of experimental data from animals to man. *Environ Health Perspect* 13:51–58.

Lamanna, C. and Hart, E. R. (1968) Relationship of lethal toxic dose to body weight of the mouse. *Toxicol Appl Pharmacol* 13:307–315.

Lamanna, C., Jensen, W. I., and Bross, I. D. J. (1955) Body weight as a factor in the response of mice to botulinal toxins. *Am J Hyg* 62:21–28.

Lathe, G. H. and Walker, M. (1958) The synthesis of bilirubin glucuronide in animal and human liver. *Biochem J* 70:705.

Lester, R. and Schmid, R. (1964) Bilirubin metabolism. *N Engl J Med* 270(15):779.

Litchfield, J. T., Jr. (1958) Facts and fallacies in predicting drug effects in man. H. Raskov, ed., *Third International Pharmacology Meeting,* pp. 89–96. London: Pergammon Press.

Litchfield, T. J., Jr. (1961) Forecasting drug effects in man from studies in laboratory animals. *JAMA* 177:34.

Litchfield, T. J., Jr. (1962) Evaluation of the safety of new drugs by means of tests in animals. *Clin Pharmacol* 3:665.

NAS (1975) *Principles for Evaluating Chemicals in the Environment.* National Academy of Sciences, Washington, D.C. pp. 134–155.

Nyhan, W. L. (1961) Toxicology of drugs in the neonatal period *J Pediatr* 59(1):1.

Oehme, F. M. (1969) A comparative study of the biotransformation and excretion of phenol. Ph.D. Dissertation, University of Missouri.

Oehme, F. W. (1970) Species difference: the basis for and importance of comparative toxicology. *Clin Toxicol* 3(1):5–10.

Okita, G. T. (1967) Species differences in duration of action of cardiac glycosides. *Fed Proc* 26(1):1125–1130.

Owens, A. H., Jr. (1962) Predicting anticancer drug effects in man from laboratory animal studies *J Chronic Dis* 15:223.

Pallotta, A. J., Kelly, M. G., Rall, D. P., and Ward, J. W. (1962) Toxicology of acetoxycycloheximide as a function of sex and body weight. *J Pharmacol Exp Ther* 136:400–405.

Parke, D. V. and Williams, R. T. (1956) Species differences in the 0- and p- hydroxylation of aniline. *Biochem J* 63:12.

Pinkel, D. (1958) The use of body surface area as a criterion of drug dosage in cancer chemotherapy. *Cancer Res* 18:853–856.

Quinn, G. P., Axelrod, J., and Brodie, B. B. (1958) Species, strain, and sex differences in metabolism of hexobarbitone, amidopyrine, antipyrine and aniline. *Biochem Pharmacol* 1:152–159.

Rall, D. P. (1969) Difficulties in extrapolating the results of toxicity studies in laboratory animals to man. *Environ Res* 2:360–367.

Rall, D. P. (1974) Problems of low doses of carcinogens. *J Wash Acad Sci* 64(2):63–68.

Rall, D. P. and North, W. C. (1953) Consideration of dose-weight relationships. *Proc Soc Exp Biol Med* 83:825–827.

Sanotskiy, I. V. (1974) The concept of the threshold nature of the reaction of living systems of external actions and its consequences in the problem of protecting the biosphere against chemicals. *Proceedings of the 1st US/USSR Symposium On A Comprehensive Analysis of the Environment.* March 25–29, 1974.

Scharding, N. N. and Oehme, F. W. (1973) The use of animal models for comparative studies of lead poisoning. *Clin Toxicol* 6(3):419–424.

Schien, P. S., Davis, R. D., Carter, S., Newman, J., Schien, D. R., and Rall, D. P. (1970) The evaluation of anticancer drugs in dogs and monkeys for the prediction of qualitative toxicities in man. *Clin Pharmacol Therap* 11:3–40.

Schmid, R. (1972) Hyperbilirubinemia. In J. B. Stanbury, J. B. Wyngaarden, and D. S. Fredrickson, eds., *The Metabolic Basis Of Inherited Disease*, p. 1141. New York: McGraw-Hill.

Sidorenko, C. I., Koreneuskaya, Ye. I., Pinigin, M. A., and Krasovskii, G. N. (1974) Hygenic bases for protecting the environment. *Proceedings of the 1st US/USSR Symposium on the Comprehensive Analysis of the Environment.* March 25–29, 1974, pp. 63–66.

Smith, H. W. (1951) *The Kidney Structure and Function in Health and Disease,* pp. 495, 543, 562. New York: Oxford University Press.

Smith, R. L. (1970) Species differences in the biliary excretion of drugs. In *The Problems of Species Difference and Statistics in Toxicology.* Proceedings of the European Society for the study of drug toxicity, vol. XI, pp. 19–32. Amsterdam: Excerpta Medica Foundation.

Smith, R. L. and Williams, R. T. (1966) Implication of the conjugation of drugs and other exogenous compounds. In Dutton, G. J. ed., Glucuronic Acid, p. 457. New York: Academic.

Stecher, P. G., ed. (1968) *The Merck Index,* 8th ed., p. 810. Rhaveay, New Jersey: Merck and Co.

Sterling, T. D. (1971) Difficulty of evaluating the toxicity and teratogenicity of 2,4,5-T from existing animal experiments. *Science* 174:1358–1359.

Talbot, N., Crawford, J., and Butler, A. (1953) Homeostatic limits to safe parenteral fluid therapy. *N Engl. J Med* 248:1100–1108.

Ulanova, I. P., Sanotskiy, I. V., and Khalepo, A. I. (1969) Problem of the transfer of data obtained in animal experiments to hygiene practice *Gigiena i Prof Zabolevaniya* 13(7):22.

Van Nordwijk, T. (1964) Communication between the experimental animal and the pharmacologist. *Statist Neerl* 18:403.

Way, E. L. (1967) Brain uptake of morphine: pharmacologic implications. *Fed Proc* 26:1115–1118.

Weisburger, J. H. and Rall, D. P. (1972) Do animal models predict carcinogenic hazards for man? In *Environment and Cancer* pp. 437–452. Baltimore: Williams and Wilkins.

White, E. G. and Cotchin, E. (1948) Natural and experimental cases of poisoning calves by flaking lead paint. *Vet J* 104:75–91.

Wilkenson, T. T. (1968) A review of drug toxicity in the cat. *J Small Animal Pract* 9:21.

Wood, A. E. (1972) Interrelations of humans, dogs and rodents. *Science* 175:437.

Zahl, P. A., Hunter, S. H., and Cooper, F. S. (1943) Age as a factor in susceptibility of mice to the endotoxin of bacillary dysentery. *Proc Soc Exp Biol Med* 54:137–139.

Zinkham, W. H., Lenhard, R. D., and Childs, B. (1958) A deficiency of G-6-PD activity in erythrocytes from patients with favism. *Bull Johns Hopkins Hosp* 102:169–175.

3

High-Risk Groups

AN IMPORTANT AND RELATIVELY new area of environmental and occupational medicine is the identification and quantification of individuals who may be at increased risk to the toxic and/or carcinogenic effects of pollutants. Numerous studies have indicated that there is a high degree of variability in the response of humans to different levels of air pollutants (Carnow 1966, 1970; CHESS, 1974). In fact, Carnow (1976) noted that there is a highly variable dose-response relationship in a heterogeneous population such that, at each level of pollutant exposure, depending on the adaptive capacity of the specific individual considered, there is likely to be an increase in the toxic response, including prolongation of illness and, in certain cases, premature death. Thus Carnow logically concluded that the most important question is not, what is a "safe" numerical standard, but how many individuals are adversely affected at each successive level of exposure? In light of the need to establish more accurate cost/benefit relationships with regard to proposed health standards, it is imperative to provide decision makers with information that defines the responses of humans to the toxic substances in question at each progressive level of exposure.

The concept of human hypersusceptibility to environmental and occupational pollutants has been discussed in several studies (Stokinger and Scheel, 1973, Stokinger and Mountain, 1963; Carnow and Carnow, 1974), which have focused on specific biologic factors that predispose individuals to a toxic response from a diversity of pollutants in our air, water, and food environments. Within the past few years, the concept of high-risk groups has been considerably strengthened by new research findings (Calabrese et al., 1977; Calabrese, 1978), and new applications of their usefulness in addressing environmental health problems have emerged (Wadden et al., 1976). This chapter summarizes the current research findings with respect to high-risk groups and discusses their application to the process of standard derivation.

For a point of departure, it is necessary to precisely define the term high-risk group. High-risk groups are those individuals who experience toxic and/or carcinogenic effects significantly before the general population as a result of one or more biologic factors, including developmental influences, genetic factors, nutritional inadequacies, disease conditions, and behavioral or life style characteristics (Calabrese, 1978).

DEVELOPMENTAL PROCESSES

There are certain periods during an individual's ontogeny (that period from conception to death) in which he is significantly more susceptible to the adverse effects of environmental stressor agents than at other stages of development. Most people generally recognize the enhanced susceptibility of the very young and the very old to respiratory infections. There are, therefore, certain stages in the lives of each person that can be classified as high risk with respect to certain toxicants. It is important to note that these stages of development are shared by all people and that at some time everyone is considered at increased risk.

Several examples of developmental factors that predispose toxic responses include (1) immature enzyme detoxification systems in fetuses and neonates, (2) the presence of less than adult levels of immunoglobulin A in the sera of children up to the age of about 10–12 years, (3) the effects of aging on the functioning of the immune system, and (4) the enhanced susceptibility of the very young to the carcinogenic effects of ionizing radiation. A variation of the developmental characteristics that may also predispose one to the toxic effects of certain pollutants is the state of pregnancy (Calabrese, 1978). Table 3-1 lists examples of these and other developmental processes that may heighten one's sensitivity to pollutant toxicity. The table also indicates

Table 3-1. The Identification and Quantification of High Risk Groups

High-Risk Groups	Estimated Number of Individuals in U.S. Affected	Pollutant(s) to which High-Risk Group is (may be) hypersusceptible
	Developmental Processes	
Immature enzyme detoxification systems	Embryos, fetuses, and neonates to the age of approximately 3–6 months	Pesticides PCBs
Immature immune system	Infants and children do not reach adult levels of IgA until the age of 10–12	Respiratory irritants
Deficient immune system as a function of age	Progressive degeneration after adolescence	Carcinogens Respiratory irritants
Adolescents, especially carriers of porphyria genes	See population frequency for porphyrias	Chloroquine Hexachlorobenzene Lead Variety of medical drugs
Differential absorption of pollutants as a function of age	Infants and young children	Barium Lead Radium Strontium
Retention of pollutants as a function of age	Individuals above the age of 50	Fluoride
Pregnancy	Approximately 20 females per 50,000 females per year in the U.S.	Carbon monoxide Insecticides Lead
Circadian rhythms including phase shifts	All people have certain periods of the day when they are more susceptible to challenge	Carcinogens Probably most other pollutants
Infant stomach acidity	Infants	Nitrates Nitrites

Genetic Factors		
Albinism		
Tyrosinase positive	1 per 14,000 Blacks; 1 per 60,000 Caucasians; very high frequency in American Indians;	Ultraviolet radiation
Tyrosinase negative	1 per 34,000-36,000 Negroes and Caucasians; 1 per 10,000-15,000 in Ireland	
Catalase Deficiency		
Hypocatalasemia	5,000,000 heterozygotes	Ozone
Acatalasemia	16,000 homozygotes	Radiation
Cholinesterase variants	1) Highly sensitive homozygous and heterozygous individuals of European ancestry have a combined frequency of about 1 per 1250 2) Moderately sensitive genotypic variants of European ancestry have a frequency of 1 per 15,000	Anticholinesterase Insecticides
Crigler-Najjar Syndrome	Few individuals live to adulthood	PCBs
Cystic fibrosis	8,000,000 heterozygous for CF 100,000 homozygous for CF Most common among Caucasians of European ancestry	Ozone Respiratory irritants
Cystinosis	Most common among Caucasians of European ancestry; unknown frequency;	Cadmium Lead Mercury Uranium
Cystinuria	1 per 200–250 individuals, although asymptomatic, are affected; 1 per 20,000–100,000 are homozygous	Cadmium Lead Mercury Uranium
Glucose-6-phosphate dehydrogenase deficiency	1,600,000 American Black males (11–13%) Mediterranean Jews 11.0%; Greeks 1–2%; Sardinians 1–8%	Carbon monoxide Ozone Radiation
Glutathione deficiency	Only a few cases have been reported	Lead Ozone

Table 3-1. *(Continued)*

High-Risk Groups	Estimated Number of Individuals in U.S. Affected	Pollutant(s) to which High-Risk Group is (maybe) hypersusceptible
Glutathione peroxidase deficiency	Partial deficiency in some neonates; only a few adult cases reported	Lead Ozone
Glutathione reductase deficiency	Unknown, but thought to be rare	Lead Ozone
Gilbert's syndrome	6% of the normal, healthy adult population	PCBs
Immunoglobin A deficiency	500,000 homozygotes	Respiratory irritants
Immunologic hypersensitivity	2% of some worker populations	Isocyanates
Inducibility of aryl hydrocarbon hydroxylase	Approximately 45% of the general population are at high risk; 9% of the 45% are at very high risk	Polycyclic aromatic hydrocarbons
Leber's optic atrophy	Thought to be rare	Cyanide
Methemoglobin reductase deficiency	Found in many ethnic groups; gene frequency is uncertain, but thought to be rare; most infants experience a temporary deficiency	Nitrates Ozone Vanadium
Phenylketonuria	1 per 80 in the U.S. is a carrier; 1 per 25,600 has the disease	Ultraviolet light
Porphyrias	1.5 per 100,000 in Sweden, Denmark, Ireland, West Australia; 3 per 1000 in South African Whites; rare in Blacks	Chloroquine Hexachlorobenzene Lead Various drugs including barbituates, sulfonomides and others
Serum alpha$_1$ antitrypsin deficiency	1) Approximately 4.0–9.0% individuals of Northern European descent are heterozygotes 2) Approximately 160,000 homozygotes based on frequency found in Norway and Sweden	Respiratory irritants Smoking
Sickle cell		
1. Trait (Heterozygote)	1) 7–13% of American Blacks—heterozygotes	Aromatic amino and nitro compounds

2. Anemia (homozygote)	2) Homozygotic condition highly fatal	Carbon monoxide Cyanide
Sulfite oxidase	Unknown	Sulfur dioxide; sulfite
Thalassemia (Cooley's anemia)	Homozygous condition highly fatal; heterozygote frequency high (0.1–8.0%) in certain people of Italian, Greek, Syrian, and Black origin	Lead Organic chemicals such as benzene and its derivatives Ozone
Tyrosinemia	Frequency in general population unknown; frequency of heterozygous carriers in the Chicoutimi region of Northern Quebec is one carrier for every 20–31 people	Cadmium Lead Mercury Uranium
Wilson's disease	400,000 heterozygotes; 200 homozygotes—Most frequent among Jews of Eastern European ancestry and non-Jews from the Mediterranean regions, especially Sicily	Lead Vanadium
Xeroderma pigmentosum	Unknown	Certain hydrocarbon carcinogens Ultraviolet radiation
	Nutritional Deficiences	
Vitamin A	25% of children between 7 and 12 have lower than recommended dietary allowance (RDA); slightly higher percentage among the lower-income groups	Carcinogens DDT PCBs
Vitamin C	10–30% of infants, children, and adults of low-income groups receive less than the RDA	Arsenic Cadmium Carbon monoxide Chromium DDT Dieldrin Lead Mercury Nitrates Ozone

Table 3-1. *(Continued)*

High-Risk Groups	Estimated Number of Individuals in U.S. Affected	Pollutant(s) to which High-Risk Group is (maybe) hypersusceptible
		Nitrites Ozone
Vitamin E	7% of the general population are "physiologically deficient"	Cadmium Lead Ozone
Calcium	65% of children between the ages of 2 and 3 receive less than the RDA	Lead
Iron	98% of children between the ages of 2 and 3 receive less than the RDA	Hydrocarbon carcinogens Lead Manganese
Magnesium	Most U.S. males have a partial deficiency	Cadmium
Phosphorous	Deficiency in people with various kidney diseases	Lead
Selenium	Unknown deficiency, thought to be rare	Cadmium Mercury Ozone
Zinc	Deficiency present in association with various diseases, but not thought to be widespread	Cadmium
Riboflavin	30% of women and 10% of men aged 30–60 ingest less than ⅔ of the RDA	Hydrocarbon carcinogens Lead Ozone
Dietary protein	10% of women and 5% of men aged 30–60 ingest less than ⅔ of the RDA for protein	DDT
Methionine	Unknown, but thought to be more common in individuals following certain vegetarian diets	DDT
	Diseases	
Kidney disease	Relates to genetic diseases (see cystinosis, cystinuria, tyrosinemia), bacterial and virus infections, and hypertensive disease	Lead Other heavy metals Excessive sodium in diet

Liver disease	Relates to genetic diseases (Gilbert's syndrome) and virus infections	Carbon tetrachloride DDT and other insecticides PCBs
Asthmatic diseases	4,000,000–10,000,000 of the general population	Respiratory irritants: nitrogen dioxide ozone sulfates sulfur dioxide
Chronic respiratory disease	6,000,000–10,000,000 of the general population	Respiratory irritants: nitrogen dioxide ozone sulfates sulfur dioxide
Heart disease	15,000,000 of the general population	Respiratory pollutants: carbon monoxide ozone sulfur dioxide Excessive sodium in diet
	Behavioral Activities	
Smoking	Widespread	Cadmium Lead Polycylic aromatic hydrocarbons Radium 226
Alcohol consumption	Widespread	Lead Pesticides PCBs
Drug taking	Widespread	Pesticides PCBs

Source: Calabrese, E. J. (1978) *Pollutants and High Risk Groups.* Wiley-Interscience, New York. p. 187.

the toxic substance(s) to which the affected individual(s) may be predisposed.

GENETIC FACTORS

In addition to developmental factors that affect pollutant toxicity, certain segments of the population exhibit genetic factors that make them hypersusceptible to certain environmental stressors. Stokinger and Scheel (1973) and Cooper (1973) discussed the role of identification of the hypersusceptible worker in occupational health management. They also discussed the nature of the hypersusceptible worker and his/her identification and quantification. As in the case of developmental factors, there is enhanced susceptibility to stressor agents, but genetic deficiency, in marked contrast to developmental and aging processes, usually affects only minor subsegments of the population. Of course, those with genetic deficiencies must endure the developmental weaknesses as well. Thus they have one risk superimposed on another.

There are at least 150 genetic diseases that have been identified in humans (Stanbury et al., 1972). Calabrese (1978) identified 26 genetic diseases for which there is at least a theoretical basis for their causing enhanced susceptibility to toxicants. Certain genetic conditions have a long history of association with enhanced susceptibility to pollutants. Stokinger and Scheel (1973) actually identified five genetic conditions for which they recommended preemployment screening prior to entry into certain industrial jobs. These conditions include (1) serum $alpha_1$ antitrypsin deficiency, (2) carbon disulfide sensitivity, (3) glucose-6-phosphate dehydrogenase deficiency, (4) sickle cell anemia and sickle cell trait, and (5) hypersensitivity to isocynates. Table 3-1 lists the 26 genetic conditions thought to predispose afflicted individuals to pollutant toxicity. Furthermore, an assessment based on population genetics has been made that lists the estimated number of the United States population with these specific genetic conditions.

NUTRITIONAL DEFICIENCIES

Another important factor influencing pollutant toxicity is nutritional status. Considerable research has reported that nutritional deficiencies may exacerbate the toxic effects of certain pollutants. In fact, of all the factors that may enhance toxicity, this is the area most likely to be controlled and thus improved.

The presence of widespread deficiencies for vitamins A, C, and E and the minerals Ca and Fe are well known from nutritional surveys (Harris, 1961; Bogert et al., 1973). Despite the fact that these substances are needed for normal development and maintenance, research has shown that deficiencies of these nutrients are known to affect the toxicities of pollutants such as ozone, lead, hydrocarbon carcinogens, and other toxicants, depending on the particular nutrient (Shakman, 1974; Mahaffey, 1974; Calabrese, 1978). Table 3-1 lists the various vitamins and minerals known to affect pollutant toxicity, as well as indicating the affected pollutant and how widespread the deficiency is within the population.

DISEASE CONDITIONS

Perhaps the most recognized of the high-risk groups are those who are afflicted with various disease conditions such as asthma, bronchitis, and emphysema. The most important diseases with respect to pollutant toxicity are heart and lung disorders. This was clearly recognized as a result of major pollutant episodes in London, Meuse Valley, Belgium, and Donora, Pennsylvania. When the national ambient air standards were derived, the only high-risk groups that were considered were those with heart and lung disease. In addition to the heart and lung diseases, individuals with kidney and liver disorders are also known to have seriously enhanced toxicity (Table 3-1).

BEHAVIORAL FACTORS

The behavior of individuals, in addition to their chosen occupation, may be an important factor affecting exposure to pollutants. It is well known that smoking behavior results in additional exposure to a variety of carcinogens, ultimately culminating in a higher incidence of lung cancer in smokers as compared to nonsmokers. The presence of metals such as lead, cadmium, and nickel in appreciable amounts in both main and sidestream smoke also results in extra metal exposure to smokers (Menden et al., 1973). In addition to smokers, people who drink alcohol or take drugs that affect the liver microsomal enzyme systems may also be at increased risk to pollutant exposure (Rothman, 1975). Furthermore, the type of diet people select, such as a high fish diet, may result in greater than normal exposure to environmental pollutants (EHRC-Mercury, 1977) (Table 3-1).

HIGH-RISK GROUPS IN PERSPECTIVE

Knowing which individuals are at high risk with respect to pollutants is very important, because these are the people who will be the first to experience morbidity and mortality as pollutant levels increase. If the high-risk segments are protected, the entire population is protected. Consequently, information concerning both the identification and quantification of high-risk groups should play an integral role in the derivation of standards for pollutants in both ambient and industrial air as well as in drinking water.

The Clean Air Act Amendments of 1970 specifically require that primary air quality standards completely protect the public's health and that the standards incorporate sufficient safety margins. According to Finklea et al. (1974b), the Clean Air Act Amendments actually assume that there exists a "no-effects" level for every pollutant and for each adverse health effect. Despite their enhanced susceptibility to the toxic effects of pollutants, high-risk individuals, according to Finklea et al., have not been specifically and separately considered in setting environmental health standards. The rationale is that adequate protection for the larger general population segments and the margins of safety included will also ensure protection for the "large number of relatively small susceptible segments of the population for which we have little or no quantitative exposure information" (Finklea et al., 1974b). Thus high-risk segments of the population were not sufficiently considered in the standard derivation process because there was not enough evidence concerning them to offer a precise assessment of risk and because it was thought that they made up only a negligible percentage of the population.

Table 3-1 shows the estimated numbers of individuals at high risk to various pollutants from a variety of causes. The numbers are very large in certain cases, often encompassing significant percentages of the population of specific racial ancestries. These data undercut the presumed validity of the previously accepted assumption that states that the high-risk segments of the population are very small and constitute only insignificant numbers of the total population.

Figure 3-1 shows a theoretical comparison between normal and high-risk segments of the population with regard to the onset of toxic effects at increasing levels of pollutant exposure. Any normal adjustment or homeostatic adaptive response as well as compensatory capabilities are reduced in the high-risk group as compared to the normals. This results in the earlier onset of disease, disability, and death depending on the specific situation.

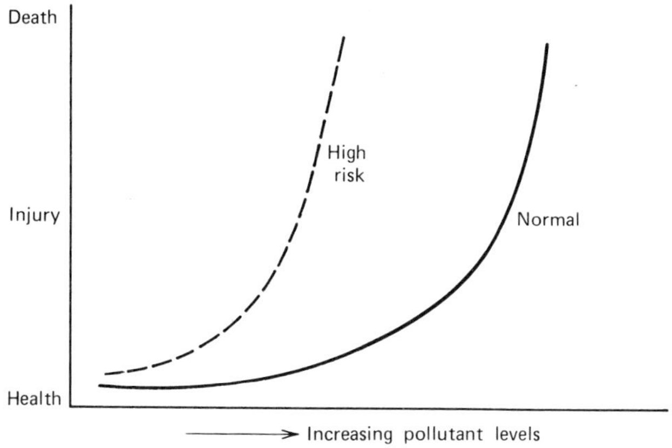

Figure 3-1. Comparison of the response of high-risk and normal individuals to increasing pollutant levels.

Experiments by Buckley et al. (1975) concerning the toxic effects of ozone on humans also clearly indicate the efficiency of the homeostatic compensatory responses of normal individuals during ozone stress. Such individuals respond to the ozone (0.5 ppm for 2¾ hours) by increasing the activity of the enzyme glucose-6-phosphate dehydrogenase (G-6-PD) by 20%. This allows the normal individual to maintain sufficient reduced glutathione (GSH) levels, thereby ensuring the integrity of red blood cell membranes. In contrast to normal individuals, those deficient in G-6-PD are known to be at extremely high risk to oxidant chemicals with regard to the development of hemolytic anemia. Such information clearly demonstrates that there are variable thresholds for pollutant effects in a population with a diverse gene pool such as ours (Calabrese, 1978).

The precise difference in sensitivity between a statistically "normal" individual and one in the high-risk subpopulation varies with respect to the specific causes of the high-risk condition. In any case, it is an illusion to assume that a threshold exists in our highly diverse heterogeneous human population, although such an assumption may be of practical significance in many cost/benefit analyses. It cannot even be said that there are separate thresholds for the normal as well as the high-risk segments. Even among such a high-risk group as those with a G-6-PD deficiency, there are now 80 recognized genetic variants (Beutler, 1972). Ultimately, each individual has his or her own unique genetic composition and, consequently, an individual threshold (Calabrese, 1978).

The identification of hypersusceptible workers has developed into a significant issue between management and labor. Ashford (1976) indicated that industrial physicians may continue the "blame the worker" attitude by management in their attempts to identify hypersusceptible workers. He suggests that numerous industrial physicians conceive of preventive medicine as identifying the high-risk individual and suggesting a job transfer or denial of a job offer. Ashford concluded that the goal of preventive medicine should be the removal of the hazardous conditions and not the hypersusceptible individual.

However, according to Kotin (1977), the enhanced risk of the hypersusceptible worker must be a shared employer-employee concern. Kotin outlined four options of management when dealing with the issue of hypersusceptibility: (1) restrict activities and products to those which offer no obvious risk, (2) more emphasis on environmental controls, (3) nonselection of those at increased risk (see the Ashford criticism above), (4) elimination of all carcinogens, mutagens, and teratogens from the workplace. Certainly options l, 2, and 4 will have a potentially adverse economic impact on the companies in question. The implementation of option 4, according to Kotin, would even criple industries such as steel, petrochemicals, and so on and is most likely nonviable.

The implementation of option 3, that is, not hiring the hypersusceptible worker, would seem to be the course of action most likely for industry to actively endorse. This is especially true in light of the growing number of workers turning to litigation or seeking redress for injury. In fact, Kotin reported that Johns-Manville, a major user of asbestos, has decided not to hire people who smoke, since exposure to tobacco smoke markedly increases the risk of both lung cancer and asbestosis. Although this action has been challenged by the unions in two plants in which trial programs were instituted, such action by Johns-Manville reflects an enlighted and reasonable application of biomedical knowledge of high-risk groups to a very real occupational health issue.

According to Ashford (1976) and Page and O'Brien (1973), the protection of the hypersusceptible individual has not been adequately considered by industrial health standards. "Threshold limit values" (TLVs), which have been developed for the past 30 years by the American Conference of Governmental Industrial Hygienists (ACGIH) and adopted as guidelines by the industry, are not designed to protect all workers (i.e., the hypersusceptibles). To date, TLVs have been derived for more than 450 chemical agents used in industry. In contrast to the content of TLVs, the Occupational Safety and Health Act of 1970 demands that no worker will acquire any impairment of health (OSHAct, 1970). With the expected implementation of the OSHAct mandate to protect the health of all workers, the need for a greater knowledge of hypersusceptible segments of the population becomes more urgent.

HIGH-RISK GROUPS AND AIR STANDARDS

Nitrogen Dioxide

The nitrogen dioxide (NO_2) standard is based to a considerable degree on a major epidemiologic research project conducted in the vicinity of a TNT plant situated near Chattanooga, Tennessee. The famous Chattanooga Study investigated the health effects of NO_2 exposure resulting from a large stationary source (i.e., the TNT plant). Among the diversity of epidemiologic studies attempted was an analysis of the health status of school children with respect to ventilatory function, the occurrence of acute respiratory illness (colds, sore throat) within recent weeks, and the retrospective occurrence of respiratory infections (pneumonia, croup, bronchitis) for periods of up to 3 years. Since young children are generally considered more susceptible to respiratory disease than young and middle-aged adults, these children represented a potential high-risk group (Shy et al., 1970a, 1970b; Pearlman et al., 1971).

The results indicated that the ventilatory performance (FEV 0.75) of the children in an elevated NO_2 area was significantly reduced as compared to an equivalent control group. Furthermore, there was an 18.8% relative excess of respiratory illness within families living in the high NO_2 area. The retrospective study alluded to above revealed that elevated NO_2 concentrations in ambient air were positively correlated with a significant increase in the incidence of acute bronchitis in infants exposed for 3 years and school children exposed for 2 and 3 years (Shy et al., 1970a, 1970b; Pearlman et al., 1971).

This major epidemiologic project represented the only human study in which a true high-risk group was considered. The study was a logical followup of previous experiments with mice that noted enhanced susceptibility to respiratory infection by *K. pneumoniae* following 3 months of constant exposure at 0.5 ppm (Ehrlich and Henry, 1968). Since these studies were conducted during the late 1960s, they were incorporated into the EPA's Air Quality Criteria (1971a) for nitrogen oxides and played a significant role in the adoption of the present NO_2 standard.

Carbon Monoxide

The ambient air quality standards for carbon monoxide (CO) were designed to "protect against the occurrence of carboxyhemoglobin (COHb) levels above 2 percent." At the time of the promulgation of the CO standards the data were such that the primary basis for the CO standards rested on a short-term human study by Beard and Wertheim (1967), which indicated impaired time interval discrimination after 90 minutes exposure to 50 ppm of CO (this is the equivalent of approximately 2.5% COHb). Another study by Beard and Wertheim (1969) revealed that at approximately 5% COHb, impairment in the performance of several psychomotor tests and in visual discrimination occurred. These studies were not based on high-risk groups per se. However, studies with coronary heart disease patients experiencing from 5 to 9% COHb revealed a lack of capacity to develop a compensatory increase in coronary blood flow during a time of increased myocardial oxygen requirement in contrast to patients with noncoronary heart disease, who demonstrated a compensatory increase in coronary blood flow following the increase in myocardial oxygen requirements (Ayres et al, 1965). The 1971 criteria document that served as the data base from which the CO standard was derived concluded "that persons with certain forms of heart disease may be particularly susceptible to exposures of CO that could lead to blood COHb levels in excess of 5 percent."

Thus it can be seen that the primary basis of the CO standard was not the protection of high-risk groups, but actually the prevention of impairment to psychomotor performance in normal subjects (see the Beard and Wertheim study, 1967). This was not because the EPA ignored the high-risk groups, but because the impaired psychomotor performance was demonstrated to occur at COHb levels of about 2.5%, whereas the first sign of functional impairment in patients with coronary artery disease was reported at approximately 5% COHb. Consequently, since the standard was designed to prevent COHb levels above 2% COHb, protection would be apparently offered to all the known risk groups.

Since the promulgation of the original CO standard, the validity of the Beard and Wertheim (1967) study has been challenged by Heuss et al. (1971) because (1) it was a single-blind study rather than the methodologically more proper double-blind study, and (2) other studies (Stewart et al., 1970; Theodore et al., 1971) have not supported the original Beard and Wertheim (1967) report.

However, despite the fact that the original Beard and Wertheim (1967) study has been challenged, evidence has continued to emerge that CO may be harmful at concentrations much lower than originally thought. Stewart (1976) reported that every molecule of CO that enters the body displaces a molecule of O_2, thereby diminishing the O_2-carrying capacity of the blood. He reasoned that there is no dose of CO that is without an effect on the body. Whether this effect results in any clinical response is of course dependent on the dose of CO and the state of health of the exposed individual. The body usually adapts, as implied earlier, by increasing cardiac output or by increasing blood flow to a specific organ. Adaptive responses as low as 2–3% COHb in the blood have been measured. If the capacity to adapt is exceeded, tissue hypoxia ensues.

Research subsequent to the 1971 CO standard has verified the earlier data of Ayres et al. (1965, 1970). Aronow et al. (1972), Aronow and Isbell (1973), and Anderson et al. (1973) have shown that patients with advanced coronary artery disease and angina pectoris experience a significant decrease in exercise tolerance after exposure to low concentrations of CO (i.e., enough to increase the COHb saturation to 5%). Furthermore, Radford (1976) reported that detectable effects can be measured on susceptible individuals when COHb levels are as low as 2.8%. According to Stewart (1976), this apparent 2.8% COHb "threshold" may be entirely due to the fact that the methodology to monitor effects below such a level does not yet exist. In fact, Stewart suggested that there may be no level of CO exposure that does not exert a significant stress on patients with advanced cardiovascular disease. Thus, during the past decade, the effects of CO on high-risk groups has become much more established. Any official reevaluation of CO by the EPA would be expected to place a higher priority on the effects of CO on people with preexisting coronary disease. According to Stewart, the major high-risk group is not really those who have already had heart attacks, but males over 45 years of age who have preexisting coronary disease that may be severe and who are unaware of it. This may include upwards of 25% of the males in this age group.

Ozone

Epidemiologic studies on high-risk groups have played an important role in the derivation of the ambient oxidant standard. Schoettlin and Landau (1961) studied 137 asthmatics in the Los Angeles basin during a 3-month period when high oxidant concentrations were anticipated. The results indicated a statistically significant increase in the number of mild attacks that occurred as the peak oxidant levels exceeded 0.25 ppm. More specifically, based on 3435 separate attacks noted, approximately 5% were related to smog by the patients. Heuss et al. (1971), commenting on the Air Quality Criteria Document for Oxidants (1971), related this adverse health effect with hourly average concentrations as low as 0.15 ppm by using the 99th percentile of the relation between instantaneous and hourly average concentrations. Thus, when the oxidant level is 0.15 ppm, there is a 1% chance of a 5% increase in asthmatic attacks. Barth et al. (1971) extrapolated this value in a more conservative manner and indicated that there "is a likelihood of an increased asthmatic attack incidence for very sensitive patients at levels well below 0.15 ppm rather than just a chance of a small increase in attacks at the 0.15 ppm level." Be that as it may, with the exception of the study of Schoettlin and Landau (1961), dose-response relationships of oxidant pollution on high-risk groups such as those with chronic respiratory illness have not been adequately developed (Barth et al., 1971). Several other studies, however, have been directed toward the characterization of the effects of ozone on people with bronchitis and those with obstructive pulmonary disease.

Schoettlin (1962) investigated the effects of air pollution on respiratory symptoms and/or function in groups of bronchitic men. Despite the fact that statistically significant differences between the groups were not noted, oxidant and oxidant precursor values consistently explained considerably more of the variation in the frequency of respiratory symptoms in the bronchitics than in the group of normal controls. Also, Balchum (1973) indicated that patients with obstructive pulmonary disease noticed that as oxidants in normal urban (Los Angeles) air were removed by filtering, difficulty in breathing was significantly lessened. Thus these two studies, although providing support for the hypothesis that individuals with respiratory disease should be considered at high risk with respect to ozone toxicity, do not substantially add to the development of an accurate dose-response relationship.

Since the promulgation of the oxidant standard in 1971, certain research has been directed to the potential systemic effects of ozone. For example, ozone has been found to be a mutagen in human lymphocytes during in vivo studies at 0.5 ppm (Merz et al., 1975). Ozone is now also known to adversely affect the integrity of the red blood membrane via a fatty acid ozonide (Menzel, 1976).

Calabrese et al. (1977) and Calabrese (1978) presented a theoretical study that indicates that individuals with a G-6-PD deficiency should be considered at increased risk with respect to the development of acute hemolysis following a 0.4–0.5 ppm O_3 exposure for approximately 2¾ hours. Since 13% of the Black male population is known to possess the G-6-PD deficiency trait, this potential high-risk group represents a rather large number of individuals who live in large cities known to have elevated levels of ozone.

What role should such theoretical evidence play in the standard-setting process? This question became a practical reality in 1976 during hearings held by the Illinois Pollution Control Board for a statewide ambient ozone standard.

In testimony offered at the Illinois hearings concerning the effects of ozone on human health, perhaps the most controversial data concerned the relationship of the enhanced susceptibility of G-6-PD-deficient individuals to ozone toxicity. Although the data did not have a significant influence on the promulgation of the 0.07 ppm state standard, it did provide a difficult situation for decision makers with respect to the derivation of subsequent episode standards (Calabrese, 1976). Table 3-2 lists the relationship of ozone to human health effects and Illinois' Environmental Health Resource Center's recommended alert and warning system levels for ozone. A major significance of the subsequent ozone standard concerned its associated implementation plans; that is, as each ozone episode level is reached, a series of implementation procedures (e.g., closing parking lots with greater than 200 parking spaces, restricting incineration activity, closing state and federal buildings, closing O'Hare International Airport, etc.) may come into effect. Thus the proposed ozone episode standards were highly controversial, not so much for the actual concentrations, but for what the Illinois EPA was prepared to do to prevent higher levels from being realized. It should be noted that the Illinois Pollution Control Board, after due consideration, decided that it could not give the same weight to a theoretical study as to one based on actual data. Consequently, the data presented above did not significantly affect the actual derivation of the Illinois ozone standard (Farley, 1976, personal communication). One can certainly sympathize with the decision maker who is trying to determine a "reasonably safe" level of a pollutant to which the general public can be permitted exposure. However, the issue of the role theoretical studies may play in environmental decision making should not be dismissed easily. The theoretical studies that have predicted the depletion of the stratospheric ozone layer as a result of both fluo-

Table 3-2. Relationship of Ozone and Photochemical Oxidant Exposure to Human Health Effects and a Recommended Alert and Warning System Level

Recommended Episode Levels	Ozone/photo-chemical oxidants (ppm)	Duration of Exposure	Health Effects
	0.70	2.0 hours	Soreness of upper respiratory tract, tendency to cough while taking deep breaths, significant increase in breathing difficulty. These conditions were made worse by 15 minutes of light exercise.
Emergency	0.50[a]	2.75 hours	Measurable biochemical changes in blood sera enzyme levels and red blood cell membrane integrity; some subjects became physically ill and unable to perform normal jobs for several hours.
	0.37	2.0 hours	Impairment of pulmonary function in young adults, probably due to a decreased lung elastic recoil; increased airway resistance and small airway obstruction.
	0.37[a]	2.75 hours	Significant biochemical changes in blood sera enzyme levels and red blood cell membrane integrity, but less severe than at 0.50 ppm; some subjects became physically ill and unable to perform normal jobs for several hours.
Red alert	0.30		A precipitous increase in the rates of cough and chest discomfort in young adults.

Table 3-2. *(Continued)*

Recommended Episode Levels	Ozone/photo-chemical oxidants (ppm)	Duration of Exposure	Health Effects
	0.25		Greater number of asthma attacks in patients on days when daily maxima equaled or exceeded 0.25 ppm during a 14-week period.
	0.25[a]	2.75 hours	Biochemical changes in blood sera enzyme levels.
Yellow alert	0.10	1 hour	Breathing impaired.
	0.10		Tokyo elementary school children had significantly reduced respiratory function associated with ozone levels less than 0.1 ppm during a long-term epidemiologic study. Beginning of headache without fever in young adults—median age 18.6 years.
Watch	0.07	2 hour average	
	0.065		Impairment of performance of student athletes during running competition.
	0.05	15–30 min	Threshold of respiratory irritation.
	0.02		Odor perception.
	0.005		Decreased electrical activity of the brain.

[a] These values were derived by standard EPA analytic procedures. They exceed the "absolute" value of ozone determined by the ultraviolet photometer method by approximately 25%. Thus the stated values may be approximately 25% lower than reported here. For example, 0.50 ppm would be approximately 0.40 ppm, 0.37 ppm would be approximately 0.30 ppm, and 0.25 ppm would be approximately 0.19 ppm.

Source: EHRC (1975) *Health Effects and Recommended Alert and Warning System for Ozone.* Chicago: Illinois Institute for Environmental Quality.

rocarbons and supersonic aircraft are certainly examples of how "unproven" yet highly plausible theories may project themselves into highly sensitive health-economic issues. The major problem is, what regulation or standard (if any) should be set in the interim between the initial explanation of the theory and the actual testing and verification of the theory? Since each theory must be viewed on its individual merits, it appears that the only reasonable way to deal with this issue is professional judgment. The option should be available for the decision makers to act forthrightly if the risk is judged as dangerous, yet they should also be free to reject the theory until its verification.

Another potential high-risk group with respect to oxidant toxicity are those with a dietary deficiency of vitamin E. Animal studies using rats have demonstrated that vitamin E deficiency increases the susceptibility to ozone toxicity, whereas vitamin E supplementation decreases ozone toxicity. Furthermore, the preventive action of supplemental vitamin E has been demonstrated in rats at levels of ozone that are often exceeded on a smoggy day in southern California (Shakman, 1974). In addition, recent studies by Menzel (1976) have verified the protective effect of vitamin E on humans exposed to elevated levels of ozone. Although the present oxidant standard does not take nutritional status into account, it would appear that future reevaluations of the standard should consider the extent of vitamin E deficiency in the population and the degree to which it may enhance ozone toxicity. In light of surveys that have indicated that 7% of the general population of the United States has a "functional" deficiency of vitamin E (Harris, 1961), it would seem that this pollutant–nutrient deficiency association may be more serious than previously thought.

Sulfur Dioxide

There has been considerable research conducted on the effects of sulfur dioxide on animal and human health. The Air Quality Criteria Document for Sulfur Oxides (EPA, 1971b), itself a summary of research efforts up to that time, contains more than 150 pages and approximately 400 references. Since 1971, considerably more research has focused on the sulfur oxides (note the CHESS studies), especially in light of the problem of sulfur acid mist emissions from automobiles equipped with catalytic converters. However, despite this vast array of literature concerning the effects of sulfur dioxide on human health, it is possible to focus on several significant studies that played a major role in the derivation of the present ambient air standard for sulfur dioxide and their relationship to the concept of high-risk groups.

Since sulfur dioxide is considered a respiratory irritant, much of the relevant research in this area has focused on those segments of the population susceptible to respiratory disease. Such hypersusceptible segments include young children, smokers, the aged, and people with asthma, bronchitis, and chronic bronchopulmonary diseases.

According to Nelson et al. (1974), children are an excellent group with which to study the influence of air pollutants on the respiratory system. The young are known to be very susceptible to acute respiratory disease, and they are a study group with considerably fewer confounding variables (i.e., minimum cigarette smoking, no additional industrial exposures, etc.) than adults. Several British epidemiologic studies of children have revealed a direct correlation between the degree of particulate and sulfur dioxide air pollution and the frequency of respiratory disease. Furthermore, not only the frequency, but also the severity of acute respiratory tract infections showed increases directly associated with increased ambient concentrations of sulfur dioxide and particulates (Douglas and Waller, 1966; Lunn et al., 1967; Holland et al., 1969). Corroborating the British studies are reports of similar studies on children in other foreign countries including Japan (Toyama, 1964) and the Soviet Union (Manzhenko, 1966). For example, school children aged 10–11 who resided in a rather heavily polluted area of Japan experienced an elevated incidence of nonproductive cough, irritation of the upper respiratory tract, and enhanced mucus secretion compared to matched children in a less polluted area (Toyama, 1964). Although these studies were highly consistent, it has been pointed out by Nelson et al. (1974) that they are not capable of distinguishing the effects of sulfur dioxide from those of the particulate. In studies of children living in communities of the Salt Lake basin, there was a highly positive association of sulfur dioxide and/or suspended sulfate levels with the incidence of singular or repeated cases of acute lower respiratory disease. Significant morbidity (croup and bronchitis) excesses occurred following only 1 year of exposure to the increased levels of sulfur dioxide (annual average levels of 91 mg/m^3), along with reported increased morbidity following prolonged exposure (5–9 years) to annual average suspended sulfate concentration of 9 $\mu g/m^3$ in the absence of increased levels of other air pollutants. The authors suggested that a decrease in the present ambient sulfur dioxide levels as required by the primary ambient air quality standard as well as a concomitant decrease in suspended sulfates could significantly decrease acute lower respiratory illness in children exposed to elevated pollution levels (Nelson et al., 1974).

It has long been known that one of the major high-risk groups with respect to respiratory irritants is asthmatics (Gersh, 1967; Schoettlin and Landau, 1961; Zeidberg et al., 1961; Yoshida et al., 1964). Despite the large number of qualitative-type studies concerning the adverse effects of sulfur dioxide on the health of asthmatics, such studies have not provided the precise dose-response relationship on which a truly adequate sulfur dioxide standard must be based. To provide more quantitative data to assist in the development of a more adequate dose-response relationship, Finklea et al. (1974a) conducted a diary study for 6 months with asthmatics residing in Utah. The results indicated that the occurrence of increased asthmatic episodes was associated with 24-hour exposures to somewhat elevated concentrations of suspended particulate matter (71 $\mu g/m^3$) when the temperature equaled or exceeded 50°F. The threshold level for sulfur dioxide ranged from 23 to 54 $\mu g/m^3$ when the temperature equaled or exceeded 40°F. The threshold for the irritation of the asthmatic response via the effects of suspended sulfates was 1.4 $\mu g/m^3$ on days when the temperature equaled or exceeded 50°F. It should be noted that the thresholds of response were shifted upward as the temperature decreased.

Finklea et al. inferred from their data that the present primary air quality standards will most likely not be able to protect asthmatics from attacks induced by air pollution. Their results suggested that greater than normal asthmatic attacks associated with sulfur dioxide may occur on 5–10% of summer days; however, greater than normal asthmatic attacks due to elevated suspended sulfates may occur on 90% of summer days. Excess asthmatic attacks associated with elevated total suspended particulates would occur on up to 5% of summer days.

The enhanced susceptibility of chronic bronchitics to varying levels of sulfur dioxide was reported for more than 500 patients by Carnow et al. (1969, 1970). The patients were separated into groups based on the severity of the disease condition. Patients who were 55 years of age or older and had severe bronchitis experienced a significant relationship between the level of sulfur dioxide and person-days of illness. The data indicated that, as the sulfur dioxide levels increased from 0.25 to 0.30 ppm, a highly significant increase in the illness rate occurred. In contrast, for the patients above 55 years of age with mild bronchitis, illness was not associated with air pollutant levels. Carnow et al. (1969) concluded that the elderly individuals with severe chronic bronchitis are clearly a hypersusceptible subgroup of the population with respect to the development of sulfur dioxide induced respiratory symptoms.

REFERENCES

Anderson, E. W., Andelman, R. J., Strauch, J. M., Fortuin, N. J. and Knelson, J. H. (1973) Effect of low level carbon monoxide exposure on onset and duration of angina pectoris. *Ann Intern Med* 79:46–50.

Aronow, W. S. et al. (1972) Effect of freeway travel on angina pectoris. *Ann Intern Med* 77:669–676.

Aronow, W. S. and Isbell, M. W. (1973) Carbon monoxide effect on exercise-induced angina pectoris. *Ann Intern Med* 79:392–395.

Ashford, N. A. (1976) *Crisis in the Workplace: Occupational Diseases and Injury,* pp. 69–92. Cambridge, Mass.: MIT Press.

Ayres, S. M., Gianelli, S., Jr., and Armstrong, R. G. (1965). Carboxyhemoglobin: hemodynamic and respiratory responses to small concentrations. *Science* 149(3680): 193–194.

Ayres, S. M., Giannelli, S., Jr., and Mueller, H. (1970) Myocardial and systemic responses to carboxyhemoglobin. *Ann NY Acad Sci* 174:268–293.

Balchum, O. J. (1973) Toxicological effects of ozone, oxidant, and hydrocarbons. U.S. Government Proceedings of the Conference on Health Effects of Air Pollution, Serial #93-15, pp. 489–503. Prepared for U.S. Senate, Committee on Public Works.

Barth, D. S., Romanovsky, J. C., Knelson, J. H., Altshuller, A. P., and Horton, R. J. M. (1971) Discussion of the article by Heuss et al. JAPCA 21(9): 535–544. *J Air Pollution Control Assoc* 21(9):544–548.

Beard, R. R. and Wertheim, G. A. (1967) Behavioral impairment associated with small doses of carbon monoxide. *Am J. Publ Health* 55:2012–2022.

Beard, R. R. and Wertheim, G. A. (1969) Behavioral manifestations of carbon monoxide absorption. Presented at 16th Int. Congr. Occupational Health, Tokyo.

Beutler, E. (1972) Glucose-6-phosphate dehydrogenase deficiency. In Standbury, J. B., Wyngaarden, J. B. and Fredrickson, D. B., eds. *The Metabolic Basis of Inherited Disease,* pp. 1358–1388. New York: McGraw Hill.

Bogert, L. J., Briggs, G. M., and Calloway, D. H. (1973) *Nutrition and Physical Fittness,* 9th ed. Philadelphia: Saunders.

Buckley, R. D., Hackney, J. D., Clark, K., and Posin, C. (1975) Ozone and human blood. *Arch Environ Health* 30:40.

Calabrese, E. J. (1976) Testimony at State of Illinois Pollution Control Board Hearings on ozone episode regulations. Chicago, Illinois.

Calabrese, E. J. (1978) *Pollutants and High Risk Groups,* p. 266. New York Wiley-Interscience.

Calabrese, E. J., Kojola, W. and Carnow, B. W. (1977) Ozone: a possible cause of hemolytic anemia in glucose-6-phosphate dehydrogenase deficiency. *J Toxicol Environ Health* 2:709–712.

Carnow, B. W. (1966) Air pollution and respiratory diseases. *Scientist and Citizen,* May, p. 1.

Carnow, B. W. (1970) Relationship of SO_2 levels to morbidity and mortality in (high risk) populations. Air Pollution Medical Research Conference. Oct. 5, 1970, New Orleans.

Carnow, B. W. (1976) Panel discussion on TLV's-Lead. In B. W. Carnow, ed., *Health*

Effects of Occupational Lead and Arsenic Exposure: A Symposium, p. 197. U.S. HEW, PHS. NIOSH.

Carnow, B. W. and Carnow, V. (1974) Air pollution, morbidity, and mortality and the concept of no threshold. In J. W. Pitts and R. L. Metcalf, eds, *Advances of Environmental Sciences and Technology*, vol. 3, pp. 127–156. New York: Wiley.

Carnow, B. W., Lepper, M. H., Schekelle, R. B., and Stamler, J. (1969) Chicago air pollution study: SO_2 levels and acute illness in patients with chronic bronchopulmonary diseases. *Arch Environ Health* 18:768–776.

CHESS (1974) A Report from CHESS, 1970–1971: Health consequences of sulfur oxides. Washington, D.C.: U.S. EPA.

Cooper, W. C. (1973) Indicators of susceptibility to industrial chemicals. *J Occup Med* 15(4):355.

Criteria Document for Photochemical Oxidants (1971) U.S. Dept. HEW, PHS; Environmental Health Service, National Air Pollution Control Administration, Washington, DC.

Douglas, J. W. B. and Waller, R. E. (1966) Air pollution and respiratory infection in children. *Br J Prevent Soc Med* 20:1–8.

EHRC (1975) Health effects and recommended alert and warning systems for ozone. Chicago: Illinois Institute for Environmental Quality.

EHRC (1977) Recommended ambient air standard for total mercury. Environmental Health Resource Center. Chicago: Illinois Institute for Environmental Quality.

Ehrlich, R. and Henry, M. C. (1968) Chronic toxicity of nitrogen dioxide. I. Effects on resistance to bacterial pneumonia. *Arch Environ Health* 17:860–865.

EPA (1971a) Air quality criteria for nitrogen oxides. Environmental Protection Agency. Washington, D.C.: Air Pollution Control Office.

EPA (1971b) Air quality criteria for sulfur dioxide. Washington, D.C.: Environmental Protection Agency.

Farley, D. (June, 1976). Illinois Pollution Control Board. Chicago, Illinois. Personal Communication.

Finklea, J. F., Calafiore, D. C., Nelson, C. J., Riggan, W. B., and Hayes, C. G. (1974a) Aggravation of asthma by air pollutants: 1971 Salt Lake Basin studies. In *Health Consequences of Sulfur Oxides: A Report for CHESS*, 1970–1971. U.S. E.P.A. Office of Research and Development, National Environmental Research Center, Research Triangle Park, NC.

Finklea, J. F., Shy, C. M., Moran, S. B., Nelson, W. C., Larsen, R. I., and Akland, G. G. (1974b) The role of environmental health assessment in the control of air pollution. In J. N. Pitts and R. L. Metcalf, eds, *Advances in Environmental Science and Technology*, vol. 7, pp. 315–389. New York: Wiley.

Gersh, L. S., Shubin, E., Dick, C. and Schulaner, F. A. (1967) A study of the epidemiology of asthma in children in Philadelphia. *J Allergy* 39:347–357.

Harris, P. L., Hardenbrook, E. G., Dean, F. P., Cusack, E. R., and Jensen, J. L. (1961) Blood tocopherol values in normal adults and incidence of vitamin E deficiency. *Proc Soc Exp Biol Med* 107:381.

Heuss, J. M., Nebel, G. J., and Colucci, J. M. (1971) National air quality standards for automotive pollutants—A critical review. *J Air Pollut Control Assoc* 21(9):535–544.

Holland, W. W., Halil, T., Bennett, A. E., and Elliot, A. (1969) Factors influencing the onset of chronic respiratory disease. *Br Med J* 2:205–208.

Kotin, P. (Oct. 7, 1977) Hypersusceptibility—Role in worker selection. Presentation to Joint Conference on Occupational Health, sponsored by American Academy of Occupational Medicine. Denver, Colorado.

Lunn, J. E., Knowelden, J., and Handyside, A. J. (1967) Patterns of respiratory illness in Sheffield infant school children. *Br J Prevent Soc Med* 21:7–16.

Mahaffey, K. R. (1974) Nutritional factors and susceptibility to lead toxicity. *Environ Health Perspect Exp* Issue No. 7, p. 107.

Manzhenko, E. G. (1966) The effect of atmospheric pollution on the health of children. *Hyg Sanit* (Moscow):31:126–128.

Menden, E. E., Elia, V. J., Michael, L. W., and Petering, H. G. (1973) Distribution of cadmium and nickel of tobacco during cigarette smoking. *Environ Sci Technol* 6(9):830–832.

Menzel, D. B. (1976) Oxidants and human health. *J Occup Med* 18(5):342–345.

Merz, T., Bender, M. A., Kerv, H. D., and Kuller, T. J. (1975) Observations of aberrations in chromosomes of lymphocytes from human subjects exposed to ozone at a concentration of 0.5 ppm for 6 hours and 10 hours. *Mut Res* 31:299.

Nelson, W. C., Finklea, J. F., House, D. E., Calafiore, D. C., Hertz, M. B., and Swanson, D. H. (1974) Frequency of acute lower respiratory disease in children: retrospective survey of Salt Lake Basin communities, 1967–1970. In *Health Consequences of Sulfur Oxides: A Report for CHESS, 1970–1971.* U.S. E.P.A., Office of Research and Development, National Environmental Research Center, Research Triangle Park, NC.

O'Donnell, R., Mikulka, P., Heinig, P., and Theodore, J. (1971) Low level carbon monoxide exposure and human performance. *Toxicol Appl Pharmacol* 18:593.

OSHAct of 1970. Public Law 91-596.

Page, J. and O'Brien, M. V. (1973) *Bitter Wages.* New York: Grossman.

Pearlman, M. E., Finklea, J. F., Creason, J. P., Shy, C. M., Young, M. M., and Horton, R. J. M. (1971) Nitrogen dioxide and lower respiratory tract illness. *Pediatrics* 47(2):391–398.

Radford, E. P. (1976) Carbon monoxide and human health. *J. Occup Med* 18(5):310–315.

Rothman, K. J. (1975) Alcohol. In J. F. Fraumeni, Jr., ed., *Persons At High Risk of Cancer: An Approach To Cancer Etiology and Control,* pp. 139–150. New York: Academic.

Schoettlin, C. E. (1962) The health effects of air pollution on elderly males. *Am Rev Resp Dis* 86:878.

Schoettlin, C. E. and Landau, E. (1961) Air pollution and asthmatic attacks in the Los Angeles area. *Public Health Rep* 76:545–548.

Shakman, R. A. (1974) Nutritional influences on the toxicity of environmental pollutants. *Arch Environ Health* 28:105.

Shy, C. M., Creason, J. P., Pearlman, M. E., McClain, K. E., Benson, F. B., and Young, M. M. (1970a) The Chattanooga school children study: effects of community exposure to nitrogen dioxide. I. Methods, description of pollutant exposure and results of ventilatory function testing. *J Air Pollut Control Assoc* 20(8):539–545.

Shy, C. M., Creason, J. P., Pearlman, M. E., McClain, K. E., Benson, F. B., and Young, M. M. (1970b) The Chattanooga school children study: effects of community exposure to nitrogen dioxide. II. Incidence of acute respiratory illness. *J Air Pollut Control Assoc* 20(9):582–588.

Standbury, J. B., Wyngaarden, J. B., and Fredrickson, D. S., eds. (1972) *The Metabolic Basis of Inherited Disease.* New York: McGraw-Hill.

Stewart, R. D. (1976) The effect of carbon monoxide on humans. *J Occup Med* 18(5):304–309.

Stewart, R. D., Peterson, J. E., Baretta, E. D. et al. (1970) Experimental human exposure to high concentrations of carbon monoxide. *Arch Environ Health* 21:154–164.

Stokinger, H. E. and Mountain, J. T. (1963) Test for hypersusceptibility to hemolytic chemicals. *Arch Environ Health* 6:57.

Stokinger, H. E. and Scheel, L. D. (1973) Hypersusceptibility and genetic problems in occupational medicine—A consensus report. *J Occup Med* 15:564–573.

Theodore, J., O'Donnell, R., and Back, K. C. (1971) Toxicological evaluation on carbon monoxide in humans and other mammalian species. *J Occup Med* 13:242.

Toyama, T. (1964) Air pollution and its health effects in Japan. *Arch Environ Health* 8:153–173.

Wadden, R. A., Farley, D. O., and Carnow, B. W. (1976) Transportation emissions and environmental health: an evaluatory planning methodology. *Environ Planning* A 8:3–21.

Yoshida, K., Oshima, H., and Swa, M. (1964) Air pollution and asthma in Yokkaichi. *Arch Environ Health* 13:763–768.

Zeidberg, L. D., Prindle, R. A., and Landau, E. (1961) The Nashville air pollution study; I. Sulfur dioxide and bronchial asthma: a preliminary study. *Am Rev Resp Dis* 84:489–503.

4

Chemical Interactions in Standard Derivation

ONE OF THE MOST IMPORTANT and extremely difficult factors to deal with in the development of environmental and occupational health standards is chemical interactions and how they affect toxicity. Air and water are known to have multiple impurities constantly present. Consequently, it has been long recognized that environmental and industrial health practices should take into consideration the presence of mixtures (Brieger, 1957; Stokinger, 1960). Yet all the United States national ambient air quality standards do not specifically incorporate the concept of synergism or additivity. However, the National Institute for Occupational Safety and Health (NIOSH) (1976) has considered the phenomenon of additivity on a very limited basis in recommending a standard to the Occupational Safety and Health Administration (OSHA) for methylene chloride. This chapter considers the occurrence of chemical interactions as factors affecting toxicity and their potential role in standard development. Special consideration is directed toward the implementation of chemical interactions in the derivation of American Conference of Governmental Industrial Hygienists (ACGIH) TLVs and in Soviet standards.

HISTORY AND THEORY

To establish a framework on which subsequent discussion may be based, it is necessary to define several terms that describe the principal types of chemical interactions relevant to toxicologic studies according to Shy et al. (1974).

Chemical Interactions

Synergism. The effect produced by combination is greater than the sum of the effects of the individual components.

Additive reaction (additivity). The effect produced by combination is equal to the sum of the effects of individual components.

Indifference. The effect produced by combination equals the effect of the single most active component; other components do not add, enhance, or diminish the effect of the most active component.

Antagonism. The effect produced by combination is less than the effect of the single most active component.

Tolerance. An increased capacity of the host to resist the effects of subsequent acute exposures to the same agent or different agents (cross tolerance). Thus tolerance has the quality of persistence for varying periods of time, depending on the degree of development of the tolerance. This differentiates tolerance from antagonism which may result from the simultaneous and nonpersistent interaction of two or more agents in the body.

Table 4-1 summarizes many of the studies that have been directed toward trying to elucidate the influence of the various types of chemical interactions on animal models and humans. Although this table is not claimed as an exhaustive survey of the literature on chemical interaction, it does represent a fairly comprehensive summation of the types of research attempted in this area. Attempts to track down all articles in this research area are, at times, quite difficult, because many subject titles are often devoid of words that would even suggest the occurrence of chemical interaction.

The interaction of chemical pollutants has been noted for over 30 years, with some early research findings by Von Oettingen (1944), Hough et al. (1944), Smith and Mayers (1944), Quadland (1943a, 1943b, 1944), Parmeggiani and Sassi (1954), and McGowan (1955). Several attempts to summarize the early published literature were undertaken by Brieger (1957), Ball (1959), and Stokinger (1960). Perhaps the most important factor that helped focus research efforts on the nature of chemical interactions and their effects on health evolved from the pollutant episode disasters at Meuse Valley, Belgium in 1930 and especially London in 1952. Since the average concentrations of individual pollutants during these episodes were not considered to be especially dangerous, it has been hypothesized that the adverse health effects may have resulted from an interaction of a variety of toxic agents present. More specifically, since the average sulfur dioxide concentration was only 1.7 ppm (the TLV for sulfur dioxide is 5 ppm) during the 1952 London episode, it was suggested that other contributing factors, acting according to some undefined interaction with sulfur dioxide, may have been the cause of the widespread respiratory disorders. Subsequent research (Kellogg et al., 1972; Amdur and Underhill, 1968; Amdur, 1959; Toyama and Nakamura, 1964; and Frank et al., 1966) has established the occurrence of an interaction of sulfur dioxide with a variety of atmospheric aerosals (e.g., soluble salts of iron, manganese, and vanadium) which could contribute to increased respiratory irritation. One of the difficulties in this area is the immensity of the task. The number of possible chemical interactions, from a mathematical probability perspective, almost defies the physical (and financial) capabilities of comprehensively testing each possible chemical interaction.

Finney (1952) developed a theoretical mathematical approach for predicting the degree of toxicity derived from various types of chemical interactions. Pozzani et al. (1959) indicated that only two of the 36 pairs of mixtures of industrial vapors tested for acute toxicity in rats deviated by greater than 1.96 SE from the calculations of Finney's theoretical approach for additive joint toxicity. According to Smyth et al. (1969), the study by Pozzani et al. supported the hypothesis that the acute toxicity of chemical mixtures randomly chosen has a high likelihood of being accurately predicted by Finney's theoretical formula for additive joint toxicity. In an attempt "to evaluate the overall confidence that can be placed on the prediction of the joint toxicity of many chemical pairs," Smyth et al. studied the toxicity of 27 industrial chemicals in all possible pairs to rats. Their results were consistent with the prediction of Finney that most interactions should be considered as additive until proven otherwise.

Based on Finney's mathematical model for joint toxicity, the following equation was employed to predict the toxic concentrations:

$$1/\text{predicted LD}_{50} = P_a/\text{LD}_{50} \text{ of component } A + \text{P}_b\text{LD}_{50} \text{ of component } B$$

P_a and P_b are the proportions of component A and component B in the

Table 4-1. Chemical Interactions of Gases, Heavy Metals, and Organic Compounds: Synergism, Additivity, Indifference, Tolerance, Cross Tolerance, and Antagonism

I. Chemical Interactions of Gaseous Pollutants	
A. Synergism	
1. SO_2 and H_2SO_4	H_2SO_4 mist (8 mg/cu m) and SO_2 (89 ppm) in combination lead to enhanced slowing of growth, damage to the lung, and enhanced impairment of respiration. (Amdur, 1959)
2. SO_2 and H_2O_2	Produced a synergistic effect when concentrations used included H_2O_2 at 0.29 mg/cu m for 5 minutes with particle size of 4.7 μ, but when the particle size increased to 11.8 μ and all other factors remained constant, no synergism was noted. (Toyama and Nakamura, 1964)
3. SO_2 and NaCl aerosol	As in the case of the SO_2 and H_2O_2 interaction, there is a synergistic interaction between SO_2 and NaCl, depending in part on the diameter of the NaCl. Experiments designed to determine the occurrence of an SO_2–NaCl synergistic interaction in humans have not been successful. (Frank et al., 1966)
4. SO_2 and other aerosols	Sodium chloride, potassium chloride, and ammonium thiocyanate aerosols potentiated the adverse respiratory effects of SO_2; furthermore, soluble salts of manganese, ferric iron, and vanadium were more effective than sodium chloride in potentiating the adverse effect of SO_2 on respiratory function (pulmonary flow resistance); insoluble aerosol salts were not able to cause a synergistic effect with SO_2 exposure. (Amdur, 1960; Amdur and Underhill, 1968; Shy et al., 1974)
5. SO_2 and arsenic trioxide	Enhanced the occurrence of respiratory cancers among male smelter workers; SO_2 is thought to act as a cocarcinogen in this case. (Landau, 1977)
6. SO_2 and manganese dioxide aerosol	SO_2 (5.0 ppm) and MnO_2 (5.9 mg/cu m) in combination caused a synergistic toxic

Table 4-1. *(Continued)*

	effect on the pulmonary clearance of nonpathogenic bacteria. (Rylander et al., 1971)
7. SO_2, CO, phenol, and dust	A mixture of SO_2, CO, phenol, and dust at 0.05 mg/cu m, 1 mg/cu m, 0.01 mg/cu m, and 0.15 mg/cu m, respectively, resulted in decreased red blood cell counts, decreased ACHase activity and lower levels of nucleic acids in whole blood; however, increases in catalase activity and cell permeability in the spleen, liver, kidney, and ovaries also occurred. (Yelfimora et al., 1972)
8. SO_2 and O_3	Human subjects exposed to SO_2 (0.37 ppm) and O_3 (0.37 ppm) exhibited a marked reduction (20–40%) in respiratory function as compared to a 0% decrease for SO_2 alone and a 10% reduction with only O_3 exposure. It took O_3 2 hr to cause its reduction; in contrast, when O_3 and SO_2 were combined, the reduction was noted in only 30 minutes. (Bates and Hazucha, 1973)
9. SO_2, aerosols, and relative humidity	When the relative humidity exceeded 80%, SO_2 (2.86 mg/cu m) and NaCl (1 mg/cu m) exhibited a synergistic toxic effect on pulmonary flow resistance in guinea pigs. The synergistic response did not occur at 40% relative humidity. (McJilton et al., 1973)
10. NO_2 and CO	All experimental mice (10 of 10) exposed to 0.5–0.8 ppm NO_2 and 50 ppm CO developed epithelial hyperplasia of the terminal bronchioles; with NO_2 exposure only 60% (6 of 10) of the mice developed the hyperplasia; at 50 ppm CO alone, no hyperplasia occurred; when the NO_2 concentration was decreased to 0.2–0.5 ppm and the CO level remained at 50 ppm, no hyperplasia developed. (Nakajama et al., 1972)
11. NO_2 and NaCl aerosol	The synergistic effect (decreased lung function in healthy young men) of the gas-aerosol mixture occurred when the

Table 4-1. *(Continued)*

	particle diameter was 0.95 mm; no synergism occurred when the particle size was 0.22 μm in diameter. (Nakamura, 1964)
12. Air pollutants (CO, O_3, NO_2, NO) and microbial infections	Mice exposed for 4 hours to CO (100 ppm), O_3 (0.35–0.67 ppm), NO_2 (0.5–1 ppm), and NO (0.03–1.96 ppm) had a fivefold increase in mortality as compared to controls. (Coffin and Blommer, 1965, 1967; Coffin et al., 1968)
13. Artificial smog and influenza viral infections	Squamous cell carcinoma of the lung occurred only in the treatment group continuously exposed to a combination of smog and the three virus intubations. (NAS, 1976; Kotin, 1966)
14. O_3 and H_2O_2	Lethal responses in animal models occurred when H_2O_2 was 1.5 ppm and O_3 was 1 ppm; however, when comparable animals were exposed to only 200 ppm of H_2O_2, a minor toxic effect was noted. (Svirbely et al., 1961)
15. O_3 and CO_2	Increased rate of respiration in guinea pigs. (Mittler et al., 1957)
16. NO and CO	Found to be highly toxic (lethal) to cats when given in combination. (Stokinger, 1962)
17. O_3 and NO_2	At low doses a synergistic effect was demonstrated with regard to the formation of lipid peroxides when human blood was exposed in an in vivo experimental setting. (Goldstein, 1976)
18. O_3, NO_2, CO, and SO_2 (synthetic smog)	Initiated significant changes in the lungs of mice, such as marked hyperplasia of the bronchial lining membranes and a thickening of the alveolar wall tissues; since precise studies of the same bioindicators were not conducted, Shy et al. (1974) could not determine whether this was an additive, synergistic, or indifferent reaction. (Loosli et al., 1972)
19. Particulate matter, sodium chloride, silica and volatile irritants (sulfur dioxide, formaldehyde, and nitric acid vapor)	Enhanced eye irritations. (Dautrebande et al., 1951)

Table 4-1. *(Continued)*

	B. Additivity
1. SO_2 and NO_2	Effects of $NO_2 - SO_2$ mixture had additive effects on human pulmonary functions. (Abe, 1967)
2. SO_2 and NO_2	Effects considered were the human thresholds for odor perception and dark adaptation. (Shalamberidze, 1967)
3. O_3 and NO_2	Affected blood parameters such as osmotic fragility, ACHase activity, GSH levels, etc.; at lower levels of exposure synergism for the formation lipid peroxides was noted. (Goldstein, 1976)
4. O_3 and NO_2	Nitrogen dioxide and ozone in combination cause a decrease in bactericidal activity within lungs of mice; effects were below the threshold concentration for the same effect caused by nitrogen dioxide and ozone acting separately. (Goldstein et al., 1974)
5. SO_2, H_2SO_4, NO_x, and NH_3	Effects considered were human odor thresholds and alpha rhythms. (Korniyenko, 1972)
6. H_2SO_4, HCl, and HNO_3	Effects considered were human odor thresholds and dark adaptation. (Melekhina, 1966)
7. CO and heat stress	Resulted in a significant decrease in the time that smokers were capable of exercising; this effect did not occur with either CO or heat alone. (Drinkwater et al., 1974)
	C. Indifference
1. SO_2 and NO_2	Six months exposure; 10 mg/m^3 of NO_2 and SO_2 each; rats were used as the experimental model. (Antweiler and Brockhaus, 1976)
2. SO_2 and NO_2	Two years exposure at 23.5 hr/day, 7 days per week, NO_2 at 0.5 and 7.5 ppm, SO_2 at 1 and 10 ppm. Suggested reasons for the lack of agreement between Amdur work and these studies is that Amdur completed her measurements during a maximum exposure of 1–2 hr. Whereas these two studies took measurements following a time lapse (3–5 hours) after expo-

Table 4-1. *(Continued)*

	sure was completed. (MacFarland, H.N., 1976; see Antweiler and Brockhaus, 1976)
3. SO_2 and NO_2	Low-level exposure did not produce synergistic response. (Mitina, 1962)
4. Peroxyacetylnitrate and CO	No greater effect on work capability of human males than CO alone. (Drinkwater et al., 1974)
5. SO_2, fly ash, and sulfuric acid mist	Long-term exposure ($\geq$ 1 year) of monkeys and guinea pigs to SO_2 (0.1–5 ppm), H_2SO_4 mist (0.1–1 mg/cu m) and fly ash (0.5 mg/cu m) based on pulmonary function tests; no synergistic relationships were noted. (Alarie et al., 1975)
6. SO_2, NO_2, and CO	104-week exposure to cyanomolgus monkeys; NO_2 at 6.78 ppm resulted in decreased pulmonary functions. Additional exposure to SO_2 at 0.48 ppm or CO at 20.3 ppm did not affect influence of the NO_2. (Busey, 1972)
	D. Tolerance
1. O_3	Previous low-level exposure to O_3 will protect against subsequent lethal doses and development of pulmonary edema, but not all toxic effects of ozone, such as labored breathing patterns, are prevented; prior exposure protects against acute effects, not against the low-level, chronic exposure. (Stokinger, 1960)
	E. Cross Tolerance
1. O_3 and ketene and H_2O_2 and cumene	Pretreatment of ozone caused tolerance to be temporarily developed for ketene in mice; the reverse is also true. (NAS, 1976)
2. Oil mists and ozone 3. Oil mists and NO_2	Oil mists provide protection in mice against acute toxic effects of O_3 and NO_2; Stokinger (1960) did not consider this interaction a true form of antagonism, since modifications of the animal's defense system may have induced a tolerance. (Waldbott, 1973)
4. Ozone and radiation	If O_3 exposure is longer than 1 hour and there is at least a 1-day lag before radia-

Table 4-1. *(Continued)*

	tion exposure, the ozone exposure is protective; if the O_3 exposure is of short duration and the radiation exposure immediately follows, the O_3 becomes a radiosensitizer. (Zelac et al., 1971)
	F. Antagonism
1. Nitrous oxide fumes and particulates (iron oxide)	Lethal effects (in mice) of nitrous oxide fumes are reduced when particulates are concurrently inhaled. (LaBelle et al., 1955)
2. O_3 and NO_2	Reduced acute toxicity for O_3 in combination with NO_2 as compared to O_3 alone. (Stokinger, 1957)
3. Aldehydes (acetaldehyde or acrolein) acid kerosene smoke	Multiple species used (guinea pigs, mice, and rabbits); concentrations used were much higher than experienced in the environment. (Salem and Cullumbine, 1961)
4. H_2S and O_3	Decreases acute O_3 toxicity when this combination is concurrently inhaled. (Stokinger, 1965)
II. Chemical Interactions of Heavy Metals	
	A. Synergism
1. Beryllium and fluoride	Enhanced deposition of fluoride in teeth and bones; exaggerated pulmonary lesions and cardiovascular changes. (Stokinger et al., 1950, 1953; Vorwald et al., 1964)
2. Arsenic and selenium	Arsenic accentuates toxocity of selenium in drinking water; this contrasts with the protective effect of arsenic when selenium was administered via the diet. (Frost, 1967)
3. Methylmercury and parathion	MetHg potentiates the toxicity of parathion in the caturina quail; first report of a synergistic relationship between a heavy metal and an organophosphate. (Dieter and Ludke, 1975)
4. Lead and Benzo (a) pyrene	Increased lung cancer in Syrian hamsters. (Kobayaski and Okamoto, 1974)
5. Lead and heavy hydrogen	Effects behavioral changes in animal models. (Bridbord, 1976)

Table 4-1. *(Continued)*

6. Uranium and smoking	Enhanced development of lung cancer in humans. (Lundin et al., 1969)
7. Mercuric chloride and temperature and air velocity	After heating, exposed juvenile chickens experienced an enhanced hyperthermia as compared to controls; high air velocity enhanced the toxic effect. (Thaxton et al., 1975)
8. Lead and ethanol	Even though both lead and ethanol separately inhibit ALAD, together they stimulate it. (Moore, 1972)
9. Cadmium and cyanide	At low concentrations (in the ppm range) cadmium toxicity to fish is enhanced by cyanide. (Stokinger, 1969)
10. Lead and oxygen	Toxicity resulting from increased partial pressures of oxygen was markedly enhanced by concomitant exposure to lead. The lead effect was dose related. Rats were used as the experimental model; suggests that lead may interact with other oxidative agents (e.g., O_3?) in ways that are harmful. (Jones et al., 1974)
11. Lead and ethanol	Epidemiologic association; toxicologic data needed to support observations. (Cramer, 1966)
12. Sodium nitrite and chlorite	Enhanced formation of methemoglobin (Becker et al., 1943)
13. Lead, cadmium, and zinc	Chromosome aberrations in lymphocytes of workers at a zinc smelting plant. (Deknudt et al., 1973)
14. Lead, cadmium, and zinc	Confirmed Deknudt et al., 1973. (Bauchinger et al., 1976)
	B. Additivity
1. Lead, cadmium, and arsenic	Adversely affected hematologic parameters (hemoglobin, hematocrit, coproporphyrin excretion). (Mahaffey and Fowler, 1977)
2. Cadmium and mercury	Inhibition of serum $alpha_1$ antitrypsin. (Chrowdhury and Louria, 1976)
3. Zinc and aluminum	Increases the activity of ALAD in an additive fashion. (Meredith et al., 1974)
	C. Antagonism
1. Selenium and methyl mercury	Japanese quail showed significantly diminished toxicity to methylmercury fol-

Table 4-1. *(Continued)*

	lowing the addition of selenium supplements to the diet. (Ganther et al., 1972)
2. Aluminum dust and silica	Aluminum dust inhibits the development of silicosis; the aluminum dust increases phagocytosis and thereby effects the removal of silica deposits. (Westerick et al., 1957)
3. Cadmium and zinc	Zinc reduced the extent of proteinuria in rabbits exposed to cadmium. (Vigliani, 1969)
4. Cadmium and selenium	Cadmium teratogenesis in animal model (hamsters) is antagonized by selenium. (Holmberg and Ferm, 1969; Parizek et al., 1973)
5. Lead and zinc	Toxic quantities of zinc prevent the development of clinical lead poisoning symptoms (pharyngeal and laryngeal paralysis) in fowls. (Willoughby et al., 1972)
6. Fluoride and selenium	Animal and human studies suggest that selenium may counteract the beneficial functions of fluoride. (Hajimarkos, 1966)
7. Lead and zinc	Zinc treatments given to rabbits (via injection) nearly eliminated the inhibitory effect of lead on ALAD. (Haeger-Aronsen et al., 1976)
8. Selenium and arsenic	When arsenic is added to poultry or cattle feed, it suppresses selenium toxicity. (Rhian and Moxon, 1943)
III. Chemical Interactions of Organic Chemicals	
A. Synergism	
1. Formaldehyde and NaCl aerosol	Synergistic effect on increased respiratory flow resistance and decreased compliance in guinea pigs after inhalation; 0.3–47 ppm was the range of formaldehyde concentrations used; NaCl consisted of an average particle size of 0.04 μm diameter and a 10 mg/cu m concentration; no synergistic effect occurred when formaldehyde concentration decreased to 0.7 ppm. (Amdur, 1959, 1960)
2. Aldehydes (formaldehyde and acrolein) and inert aerosols or particulate materials	The most significant amount of synergism was noted for aerosols of triethylene glycol, mineral oil, celite glycerin, and sodium chloride in conjunction with for-

Table 4-1. *(Continued)*

	maldehyde; as for studies with acrolein, synergism occurred only with mineral oil, NaCl, and SantocelCF (i.e., amorphous silica preparation-Monsanto Co). When aerosol penetration exceeds vapor penetration, the toxicity is increased. (LaBelle et al., 1955)
3. Alkanes (e.g., butane) and alkenes (isobutylene)	Enhanced lethal toxicity in both mice and rats. (Shugaev, 1967)
4. Ethylene dichloride and CCl_4	Slight potentiating effects noted in the acute LC_{50} and acute vapor LC_{50} in animal studies. (McCollister et al., 1956)
5. Methylethylketone and acetone	Subtoxic concentrations of indivdual components resulted in severe intoxication, including convulsions and loss of consciousness. (Smith and Mayers, 1944)
6. Trichloroethylene and alcohol	Clinical evidence suggests a synergistic relationship. (Kleinfield and Tabershaw, 1954; Gutch et al., 1965)
7. Isopropyl alcohol and CCl_4	Isopropyl alcohol intake in large amounts 16–20 hours prior to inhalation of CCl_4 increases toxicity of CCl_4. (Cornish and Adefvin, 1967; Traiger and Plaa, 1971, 1973, 1974)
8. Various solvents and PCB and DDT	Increased mortality in mice when injected intraperitoneally. (Lewin et al., 1972)
9. Insecticide carriers and virus	Increased viral lethality along with enhanced toxic effects on the liver and CNS. (Crocker et al., 1976)
10. Malathion and EPN	Synergistic effect in mammals from these two anticholinesterase substances. (Frawley et al., 1956)
11. Urethan and x-irradiation	Enhanced leukemia incidence in rats. (Myers, 1973)
12. BAP and alpha radiation	Enhanced occurrence of respiratory cancers in the hamster. (McGandy et al., 1974)
13. BAP and SO_2	Enhancement of lung tumor development. (Skvortsova et al., 1967)
14. BAP and N-methyl-n-nitrosourea	Syrian hamsters experienced an increased respiratory cancer rate. (Kaufman and Madison, 1974)
15. BAP and ferric oxide	Enhanced respiratory cancer in hamsters. (Sellakumar et al., 1973; Saffiotti et al.,

Table 4-1. *(Continued)*

	1972a, 1972b; Harris et al., 1971; Port et al., 1973)
16. BAP and Fe_2O_3 and diethylnitrosamine	Enhanced respiratory cancer in hamsters. (Montesano et al., 1970a)
17. BAP and phenol	Effected changes in a number of cellular biochemical functions. (Skvortsova and Vysochina, 1976)
18. Diethylnitrosamine and NaCl	Enhanced occurrence of tumor incidence in lower respiratory tract. (Stenback et al., 1973)
19. 2-acetylaminofluorene and 3-methyl 4-dimethylaminoazobenzene	Increased incidence of hepatic tumors in male rats. (MacDonald et al., 1952)
20. DDT and CCl_4	DDT pretreatment enhanced the toxicity of CCl_4 in rats. (McLean and McLean, 1966)
21. PCB and benzenehexachloride	Enhanced hepatic tumorigenesis in mice. (Nobuyuki et al., 1973)
22. CCl_4 and ethanol	Enhanced hepatotoxicity of CCl_4 in rats by pretreatment with ethanol. (Maling et al., 1975)
23. CCl_4 and Aroclor 1254	The PCB, Aroclor 1254, potentiated the acute toxicity of CCl_4 in the rat. (Grant et al., 1971)
24. DDT and zinc	Elevated levels of both DDT and zinc resulted in a highly significant reduction in the hemoglobin levels of maternal rats and their fetuses. (Feaster et al., 1972)
25. Chloroform and urethane	One-half the acute toxic dose for chloroform and urethane resulted in the death of experimental rabbits. The authors referred to this as "additive synergism." More results are needed to determine if it is a true additive or synergistic response. (Lushbaugh and Storer, 1948)
26. Urethane and x-irradiation	Chick embryos exposed to 300 rad x-irradiation and a single dose of urethane developed a variety of malformations. (Hawkins and Murphy, 1925; Diwan and Batra, 1968)
27. 4-nitroquinoline 1-oxide and cigarette smoke	Enhanced occurrence of lung tumors in rats. (Mori, 1964)

Table 4-1. *(Continued)*

28. 2-napthylamine and 4-nitrobiphenyl	Enhanced urinary bladder cancers in dogs; response is difficult to classify in terms of additivity or synergism. (Deichmann et al., 1965)
29. Sodium nitrite and dimethylamine	Enhanced toxic responses such as relative weight loss, mortality, and liver necrosis in mice; previous studies revealed these substances interacted synergistically during the inhibition of mouse liver protein and RNA synthesis. (Aschina et al., 1971; Friedman et al., 1971)
30. Hexachlorotetrafluorolutane and methylcholanthrene	Rats injected intravenously with methylcholanthrene mixed with hexachlorotetrafluorolutane formed respiratory squamous cell carcinomas in the reactive epithelium; neither substance induced tumor development by itself. (Stanton and Blackwell, 1961)
31. BAP and N-dodecane; benzo(a) anthracene and N-dodecane	Carcinogenic effect of benzo(a)pyrene and benzo(a)anthracene is markedly enhanced by N-dodecane activity as a cocarcinogen. (Bingham and Falk, 1969)
32. BAP and diethylnitrosamine	Enhanced incidence of respiratory cancer in hamsters. (Montesano, 1970b).
33. PCBs and alkylbenzene sulfonic acid salt	Rats developed enlarged livers (by weight). (Kamohara and Fujiwara, 1974)
34. Smoking and asbestos	Enhanced lung cancer in workers. (Selikoff et al., 1968)
35. Smoking and dust	Decreased lung function in workers. (Rossiter and Weill, 1974)
	B. Additivity
1. PCBs and dieldrin; PCBs and DDE	PCBs and either dieldrin or DDE adversely affect egg production, hatchability, and viability of the pheasant embryo at the time of hatching in an additive fashion. (Dahlgren et al., 1972; Health et al., 1970)
2. 1,2-dichloropropane and 1,2,3-trichloropropane perchloroethylene	At high concentrations these substances act in an additive fashion with respect to the loss of righting reflex. (Sidorenko et al., 1976)

Table 4-1. *(Continued)*

3. Methylene chloride and CO	Levels of COHb in the blood increased in an additive fashion. (Fodor et al., 1973; NIOSH methylene chloride, 1976)
4. Acetone Acetonitride Acetophenone Acrylonitrile Aniline Butyl ether Carbon tetrachloride Diethanolamine Dioxane Ethyl acetate Ethyl acrylate Ethyl alcohol Ethylene glycol Formalin Isophorone Morpholine Nitrobenzene Polyethylene glycol 400 Propylene glycol Propylene oxide Toluene Tetrachloroethylene	Rat peroral LD_{50}'s were determined for all possible pairs (about 300) of the 22 industrial chemicals listed here; on the whole, the responses of the rats to most mixtures reflected additivity. The 5% of pairs tested that differ most (positively and negatively) from the additivity response are listed here. (Smyth et al., 1969)
Less than Additive Diethanolamine and acrylonitrile Ethyl alcohol and ethylene glycol Diethanolamine and ethyl acrylate Aniline and formalin Ethyl alcohol and propylene oxide Propylene glycol and propylene oxide	
More than Additive Ethyl acetate and formalin Tetrachloroethylene and polyethylene glycol 400	

Table 4-1. *(Continued)*

Tetrachloroethylene and butyl ether	
Tetrachloroethylene and dioxane	
Acetonitrile and dioxane	
Tetrachloroethylene and acetophenone	
Acetonitrile and acetophenone	
Acetonitrile and acetone	
Morpholine and toluene	
C. Indifference	
1. Methylethylketone, propylene formal, 1,3-amylene oxide, tetrahydrofuran, 1,3-butylene oxide, dimethyl acetal, propylene acetal, acetone	Result of acute and extended animal studies of exposure to this did not differ from the effect of the methylethylketone. (LaBelle and Brieger, 1955)
2. Parathion and lead	Could not show a biologic interaction. (Phillips et al., 1973)

mixture. Smyth et al., in agreement with the general findings of Pozzani et al., concluded that approximately 5% of the various combinations tested exhibited more or less than additive effects. The other 95% produced toxicity effects quantifiable in terms of additivity.

Although considerably more research must be directed to the area of low-dose interactions, the results of Bates and Hazucha (1973) are worthy of comment. They exposed humans to "fairly" realistic concentrations of SO_2 (0.37 ppm) and 0_3 (0.37 ppm) and found a marked decrease in respiratory function when compared to the sum of the adverse effects of the two gases. In light of the data of Bates and Hazucha, it is clear that in the derivation of standard setting, chemical interactions should be strongly considered. Most present studies have considered synergism at only very high or acute levels of exposure (Ashford, 1976). The question of whether synergisms would occur at lower (more realistic) concentrations is a matter of debate. Ashford (1976), reporting on a personal communication with Dr. Herbert Stokinger (1973), then Chief of NIOSH's Toxicology Division, noted that

Stokinger felt that the interactions of substances at low levels of exposure would be "physiologically inconsequential."

Ashford tried to place the important research findings of Smyth et al. in a clear perspective. For instance, if there are N potentially toxic chemicals in commerce today (estimates are approximately 12,000), then the number of potential pairs is $N(N - 1)/2$. Consequently, of the $12{,}000 \times 11{,}999/2$ pairs, 2.5% or 1.8 million pairs may act synergistically. Ashford mistakenly used the value of 5% for the percentage of synergistic responses. Still, his point is well taken that the number of potential synergistic situations is enormous. Even though we may focus on synergistic responses, the vast majority are thought to be additive. This should not be ignored because of the added toxicity of such interactions.

ACGIH APPROACH TO TLVs FOR MIXTURES

Although the scope of United States industrial standards has markedly expanded in the 35 years since the adoption of the TLV by the ACGIH, their recommendations have not specifically listed standards for chemical combinations. However, the ACGIH has provided the general framework on which such standards may be based by including a section concerning workplace TLVs for mixtures in their annual booklet. In light of all the possible combinations of substances that may occur in the workplace, along with novel production processes in the future, the general framework as set forth by the ACGIH seems to be a very reasonable approach to the matter. The ACGIH (1976) approach used for setting TLVs for mixtures is now reprinted with their permission.

When two or more hazardous substances are present, their combined effect, rather than that of either individually, should be given primary consideration. In the absence of information to the contrary, the effects of the different hazards should be considered as additive. That is, if the sum of the fractions $C1/T1 + C2/T2 + \cdots Cn/Tn$ exceeds unity, the threshold limit of the mixture should be considered as exceeded. C1 indicates the observed atmospheric concentration and T1 the corresponding threshold limit [See examples 1A(a)(c)].

Exceptions to this rule may be made when there is good reason to believe that the chief effects of the different harmful substances are not in fact additive, but *independent* as when purely local effects on different organs of the body are produced by the various components of the mixture. In such cases the threshold limit ordinarily is exceeded only when at least one member of the series $C1/T1 +$ or $+ C2/T2$, etc. has a value exceeding unity.

Antagonistic action or potentiation may occur with some combinations of atmospheric contaminants. At present such cases must be determined individually. Potentiating or antagonistic agents are not necessarily harmful by themselves. The potentiating effects of exposure to such agents by routes other than inhalation is also possible, for example, imbibed alcohol and inhaled narcotic (trichloroethylene). Potentiation is characteristically exhibited at high concentrations, probably less at low concentrations.

When a given operation or process characteristically emits a number of harmful dusts, fumes, vapors, or gases, it may only be feasible to attempt to evaluate the hazard by the measurement of a single substance. In such cases, the threshold limit used for this substance should be reduced by a suitable factor, the magnitude of which depends on the number, toxicity, and relative quantity of the other contaminants ordinarily present.

Examples of processes that are typically associated with two or more harmful atmospheric contaminants are welding, automobile repair, blasting, painting, lacquering, certain foundry operations, diesel exhausts, and so on.

The following formulae apply only when the components in a mixture have similar toxicologic effects; they should not be used for mixtures with widely differing reactivities, for example, hydrogen cyanide and sulfur dioxide. In such cases the formula for independent effects should be used (1A.c.).

1A(a) General case, in which air is analyzed for each component.

Additive effects. (Note: It is essential that the atmosphere be analyzed both qualitatively and quantitatively for each component present to evaluate compliance or noncompliance with this calculated TLV).

$$\frac{C1}{T1} + \frac{C2}{T2} + \frac{C3}{T3} + \cdots = 1$$

Example Air contains 5 ppm of carbon tetrachloride (TLV = 10 ppm), 20 ppm of 1,2-dichloroethane (TLV = 50 ppm), and 10 ppm of 1,2-dibromoethane (TLV = 20 ppm).

Atmospheric concentration of mixture = 5 + 20 + 10 = 35 ppm of mixture

$$\frac{5}{10} + \frac{20}{50} + \frac{10}{20} = \frac{25 + 20 + 25}{50} = 1.4$$

The threshold limit is exceeded. Furthermore, the TLV of this mixture may be calculated by reducing the total fraction to 1.0, that is,

$$\text{TLV of mixture} = \frac{35}{1.4} = 25 \text{ ppm}$$

1A(b) Special case, the source of contaminant is a liquid mixture and the atmospheric composition is assumed to be similar to that of the original material, for example, on a time-weighted averaged exposure basis, all the liquid (solvent) mixture eventually evaporates.

Additive effects (approximate solution)

The percentage composition (by weight) of the liquid mixture is known; the TLVs of the constituents must be listed in mg/m^3.

(Note: To evaluate compliance with this TLV, field sampling instruments should be calibrated in the laboratory for response to this specific quantitative and qualitative air vapor mixture and also to fractional concentrations of this mixture, e.g., ½ the TLV, $\frac{1}{10}$ the TLV, 2 times the TLV, 10 times the TLV, etc.)

$$\text{TLV of mixture} = \frac{1}{\dfrac{\text{fa}}{\text{TLVa}} + \dfrac{\text{fb}}{\text{TLVb}} + \dfrac{\text{fc}}{\text{TLVc}} + \cdots \dfrac{\text{fn}}{\text{TLVn}}}$$

Example 1: Liquid contains (by weight)

50% heptane: TLV = 400 ppm or 1600 mg/m^3 (1 mg/m^3 = 0.25 ppm)

30% methylene chloride: TLV = 200 ppm or 720 mg/m^3 (1 mg/m^3 = 0.28 ppm)

20% perchloroethylene: TLV = 100 ppm or 670 mg/m^3 (1 mg/m^3 = 0.15 ppm)

$$\text{TLV of Mixture} = \frac{1}{\dfrac{0.5}{1600} + \dfrac{0.3}{720} + \dfrac{0.2}{670}}$$

$$= \frac{1}{0.00031 + 0.00042 + 0.00030}$$

$$= \frac{1}{0.00103} = 970 \text{ mg/m}^3$$

Of this mixture,

50%, or (970) (0.5) = 485 mg/m^3, is heptane
30%, or (970) (0.3) = 291 mg/m^3, is methylene chloride
20%, or (970) (0.2) = 194 mg/m^3, is perchloroethylene

These values can be converted to ppm as follows:

Heptane: 485 mg/m^3 × 0.25 ppm (= 1 mg/m^3) = 121 ppm
Methylene chloride: 291 mg/m^3 × 0.28 ppm (= 1 mg/m^3) = 81 ppm
Perchloroethylene: 194 mg/m^3 × 0.15 ppm (= 1 mg/m^3) = 29 ppm

TLV of mixture = 121 + 81 + 29 = 231 ppm, or 970 mg/m^3

Example 2. Liquid solvent contains (by weight):

50% isopropyl alcohol: TLV = 400 ppm or 980 mg/m^3 (1 mg/m^3 = 0.41 ppm)

30% dichloroethane: TLV = 50 ppm or 200 mg/m^3 (1 mg/m^3 = 0.25 ppm)

20% perchloroethylene: TLV = 100 ppm or 670 mg/m^3 (1 mg/m^3 = 0.15 ppm)

$$\text{TLV of Mixture} = \frac{1}{\dfrac{0.5}{980} + \dfrac{0.3}{200} + \dfrac{0.2}{670}}$$

$$= \frac{1}{0.00051 + 0.0015 + 0.000298}$$

$$= \frac{1}{0.002308} = 433\ \text{mg/m}^3$$

of this mixture

50%, or (433) (0.5) = 216 mg/m^3, is isopropyl alcohol
30%, or (433) (0.3) = 130 mg/m^3, is dichloroethane
20%, or (433) (0.2) = 87 mg/m^3, is perchloroethylene

These values can be converted to ppm as follows:

$$\text{Isopropyl alcohol: } 216\ \frac{\text{mg}}{\text{m}^3} \times 0.41\ \text{ppm}\ (= \text{mg/m}^3) = 89\ \text{ppm}$$

$$\text{Dichloroethane: } 130\ \frac{\text{mg}}{\text{m}^3} \times 0.25\ \text{ppm}\ (= \text{mg/m}^3) = 33\ \text{ppm}$$

$$\text{Perchloroethylene: } 87\ \frac{\text{mg}}{\text{m}^3} \times 0.15\ \text{ppm}\ (= \text{mg/m}^3) = 13\ \text{ppm}$$

TLV of mixture = 89 + 33 + 13 = 135 ppm or 433 mg/m^3

1A(c) Independent effects. Air contains 0.15 mg/m³ of lead (TLV, 0.15) and 0.7 mg/m³ of sulfuric acid (TLV, 1); 0.15/0.15 = 1 0.7/1 = 0.7—threshold limit is not exceeded.

1B. TLV for mixtures of mineral dusts. For mixtures of biologically active mineral dusts the general formula for mixtures may be used.

For mixture containing 80% nonasbestiform talc and 20% quartz, the TLV for 100% of the mixture is given by:

$$\text{TLV of mixture} = \frac{1}{\dfrac{0.8}{20} + \dfrac{0.2}{2.7}} = 9 \text{ mppcf}$$

TLV of asbestiform talc (pure) = 20 mppcf

$$\text{TLV of quartz (pure)} = \frac{300}{100 + 10} = \frac{300}{110} = \text{2.7 mppcf (million particles per cubic foot)}$$

Essentially the same results are obtained if the limit of the more (most) toxic component is used, provided the effects are additive. In the above example the limit for 20% quartz is 10 mppcf.

For another mixture of 25% quartz, 25% amorphous silica, and 50% talc.

25% quartz − TLV (pure) = 2.7 mppcf
25% amorphous silica − TLV (pure) = 20 mppcf
50% talc TLV (pure) = 20 mppcf

$$\text{TLV} = \frac{1}{\dfrac{0.25}{2.7} + \dfrac{0.25}{20} + \dfrac{0.5}{20}} = 8 \text{ mppcf}$$

The limit for 25% quartz approximates 9 mppcf.

NIOSH APPROACH FOR METHYLENE CHLORIDE AND CO

Research on rats by Foder et al. (1973) indicated that inhaled methylene chloride and carbon monoxide exposure both contribute to the levels of carboxyhemoglobin (COHb). In fact, the influence of methylene chloride and carbon monoxide on the formation of COHb was additive. As a result of this study, NIOSH proposed that if methylene chloride and carbon monoxide are present in the workplace atmosphere, their respective exposure limits should be decreased.

Table 4-2. TWA Exposure Limits For Methylene Chloride When CO is Jointly Present in the Occupational Environment

	CH_2Cl_2 (ppm)	
CO TWA	TWA	Action
0–9	75	37.5
10	54	27
15	43	21.5
20	32	16
25	21	10.5
30	11	5.5
35	0	0.0

Source: NIOSH Criteria for a recommended standard for methylene chloride. 1976, p. 97.

Accordingly, NIOSH (1976) recommended the following special considerations for methylene chloride.

$$\frac{C(CO)}{L(CO)} + \frac{C(CH_2Cl_2)}{L(CH_2Cl_2)} \leq 1$$

$C(CO)$ = the TWA exposure concentration of CO
$L(CO)$ = the recommended TWA exposure limit of CO = 35 ppm
$C(CH_2Cl_2)$ = the TWA exposure concentration of methylene chloride
$L(CH_2Cl_2)$ = the recommended TWA exposure limit of methylene chloride = 75.

They suggested that no adjustment in the recommended standard for methylene chloride become operational until TWA carbon monoxide levels in the workplace are greater than 9 ppm.

Table 4-2 shows the adjustment in permissible exposure limits for methylene chloride at various levels of carbon monoxide.

SOVIET APPROACHES TO REGULATING CHEMICAL MIXTURES

The Soviets have made attempts to develop standards for combinations of various pollutants. Table 4-3 shows the Soviet approach for the development of a standard for a combination of sulfur dioxide and nitrogen oxides. As in the case of standard derivation for individual

Table 4-3. Soviet Approaches for Deriving Standards for Mixtures of Gases (mg/cu m)

Substance	Approved MPC	Odor		Reflex Effect		Resorptive effect		Physiologic and Biochemical Changes[a,b]		Change in Morbidity, Physical Development, or Blood[a,b]	
		Threshold	Subthreshold	Threshold	Subthreshold	Threshold	Subthreshold	Present	Absent	Present	Absent
Sulfur dioxide and nitrogen oxides	Given by the formula for a simple sum	1.6	1.3	0.6	0.5	0.15	0.078	–/0.32	–/0.19	–/0.32	
						+	+	+	+	+	
		0.23	0.11	0.14	0.09	0.10	0.052	–/0.099	–/0.013	–/0.099	

[a] Values refer to the mean 24-hour concentration.

[b] Carried out on groups of children selected on the basis of Socioeconomic background, housing and living conditions.

Source: Izmerov, N.F. (1973) *Control of Air Pollution in the U.S.S.R.* WHO, Geneva, p. 56 (based on Salamberidze, 1969).

Table 4-4. Overall Indices of the Combined Action of Toxic Substances in the Air

Substances in Combination	Way in Which the Effects Combine	Standard Recommended for the Given Combinations	Author
Chlorine Hydrogen chloride	Partial summation	Apply MPC for each substance	V.M. Stjazkin
Sulfur dioxide Sulfuric acid	Simple summation	Apply formula ≤ 1	K.A. Bustueva
Hydrogen sulfide Carbon disulfide	Partial summation	Apply MPC for each substance	B.K. Bajkov
Sulfur dioxide Phenol	Simple summation	Apply formula ≤ 1	A.P. Mahinhja
Phenol Carbon monoxide	Simple summation	Apply formula ≤ 1	E.F. Elfimova
Phenol Acetone	Simple summation	Apply formula ≤ 1	U.G. Pogosjan
Phenol Acetophenone	Full summation	Apply formula ≤ 1.5	Ju. E. Korneev
Acetophenone Acetone	Full summation	Apply formula ≤ 1.5	N.Z. Tkac
Isopropylbenzene Isopropylbenzene hydroperoxide	Simple summation	Apply formula ≤ 1	G.I. Solomin
Ethylene Propylene Butylene	Simple summation	Apply formula ≤ 1	M.L. Krasovickaja
Hydrogen sulfide Carbon disulfide Dinyl	Simple summation	Apply formula ≤ 1	H.H. Mannanova
Isopropylbenzene Benzene	Simple summation	Apply formula ≤ 1	E.V. Elfimova
Nitric acid Sulfuric acid Hydrochloric acid	Simple summation	Apply formula ≤ 1	V.P. Melexina

Source: Izmerov, N.F. (1973) *Control of Air Pollution In The U.S.S.R.* WHO, Geneva, p. 52.

pollutants, the Soviets also consider the data from odor, reflexes, resorption, physiologic changes and morbidity studies (Ulanova and Zayeva, 1964; Stayzhkin, 1962; Sanotskij, 1969; Lyublina, 1956; Shtessel, 1958; Shugaev, 1967). In the case of sulfur dioxide and nitrogen oxides the formula is a simple summation. Examples of other cases of the Soviet standards for different combinations of toxic substances are given in Table 4-4.

The role of chemical interactions in the standard-setting process in the United States remains to be seen. Common sense demands that it be considered in the development of health criteria from which standards are to be derived. The example of methylene chloride and carbon monoxide clearly illustrates that the cumulative effects of different pollutants must be taken into account if the workplace is to be made a safe place in which to spend 8 hours each work day for 40 years.

The question naturally arises whether some of our present standards should be modified in light of recent evidence concerning synergistic interaction of various pollutants. Shy et al. (1974) were of the opinion that the present ambient standards should not be revised as a result of knowledge derived from new community studies that include various combinations of pollutants since the present standards primarily derived from studies of an epidemiologic nature actually assumed the occurrence of chemical interactions even though only single pollutants were specifically related to health effects. Similar reasoning could probably be used for industrial standards, assuming that the mix of pollutants has not changed. However, in the future, as previously noted, the emphasis is likely to be based more on toxicologic than epidemiologic studies. Consequently, chemical interactions are expected to play an important part in the derivation of future standards for industry and community.

REFERENCES

Abe, M. (1967) Effects of mixed NO_2–SO_2 gas on human pulmonary function. *Bull Tokyo Med Dent Univ* 14:415–433.

ACGIH (1976) American Conference of Governmental Industrial Hygienists. Threshold Limit Values. Cincinnati, Ohio.

Alarie, Y. C., Krumm, A. X., Busey, W. M., Virich, C. E., and Kantz, R. J. (1975) Long-term exposure to sulfur dioxide, sulfuric acid mist, fly ash, and their mixtures; Results of studies in monkeys and guinea pigs. *Arch Environ Health* 30:255–262.

Amdur, M. O. (1959) The physiologic response of guinea pigs to atmospheric pollutants. *Int J Air Pollut* 1:170–183.

Amdur, M. O. (1960) The response of guinea pigs to inhalation of formaldehyde and formic acid alone and with a sodium chloride aerosol. *Int J Air Pollut* 3:201–220.

Amdur, M. O. and Underhill, D. (1968) The effect of various aerosols on the response of guinea pigs to sulfur dioxide. *Arch Environ Health* 16:460–468.

Antweiler, H. and Brockhaus, A. (1976) Respiratory frequency, flow rate and minute volume in non-anesthetised guinea pigs during prolonged exposure to low concentrations of SO_2 and NO_2. *Ann Occup Hyg* 19:13–16.

Aschina, S., Friedman, M. A., Arnold, E., Millar, G. N., Mishkin, M., Bishop, M., and Epstein, S. S. (1971) Acute synergistic toxicity and hepatic necrosis following oral administration of sodium nitrite and secondary amines to mice. *Cancer Res* 31:1201–1205.

Ashford, N. A. (1976) *Crisis in the Workplace: Occupational Disease and Injury.* Cambridge: MIT Press.

Ball, W. L. (1959) The toxicological basis of threshold limit values. Theoretical approach to prediction of toxicity of mixtures. *Am Ind Hyg Assoc J* 20:357.

Bates, D. V. and Hazucha, M. (1973) The short-term effect of ozone on the human lung. In Assembly of Life Sciences—National Academy of Sciences—National Research Council, *Proceedings of the Conference on Health Effects of Air Pollutants,* pp. 507–540, Washington, D.C.

Bauchinger, M., Schmid, E., Einbrodt, H. J., and Dresp, J. (1976) Chromosome aberrations in lymphocytes after occupational exposure to lead and cadmium. *Mut Res* 40:57–62.

Becker, T., Dordoni, F., and Jung, F. (1943) Zur Thorie der Chloratvergiftung. II. Arch. f. exper. *Path v Pharmakol* 201:197–209.

Bingham, E. and Falk, H. L. (1969) Environmental carcinogens: the modifying effect of cocarcinogens on the threshold response. *Arch Environ Health* 19:779–783.

Bridbord, K (1976). Epidemiology and lead exposure among occupational groups. In *Health Effects of Occupational Lead and Arsenic Exposure: A Symposium, NIOSH,* pp. 110–113.

Brieger, H. (1957) Synergistic and antagonistic physiological effects of exposure to multiple air contaminants. *Med Surg* 26:315.

Busey, W. M. (1972) Summary report. Study of synergistic effects of certain airborne systems in the cyanomolgus monkey. Vienna, Virginia: Hazelton Laboratories, Inc. (Cited in Shy et al., 1974)

Chrowdhury, P. and Louria, D. B. (1976) Influence of cadmium and other trace metals on human α_1-antitrypsins: an *in vivo* study. *Science* 191:480.

Coffin, D. L. and Blommer, E. H. (1965) The influence of cold on mortality from streptococci following ozone exposure. *J Air Pollut Control Assoc* 15:523–524.

Coffin, D. L. and Blommer, E. H. (1967) Acute toxicity of irradiated auto exhaust. Its indications by enhancement of mortality from streptococcal pneumonia. *Arch Environ Health* 15:36–38.

Coffin, D. L., Gardner, D. E., Holzman, R. W., and Wolock, F. J. (1968) Influence of ozone on pulmonary cells. *Arch Environ Health* 16:633–636.

Cornish, H. H. and Adefvin, J. (1967) Potentiation of CCl_4 toxicity by aliphatic alcohols. *Arch Environ Health* 14:447–449.

Cramer, K. (1966) Predisposing factors for lead poisoning. *Acta Med Scand* 179:56 (Suppl 445).

Crocker, J. F. S., Ozene, R. L., Safe, S. H., Diyout, S. C., Rozel, K. R., and Hutzinger, O.

(1976) Lethal interaction of ubiquitous insecticide carriers with virus. *Science* 192:1351–1353.

Dahlgren, R. B., Linder, R. L., and Carlson, E. W. (1972) Polychlorinated biphenyls: their effects on penned pheasants. *Environ Health Perspect* 1:89.

Dautrebande, L., Shaver, J., and Capps, R. (1951) Studies on aerosols II. Influence of particle matter on eye irritation produced by volatile irritants. *Arch Int Pharmacodyn* 85:17 (Cited in Stokinger, 1960).

Deichmann, W. B., Scotti, T., Radomski, J., Bernal, E., Coplan, M., and Woods, F. (1965) Synergism among oral carcinogens. II. Results of the simultaneous feeding of bladder carcinogens to dogs. *Toxicol Appl Pharmacol* 7:657–659.

Deknudt, Gh., Leonard, R., and Ivanov, B. (1973) Chromosome aberrations observed in male workers occupationally exposed to lead. *Environ Physiol Biochem* 3:132–138.

Dieter, M. P. and Ludke, J. L. (1975) Studies on combined effects of organophosphates and heavy metals in birds. I. Plasma and brain cholinesterase in Catunix Quail fed methyl mercury and orally dosed with parathion. *Bull Environ Contam Toxicol* 13(3):257–262.

Diwan, B. A. and Batra, B. K. (1968) Radiation-induced abnormalities in the chick embryo treated with urethane. *Wilhelm Roux Archiv* 161:319–324.

Drinkwater, B. L., Raven, P. B., Horvath, S. M., Gliner, J. A., Ruhling, R. O., Bolduan, N. W., and Taguschi, S. (1974) Air pollution, exercise, and heat stress. *Arch Environ Health* 28:177–181.

Eckardt, R. E. (1976) Extrapolating from animals to man: carcinogenesis. *J Occup Med* 18(7):492–494.

Feaster, J. P., Van Middlelem, C. H., and Davis, G. K. (1972) Zinc-DDT interrelationships in growth and reproduction in the rat. *J Nutr* 102:523–528.

Finney, D. J. (1952) *Probit Analysis*, 2nd ed. London: Cambridge University Press.

Fodor, G. G., Prajsnar, D., and Schlipkoter, H. W. (1973) Endogenous CO formation by incorporated halogenated hydrocarbons of the methane series. *Staub-Reinhalt Luft* 33:260–261.

Frank, N. R., Amdur, M. O., and Whittenberger, J. L. (1966) A comparison of the acute effect of SO_2 administered alone or in combination with NaCl particles on the respiratory mechanisms of healthy adults. *Air Water Pollut* 8:125–133.

Frawley, J. P. et al. (1956) Marked potentiation in mammalian toxicity from the simultaneous administration of two anticholinesterase compounds. Presented, Federation of American Societies for Experimental Biology, November 9, 1956 (Cited in Brieger, 1957).

Friedman, M. A., Millar, G., Sengupta, M., and Epstein, S. S. (1971) Inhibition of mouse liver protein and nuclear RNA synthesis following combined oral treatment with sodium nitrite and dimethylamine or methylbenzylamine. *Experimentia* (Cited in Asahina et al., 1971).

Frost, D. V. (1967) Significance of the symposium. In O. H. Muth, ed., *Symposium: Selenium in Biomedicine,* pp. 7–26. Westport: AVI Publishing Co.

Ganther, H. E., Goudie, C., Sunde, M. L., Kopecky, M. J., Wagner, P., Sang-Hwan, O.H., and Hoekstra, W. G. (1972) Selenium: relation to decreased toxicity of methylmercury added to diets containing tuna. *Science* 175:1122–1124.

Goldstein, B. (1976) Combined exposure to ozone and nitrogen dioxide. *Environ Health Persp* 13:107–110.

Goldstein, B., Warshaver, D., Lippert, W., and Tarkington, B. (1974) Ozone and nitrogen dioxide exposure. *Arch Environ Health* 28:85–90.

Grant, D. L., Phillips, W. E. J., and Villeneuve, D. C. (1971) Metabolism of a polychlorinated biphenyl (Aroclor 1254) mixture in the rat. *Bull Environ Contam Toxicol* 6(2):102–112.

Gutch, G. F., Tomhave, W. G., and Stevens, S. C. (1965) Acute renal failure due to inhalation of trichloroethylene. *Ann Intern Med* 63:128–134.

Haeger-Aronsen, B., Schutz, A., and Abdulla, M. (1976) Antagonistic effect *in vivo* of zinc on inhibition of δ-Aminolevulinic acid dehydrate by lead. *Arch Environ Health* July/August. 31:215–220.

Hajimarkos, D. M. (1966) Micronutrient elements in relation to dental carries. *Borden's Rev Nutr Res* 27:1.

Harris, C. C., Sporn, M. B., Kaufman, D. G., Smith, J. M., Baker, M. S., and Saffiotti, U. (1971) Acute ultrastructural effects of benzo(a)pyrene and ferric oxide on the hamster tracheobronchial epithelium. *Cancer Res* 31:1977–1989.

Hawkins, J. A. and Murphy, J. B. (1925) The effect of ethyl urethane anesthesia on the acid-base equilibrium and cell contents of blood. *J Exp Med* 42:609.

Health, R. G. et al. (1970) Effects of polychlorinated biphenyls on birds. Proc. XV Intern. Ornithological Congress, The Hague. (Cited in Dahlgren, 1972).

Holmberg, R. E. and Ferm, V. H. (1969) Interrelationships of selenium, cadmium, and arsenic in mammalian teratogenesis. *Arch Environ Health* 18:873–877.

Hough, V. H., Gunn, F. D., and Freeman, S. (1944) Studies on the toxicity of commercial benzene and of a mixture of benzene, toluene and xylene. *J Indus Hyg Toxicol* 26:296.

Izmerov, N. F. (1973) Principles underlying the establishment of air quality standards in the U.S.S.R. In *Control of Air Pollution in the U.S.S.R.*, pp. 42–60, Geneva: WHO.

Jones, R. B., Nelson, D. P., Shapiro, S., and Kiesow, L. A. (1974) Synergism in the toxicities of lead and oxygen. *Experienta* 30(4):327–328.

Kamohara, K. and Fujiwara, K. (1974) Studies on experimental PCB poisoning: dose-response effect of PCB and synergistic effect with PCB and ABS. *Jap J Hyg* 29:321–327.

Kaufman, D. G. and Madison, R. M. (1974) Synergistic effects of benzo(a)pyrene in N-methyl-n-nitrosourea on respiratory carcinogenesis in syrian golden Hamsters. In E. Karbe and J. F. Pard, eds., *Experimental Lung Cancer: Carcinogenesis and Bioassays,* pp. 207–218. International Symposium. Seattle, Washington, June 23–26, 1974. New York: Springer.

Kellogg, W. W., Cadle, R. D., Allen, E. R., Lazrus, A. L., and Martell, E. A. (1972) The sulfur cycle. *Science* 175:587–596.

Kleinfield, M. and Tabershaw, J. R. (1954) Trichloroethylene toxicity—report of five fatal cases. *Arch Ind Hyg Occup Med* 10:134–141.

Kobayashi, N. and Okamoto, T. (1974) Effects of lead oxide in the induction of lung cancer in Syrian hamsters. *J Natl Cancer Inst* 52:1605–1608.

Korniyenko, A. P. (1972) Hygienic evaluation of a mixture of sulfuric acid aerosol, sulfurous anhydrine, nitrogen oxides, and ammonia as an atmospheric pollutant. *Gig Sanit* 37:8–10.

Kotin, P. (1966) The influence of pathogenic miases on cancers induced by inhalation. *Can Cancer Conf* 6:465.

Kotin, P. and Wiseley, D. V. (1963) Production of lung cancer in mice by inhalation exposure to influenza virus and aerosol of hydrocarbon. In Homburger, F., ed., *Progr. Exp. Tumor Res.* vol. 3, pp. 189–215. New York: Karger. N.Y.

LaBelle, C. W. and Brieger, H. (1955) The vapor toxicity of a composite and its principal components. *Arch Indust Health* 12:623.

LaBelle, C. W., Long, J. E., and Christofano, E. E. (1955) Synergistic effects of aerosols. *Arch Indust Health* 11:297–304.

Landau, E. (1977) NAS report on arsenic: a critique. Speciality Conference on Toxic Substances in the Air Environment, edited by the Air Pollution Control Association, Pittsburg, Pa., pp. 65–77.

Lewin, U., McBlain, W. A., and Wolfe, F. H. (1972) Acute intraperitoneal toxicity of DDT and PCB's in mice using two solvents. *Bull Environ Contam Toxicol* 8:245.

Loosli, G. G., Buckley, R. D., Hertweck, M. S., Hardy, J. D., Ryan, D. P., Stinson, S., and Serebrin, R. (1972) Pulmonary response of mice exposed to synthetic smog. *Ann Occup Hyg* 15:251–260.

Lundin, F. E., Jr., Lloyd, J. W., Smith, E. M., Archer, V. E., and Holaday, D. A. (1969) Mortality of uranium miners in relation to radiation exposure, hard-rock mining, and cigarette smoking—1950 through September 1967. *Health Phys* 16:571–578.

Lushbaugh, C. C. and Storer, J. B. (1948) Synergistic necrotizing action of urethane and chloroform. *Arch Pathol* pp. 494–502.

Lyublina, E. N. (1956) Two types of effects that organic solvents (narcotics) used in industry have on the nervous system. Lenigrad. (Cited in Sidorenko et al, 1976).

MacDonald, J. C., Miller, E. C., Miller, J. A., and Rusch, H. P. (1952) The synergistic action of mixtures of certain hepatic carcinogens. *Cancer Res* 12:50–54.

Mahaffey, K. R. and Fowler, B. A. (1977) Lead, cadmium and arsenic: effects on heme and porphyrin metabolism in rats. First International Conference on Toxicology. Toronto, Ont.

Maling, H. M., Stripp, B., Sipes, I. B., Highman, B., Saul, W., and Williams, M. A. (1975) Enhanced hepatotoxicity of carbon tetrachloride, thioacetamide, and dimethylnitrosamine by pretreatment of rats with ethanol and some comparisons with potentiation by isopropanol. *Toxicol Appl Pharmacol* 33:291:308.

McCollister, D. D., Hollingsworth, R. L., Oyen, F., and Rowe, V. K. (1956) Comparative inhalation toxicity of fumigant mixture individual and joint effects of ethylene dichloride, carbon tetrachloride, and ethylene dibromide. *Arch Indust Health* 13:1.

McFarland, H. N. (1976) Discussion section. *Ann Occup Hyg* 19:16.

McGandy, R. B., Kennedy, A. R., Terzaghi, M., and Little, J. B. (1974) Experimental respiratory carcinogenesis: interaction between alpha radiation and benzo(a)pyrene in the hamster. In. *Experimental Lung Cancer: Carcinogenesis and Bioassays.* International Symposium, Seattle, Washington. June 23–26, pp. 485–491. New York: Springer.

McGowan, G. C. (1955) Physically toxic chemicals and industrial hygiene. *Arch Ind Health* 11:315.

McJilton, C. J., Frank, R., and Charlson, R. (1973) Role of relative humidity in the synergistic effect of a sulfur dioxide–aerosol mixture on the lung. *Science* 182:503–504.

McLean, A. E. M. and McLean, E. K. (1966) The effect of diet and 1,1,1—trichloro-2,2-bis-(p-chlorophenyl)ethane (DDT) on microsomal hydroxylating enzymes and on sensitivity of rats to carbon tetrachloride poisoning. *Biochem J* 100:564–571.

Melekhina , V. P. (1966) The problem of combined action of three mineral acids. *USSR Lit Air Pollut Rel Occup Dis* 16:76–81.

Meredith, P. A., Moore, M. R., and Goldberg, A. (1974) The effect of aluminum, lead and zinc on δ-aminolaevulinic acid dehydratase. *Biochem Soc Tran S* 2:1243–1245.

Mitina, L. S. (1962) Problem concerned with the combined effect on the organism of nitrogen peroxide and sulfur dioxide in low concentrations. *Gigiena i Sanit* 27(10):3.

Mittler, S., Hedrick, D., and Philips, L. (1957) Toxicity of ozone. II. Effect of oxygen and carbon dioxide upon acute toxicity. *Indust Med Surg* 26:63.

Montesano, R., Saffiotti, U., and Shubik, P. (1970a) The role of topical and systemic factors in experimental respiratory carcinogenesis in inhalation carcinogenesis. In M. G. Hanna, Jr., P. Nettesheim, and J. R. Gilbert, eds., *Inhalation Carcinogenesis,* pp. 353–374. Atomic Energy Symposium Commission Series 18. Oak Ridge, Tenn.

Montesano, R., Saffiotti, U., and Shubik, P. (1970b) Synergistic effect of diethylnitrosamine, benzo(a)pyrene and ferric oxide on respiratory carcinogenesis in hamsters. *Cancer Res* (Cited in Montesano, 1970: *Tumori,* 56:335–344).

Moore, M. R. (1972) Lead, ethanol and δ-aminolaevulinate dehydratase. *Biochem J* 129:43–44.

Mori, K. (1964) Acceleration of experimental lung cancers in rats by inhalation of cigarette smoke. *Gann* 55:175–181.

Myers, D. K. (1973) Radiation-induced leukemia in rats: synergistic effect of urethane. *Experientia* 29(7):859–861.

Nakajama, T., Hattori, S., Tateisni, R., and Horai, T. (1972) Morphological changes in the bronchial alveolar system of mice following continuous exposure to low concentrations of nitrogen dioxide and carbon monoxide. *J Jap S Chest Dis* 10:16–22.

Nakamura, K. (1964) Response of pulmonary airway resistance by interaction of aerosols and gases in different physical and chemical nature. *Jap J Hyg* 19:322–333.

NAS (1976) *Vapor-Phase Organic Pollutants: volatile hydrocarbons and oxidation products,* p. 198. Washington, D.C.: National Academy of Sciences.

NIOSH (1976) Criteria for a recommended standard of occupational exposure to methylene chloride. National Institute of Occupational Safety and Health. Cincinnatti, Ohio.

Nobuyuki, I., Nagasaki, H., Arai, M., Makiura, S., Sugihara, S., and Hirao, K. (1973) Histopathologic studies on liver tumorigenesis induced in mice by technical polychlorinated biphenyls and its promoting effect on liver tumors induced by benzene hexachloride, *J Natl Cancer Inst* 51(5):1637–1642.

Parizek, J., Kalouskova, J., Babicky, A., Benes, J., and Pavlik, L. (1973) Interaction of selenium with mercury, cadmium, and other toxic metals. In W. G. Hoekstra, J. W. Suttie, H. E. Ganther, W. Merz, eds., *Trace Element Metabolism In Animals,* vol. 2, pp. 119–131. Baltimore University Park Press.

Parmeggiani, L. and Sassi, E. (1954) Occupational hazard from toluene: Environmental investigations and clinical researches in chronic poisoning. *Med Lavoro* 45:574 (Cited in Brieger, 1957).

Phillips, W. E. J., Hatina, G., Villeneuve, D. C., and Becking, G. C. (1973) Chronic ingestion of lead and the response of the immature rat to parathion: *Bull Environ Contam Toxicol* 9:28–36.

Port, C. D., Henry, M. C., Kaufman, D. G., Harris, C. D., and Ketels, K. V. (1973) Acute changes of surface morphology of hamster tracheobronchial epithelium following benzo(a)pyrene and ferric oxide administration. *Cancer Res* 33:2498–2506.

Pozzani, V. C., Weil, C. S., and Carpenter, C. P. (1959) The toxicological basis of threshold limit values: 5. The experimental inhalation of vapor mixtures by rats, with notes upon the relationship between single dose inhalation and single dose oral data. *Am Indust Hyg Assoc J* 20(5):364–369.

Quadland, H. P. (1943a). Reports of occupational injuries attributed to volatile solvents: Literature study. *Indust Med* 12:734.

Quadland, H. P. (1943b) Carbon tetrachloride: Literature study of reports of occupational injuries attributed to volatile solvents. *Indust Med* 12:821.

Quadland, H. P. (1944) Petroleum solvents and trichloroethylene: literature study of reports of occupational diseases attributed to volatile solvents. *Indust Med* 13:45.

Rhian, M. and Moxon, A. L. (1943) Chronic selenium poisoning in dogs and its prevention by arsenic. *J Pharmacol Exp Ther* 78:249.

Rossiter, C. E. and Weill, H. (1974) Synergism between dust exposure and smoking: an artefact in the statistical analysis of lung function. *Bull Physiol Pathol Resp* 10:717–725.

Rylander, R. (1969) Alteration of lung defense mechanisms against airborne bacteria. *Arch Environ Health* 18:551–555.

Rylander, R., Ohrstrom, M., Hellstrom, P. A., and Bergstrom, R. (1971) SO_2 and particles-synergistic effects on guinea pig lungs. In *Inhaled Particles*. III. 1: 535–541. Proceedings of an International Symposium Organized by the British Occupational Hygiene Society in London, England. London: Urwin Bros. Limited.

Saffiotti, U., Montesano, R., Sellakumar, A. R., Cefis, F., and Kaufman, D. G. (1972a) Respiratory tract carcinogenesis in hamsters induced by different numbers of administrations of benzo(a)pyrene and ferric oxide. *Cancer Res* 32:1073–1081.

Saffiotti, U., Montesano, R., Sellakumar, A. R., and Kaufman, D. G. (1972b) Respiratory tract carcinogenesis induced in hamsters by different dose levels of benzo(a)pyrene and ferric oxide. *J Natl Cancer Inst* 49:1199–1204.

Salem, H. and Cullumbine, H. (1961) Kerosene smoke and atmospheric pollutants. *Arch Environ Health* 2:183–187.

Sanotskij, I. V. (1969) Current state of the problem concerning the combined action of gases, vapors and aerosols. *Toksikol Novy Promyshlem Khimi Vyeshchest* No. 11,6 (Cited in Sidorenko et al., 1976).

Selikoff, I. J., Hammond, E. C., and Churg, J. (1968) Asbestos exposure, smoking and neoplasia. *JAMA* 204:106–112.

Sellakumar, A. R., Montesano, R., Saffiotti, U., and Kaufman, D. G. (1973) Hamster respiratory carcinogenesis induced by benzo(a)pyrene and different dose levels of ferric oxide. *J Natl Cancer Inst* 50:507–510.

Shalamberidze, O. P. (1967) Reflex effects of mixtures of sulfur and nitrogen dioxides. *Hyg San* 32:10–15.

Shtessel, T. A. (1958) (Experimental study of reintroduced industrial compounds). Eksperimental'nye Issledovaniya Po Toksikologi'l Vnov Vuodimykh Promyshlennykh Veshchestv. Leningrad. (Cited in Sidorenko et al., 1976)

Shugaev, B. B. (1967) Combined action of aliphatic hydrocarbons, viewed on the example of butane and isobutylene with reference to their effective concentrations in the brain tissue. *Farmakol Toksikal* 30:102–105 (Cited in NAS, 1976).

Shy, C. M., Alarie, Y., Bates, D. V., Frank, R., Hackney, J. D., Horrath, S. M., and Nadel, J. A. (1974) Synergism or antagonism of pollutants in producing health effects, pp. 483–499. Report to the U.S. Senate Committee of Public Works. National Research Council, National Academy of Science—National Academy of Engineering.

Sidorenko, G. I., Tsulaya, V. R., Korenevskaya, E. I., and Bonashevskaya, T. I. (1976) Methodological approaches to the study of the combined effect of atmospheric pollutants as illustrated by chlorinated hydrocarbons. *Environ Health Perspect* 13:111–116.

Skvortsova, N. N. et al. (1967) The combined effect of small concentrations of sulfur dioxide and benzpyrene on animal organisms in chronic experiments. Proceedings of the Conference on the Research Results for 1967, p. 9. Institute of General and Communal Hygiene imeni. A. N. Sysin of the U.S.S.R. Academy of Medical Science. Moscow, 1968 (cited in Skvortsova and Vysochina, 1976).

Skvortsova, N. N. and Vysochina, I. V. (1976) Changes in biochemical and physiological indices in animals produced by the combined effect of benzo(a)pyrene and phenol. *Environ Health Perspect* 13:101–106.

Smith, A. R. and Mayers, M. R. (1944) Poisoning and fire hazards of butane and acetone. *Indust Hyg Bull* New York State Dept. of Labor, 23:174.

Smyth, H. F., Jr., Weil, C. S., West, J. S., and Carpenter, C. P. (1969) An exploration of joint toxic action: twenty-seven industrial chemicals intubated in rats in all possible pairs. *Toxicol Appl Pharmacol* 14:340–347.

Stanton, M. F. and Blackwell, R. (1961) Induction of epidermoid carcinoma in lungs of rats: A new method based upon deposition of methylcholanthrene in areas of pulmonary infarction. *J Natl Cancer Inst* 27:375–407.

Stayzhkin, V. M. (1962) Hygienic determination of limits of allowable concentrations of chlorine and hydrochloride gases simultaneously present in atmosphere air. Translated by Levine, B. S., *USSR Lit Air Pollut Rel Occup Dis* 9:55.

Stenback, F. G., Ferrero, A., and Shubik, P. (1973) Synergistic effects of diethylnitrosamine and different dust on respiratory carcinogenesis in hamsters. *Cancer Res* 33:2209–2214.

Stokinger, H. E. (1957) Evaluation of the hazards of O_3 and oxides of nitrogen. *Arch Indust Health* 181–190.

Stokinger, H. E. (1960) Toxicologic interactions of mixtures of air pollutants: Review of recent developments. *Int J Air Pollut* 2:313–326.

Stokinger, H. E. (1962) Effects of air pollution on animals. In Stern, A. C., ed., *Air Pollution,* vol. 1, p. 325. New York: Academic.

Stokinger, H. E. (1965) Ozone toxicology: A review of research and industrial experience: 1954–1964. *Arch Environ Health* 10:719–731.

Stokinger, H. E. (1969) The spectra of today's environmental pollution—USA brand: new perspectives from an old scout. *Am Ind Hyg Assoc J* 30:195–217.

Stokinger, H. E., Ashenberg, N. J., Devoldre, J. et al. (1950) Part II. The enhancing effect of the inhalation of hydrogen fluoride vapor on beryllium sulfate poisoning in animals. *Arch Indust Hyg* 1:398.

Stokinger, H. E., Spiegl, C. J., Root, R. E., Hall, R. H., Steadman, L. T., Stroud, C. A. Scott, J. K., Smith, F. A., and Gardner, D. F. (1953) Acute inhalation toxicity of beryllium, IV. Beryllium fluoride at exposure concentrations of one and ten milligrams per cubic meter. *Arch Ind Hyg Occup Med* 8:493.

Svirbely, J. L., Dobrogorski, O. J., and Stokinger, H. E. (1961) Enhanced toxicity of ozone-hydrogen peroxide mixtures. *Am Indust Hyg Assoc* 22:21–26.

Thaxton, P., Yonushonis, W. P., and Baughman, G. R. (1975) Synergistic relationship of temperature, air velocity, and mercury in the chicken. *J Appl Physiol* 38(6):969–973.

Toyama, T. and Nakamura, K. (1964) Synergistic response of hydrogen dioxide, eye aerosols and sulfur dioxide to pulmonary airway resistance. *Indus Health* 2:34–45.

Traiger, G. J. and Plaa, G. L. (1971) Differences in the potentiation of carbon tetrachloride in rats by ethanol and isopropanol pretreatment. *Toxicol Appl Pharmacol* 20:105–112.

Traiger, G. J. and Plaa, G. L. (1973) Effect of isopropanol on CCl_4 induced changes in perfused liver hemodynamics. *Arch Int Pharmacodyn Ther* 202:102–105.

Traiger, G. J. and Plaa, G. L. (1974) Chlorinated hydrocarbon toxicity potentiation by isopropyl alcohol and acetone. *Arch Environ Health* 28:276–278.

Ulanova, N. P. and Zayeva, G. N. (1964) Problems associated with the hygiene establishment of standards for a mixture of gases and vapors. Proceedings of the 10th Moscow Scientific-Practical Conference on Hygiene Problems, pp. 91–92 (cited in Sidorenko et al, 1976).

Vigliani, E. C. (1969) The biopathology of cadmium. *Am Ind Hyg Assoc J* 30:329.

Von Oettingen, W. F. (1944) Carbon monoxide: Its hazards and the mechanisms of its action. Public Health Bulletin, No. 290, U.S. Gov. Printing Office, Washington, D.C. (cited in Stokinger, 1960).

Vorwald, A. J., Reeves, A. L., and Urban, E. C. J. (1964) Experimental beryllium toxicology. In H. E. Stokinger, ed., *Beryllium: Its Industrial Hygiene Aspects,* pp. 201–234. New York: Academic.

Wada, O., Yano, Y., Ono, T., and Toyokawa, K. (1972) Interaction of cadmium and organic mercury compounds against activity of delta-aminolevulinic acid dehydrase *in vitro*. *Indust Health* 10:59–61.

Waldbott, G. L. (1973) *Health Effects of Environmental Pollutants,* p. 94. St. Louis: Mosby.

Westerick, M. L., Gross, P., and McNerney, J. N. (1957) The topographic relation of contaminating iron to experimental modified silicosis. 19th Annual Meeting of the American Industrial Hygiene Association, St. Louis (cited in Brieger, 1957).

Willoughby, R. A., MacDonald, E., McSherry, B. J., and Brown, G. (1972) Lead and zinc poisoning and the interaction between Pb and Zn poisoning in the fowl. *Can J Comp Med* 36:348–359.

Yelfimora, Ye. V., Gusev, M. I., Novikov, Yu. V., Yudina, T. V., and Sergeyev, A. N. (1972). A study of the joint resorptive action of atmospheric pollutants (gases and dust). *Gig San* 37:11–15 (in Russian). (Translated by EPA; available from the Air Pollution Technical Center, 1 Research Triangle Park, N.C. as APTIC No. 57049.) (Cited in Shy et al., 1974.)

Zelac, R. E., Chromroy, H. L., Bolch, W. E., Jr., Donavant, B. G., and Bevis, H. A. (1971) Inhaled ozone as a mutagen. II. Effect on the frequency of chromosome aberrations observed in irradiated chinese hamsters. *Environ Res* 4:325–342.

5

Approaches To Deriving Safe Limits for Pollutant Exposure

"ONE OF THE MOST IMPORTANT principles of occupational health . . . is that exposure to a toxic agent may be permitted up to some limit of tolerance above zero, within which [one] can cope successfully with the insult with no significant threat to [human] health" (Hatch, 1971). This principle, as stated by Hatch, has been the cornerstone on which industrial health standards have been derived in the United States for some 30 years.

Despite the fact that this principle has provided the basis of industrial health standards in the United States, it has not gone without challenge. For example, toxicologists in the Soviet Union generally reject the notion that any type of biochemical adaptation following pollutant exposure should be permitted (Izmerov, 1973). What Hatch refers to as an "indication of a healthy response," the Soviets see as a "predictor of impending injury." Thus the concept of tolerance limits or thresholds of response are markedly different between those toxicologists in the United States who endorse the Hatch perspective and those in the Soviet Union, at least with respect to the interpretation of what indicates an adverse or potentially adverse health effect. A second challenge to the threshold concept emerged with the growth of radiation biology and the rapid development of mechanisms of carcinogenesis. Theoretical frameworks have been proposed that indicate that carcinogens may act according to a nonthreshold or linear fashion; that is, at each dose there is an accompanying response. Not only have carcinogens been viewed as acting according to a linear dose-response manner, but now noncarcinogens (e.g., lead) are also seen in a similar light, especially since the widespread usage of vastly more sensitive indicators of pollutant activity.

As a result of challenges to the credibility of the threshold concept, the validity of previous industrial health standards is now questioned. If the fundamental basis on which these standards has been derived is not trustworthy, how much confidence can one have in the standards they in effect helped to create?

This chapter considers the concept of the dose-response relationship in the process of standard derivation. It focuses on the controversies centered around linear (nonthreshold) versus nonlinear (threshold) dose-response curves in the development of standards for both carcinogens and substances thought to be noncarcinogens (realizing that one cannot prove a negative). Inherent in the establishment of accurate dose-response relationships are the problems of extrapolation from high-exposure doses during experimentation to minute levels as would be more typical of human exposure. The nature of this type of extrapolation is addressed, and approaches toward effectively incorporating this information in the standard derivation process are considered.

THRESHOLD LIMIT

According to Stokinger (1972), the term "threshold limits" for industrial air was derived from the work of the late Lawrence T. Fairhall, Ph.D., the former chief toxicologist for the US PHS in the Division of Industrial Hygiene, because it was more precise than the term maximum allowable concentration. The functional concept of threshold limit value, as explained by Stokinger, is that concentration at which an individual could be exposed for a considerable length of time (8 hours/day, 40 hours/week for a working lifetime) without experiencing an adverse health effect (toxicity, irritation, etc.). This concept of threshold represents a nonlinear relationship between dose and subsequent responses. As previously indicated, a linear dose response is one that exhibits a response at each level of pollutant exposure; that is, there is no dose at which there is no response. Figure 5-1 shows the difference between linear and nonlinear dose-response relationships.

When the nonlinear dose-response relationship was developed, many of today's sensitive indicators of biologic responses were not in use. Consequently, in the early years of developing tolerance or threshold limits for pollutant toxicity, the emphasis was to prevent ill health at a level to prevent fatalities from occupational disease and the elimination of acute illness. Today the emphasis is more and more directed toward preventing subclinical health effects. Thus it is vital to associate the concept of threshold with a specific response. In fact, when TLVs were established, they were intended to be a guide for employers (to achieve) so that specific adverse health effects could be prevented.

Hatch (1972) indicated that it is necessary to be aware that there are multiple types of biologic responses to environmental stressors that may tend to overlap on the dose axis of the dose-response chart. He correctly concluded that a zero response [that is, a response similar to the control group (Herman, 1971)] at a concentration greater than zero cannot be interpreted to indicate an absolute non-response, but only the degree of sensitivity of the specific type of assay used. If it is true that an effect

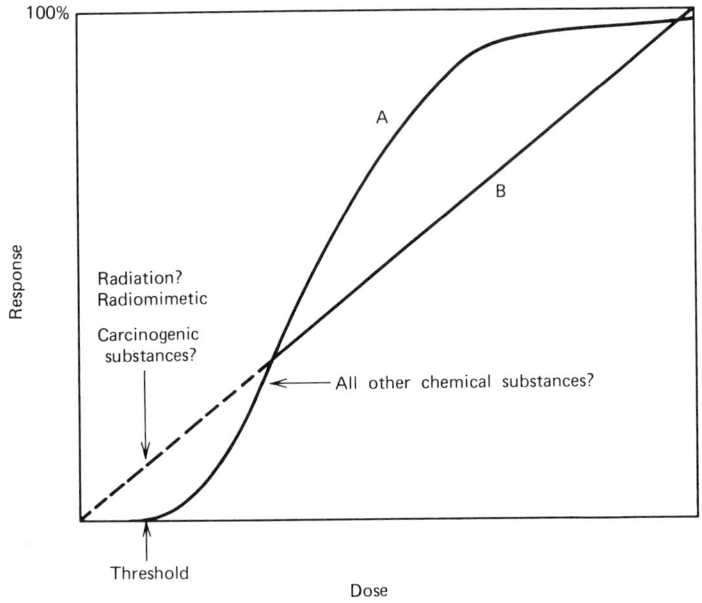

Figure 5-1. Dose-response relationship for a noncarcinogenic agent (A) and a carcinogenic (B) toxic substance. From Herbert E. Stokinger (1972) Concepts of thresholds in standard setting. Arch Environ Health 25:155.

may occur at any dose, how can one establish a permissible level of exposure?

According to Hatch (1971), the problem of establishing permissible exposure levels should be addressed by asking the question, do tolerance limits of contact with toxic agents exist below which an exposed individual will experience no threat to his health, however long the experience? If the answer is yes, as Hatch contends, the following criteria must be met.

1. The kind (type) and maximum degree of response has to be selected which serves with sufficient uniqueness and sensitivity to distinguish between health and ill health.
2. The magnitude of this response must be shown to remain below the critical level over a range of doses of the toxic agent from zero up to the established tolerance limit.

Should every response to a pollutant be considered an adverse health effect? What "kinds" of responses should be classified such that health and ill health can truly be distinguished? Is there truly a precise demarcation between these two conditions?

The answers to these questions will help establish the protocol on which environmental and occupational health standards can be derived. According to toxicologists such as Hatch (1971, 1972) and Stokinger (1972, 1977), experimental animals and humans respond to pollutant exposure with a natural process of adaptation. Adaptive responses are seen as homeostatic processes by which an organism protects itself. Stokinger (1972, 1977) states matter of factly that the existence of adaptive responses is the most convincing physiologic argument supporting the biologic reality of thresholds. Toxicity may be conceptualized as the result of two competing reactions—one is the action of the toxicant on the body, and the competing response is the adaptive (or homeostatic) action. Thus toxicity may result only when the adaptive responses have not been sufficient to overcome the action of the toxicants (Figure 5-2). This conceptual framework has been verified in numerous organisms, including humans, and with numerous substances. For example, thresholds for numerous sensory irritants are well known (Stokinger, 1972). The occurrence of biochemical mechanisms for detoxification and excretion for a wide variety of xenobiotic substances is another example of these adaptive homeostatic processes (Parke, 1968).

Unfortunately, there is no general agreement about what constitutes an adverse health effect. This is very evident in the rather large discrepancies between TLVs in the United States and industrial standards

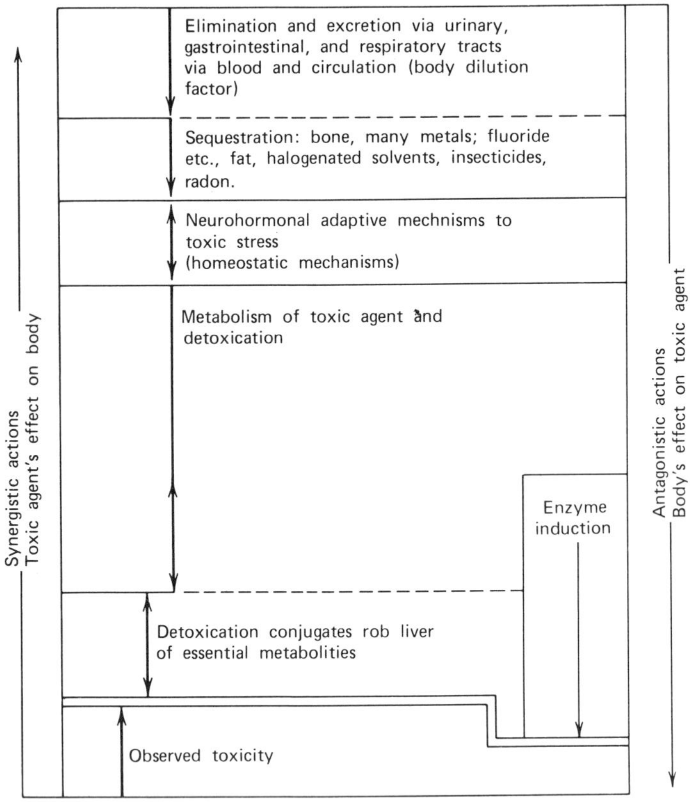

Figure 5-2. Analysis of the effects of a toxic substance. From Stokinger, Concepts of thresholds in standard setting. ***Arch Environ Health*** 25:155. Copyright 1972, American Medical Association.

in the USSR. Fundamentally, one's approach to standard setting depends in large part on the accepted concept of the term "state of health." In the case of most United States industrial standards, it is clear that "no threat to health is anticipated so long as the exposure does not induce a disturbance of a kind and degree that overloads the normal protective mechanisms of the body (Hatch, 1972)." In marked contrast, for the USSR, "a potential for ill health is said to exist as soon as the organism undergoes the first detectable change of whatever kind from its normal state" (Hatch, 1972).

Although these two perspectives are different, they do possess a fundamental agreement—that of the existence of thresholds above zero. However, there is a rather large contingent of scientists who feel that it is not possible to establish a threshold or a no-effect level for any substance. Even Stokinger (1972), who helped formulate the concept of adaptive thresholds in the Unites States, acknowledges the possibility of nonthreshold environmental agents such as ionizing radiation and some natural body metabolites like carbon monoxide and carbon dioxide.

LINEAR DOSE RESPONSE

There has been a strong attempt to rebutt the linear nonthreshold concept. The major thinking has evolved out of a paper by Hutchinson (1964) entitled "The Influence of the Environment" and has been expanded in complementary fashion by Dinman (1972), Friedman (1973), Krisko and Bolander (1974) and more recently by Stokinger (1977).

To develop the argument against a linear dose-response relationship it is necessary to accept a specific definition of the word "effect." This is in fact the same argument between United States and Soviet approaches to standard setting with respect to defining an adverse health effect. For Dinman (1972), the effect is not necessarily considered either beneficial or deleterious. In fact, the total thrust of the Dinman paper is to challenge the notion that an effect caused by an interaction of an environmental "pollutant" and a biologic system is by necessity deleterious. If the thesis of Dinman can be sustained, the concept of nonthreshold for chemical effects becomes vulnerable.

The following statements put forth by Dinman (1972), summarized by Friedman (1973), when taken collectively tend to support the threshold concept.

1. A cell (hepatocyte) is extimated to have approximately 10^{14} molecules and atoms with which a xenobiotic substance may interact.
2. A major factor influencing activity is molecular specificity as compared to the mere presence of an atom or molecule in a cell.
3. There are lower concentration limits for the occurrence of biologically significant intracellular molecular interactivity. Numerous examples of in vitro studies of specific inhibitors have demonstrated a lower concentration limit for such inhibition is 10^{-8} M.
4. The binding, or interaction with proteins or other molecules at sites when there is no resulting functional effect, may not infrequently happen.
5. The fact that all chemical components of the cells and cells themselves are in a dynamic flux. The major consideration is that the rates of loss exceed the rates of normal replacement.

6. Cells of different types of tissue have the capacity to induce normal DNA repair mechanisms to repair genetic damage caused by environmental mutagens (Cleaver, 1968). In fact, the absence of DNA repair mechanisms in persons with xeroderma pigmentosum clearly demonstrates the life-saving functional capacity of this process in normal individuals (Cleaver, 1968).

Based on the foregoing list of supportive statements and the estimates by Hutchinson (1964) and Dinman (1972) that 10^4 molecules per cell is the limiting concentration of biologic activity, Friedman (1973) attempted to calculate "safe" or threshold levels of activity for various substances. He assumed there are 6×10^{13} cells in a 70-kg human. Then he calculated that one needs 8.6×10^{15} molecules per kilogram of cells as a minimum to induce a biologic effect (or about 1×10^{-8} M). Using this scheme, he calculated that 0.12 mg of vitamin A per day or $\frac{1}{10}$ of a RDA amounts to a daily exposure of 3.6×10^{15} molecules per kilogram of cells. However, despite this seemingly large number of molecules, it is still not sufficient to prevent severe deficiency disease symptoms. Furthermore, he presented the interesting example of normal postmenopausal women and adult men having estradiol blood levels of approximately 20 ng/liter of plasma. If one conservatively estimates (as Friedman does) that the total body quantity of estradiol is in the plasma (50 ng in a 70-kg man), the adult man or postmenopausal woman has 1.5×10^{12} molecules of this hormone per kilogram of cells without notable estrogenic effect. Friedman mentioned the case of diethylstilbestrol (DES) in which the lowest level capable of producing a detectable estrogenic effect (e.g., increase in uterine weight) occurs at approximately 0.5 ppb. If mice were exposed to $\frac{1}{100}$ of the level that causes this minimum estrogenic effect, a total of 2.8×10^{12} molecules per kilogram of cells would be present. Using this identical methodology, Friedman calculated the estimated no-effect quantities of certain carcinogens and other toxic agents (Table 5-1).

Friedman's conclusion is quite clear—thresholds do indeed exist in biologic systems, and there are levels below which a response does not manifest itself. Consequently, the application of the linear nonthreshold dose-response relationship inadequately, in fact inaccurately, represents the interactions of toxic substances within biologic systems. Thus, according to Friedman, "acceptable risk doses" need not always be the "practical equivalent of zero."

The reasoning of Dinman (1972) and Friedman eloquently defends the position of the existence of thresholds. However, their position becomes less tenable when the sensitivity of the assay is considered. For

Table 5-1. Estimated No-Effect Quantities Of Some Potent Carcinogens And Toxic Agents

Compound	Molecular Weight	"Estimated" No-Effect Level (g/kg body weight)	Molarity	Molecules per Kilogram Body Weight
Aflatoxin	312	5×10^{-9}	1.6×10^{-11}	9.6×10^{12}
1,2,5,6 dibenzanthracene	278	2.5×10^{-4}	9×10^{-7}	5.4×10^{17}
1,2,5,6 dibenzanthracene (subcutaneous)	278	2.5×10^{-5}	9×10^{-8}	5.4×10^{16}
Methylcholanthrene (subcutaneous)	268	2.5×10^{-4}	9.3×10^{-8}	5.6×10^{16}
3,4 Benzpyrene (subcutaneous)	252	2.5×10^{-4}	1×10^{-6}	6×10^{17}
3,4 Benzpyrene (skin)	252	2.5×10^{-6}	1×10^{-8}	6×10^{16}
Aramite	335	1×10^{-1}	3×10^{-4}	1.8×10^{20}
Tetrachlorodibenzodioxin	320	6×10^{-8}	1.9×10^{-10}	1.1×10^{14}
Botulinum toxin (mouse)	900,000	6.5×10^{-11}	7×10^{-17}	4.2×10^{7}

Source: Leo Friedman (1973) Problems of evaluating the health significance of chemicals present in foods. *Pharmacology and the Future of Man*, vol. 2, p. 38. Proc. 5th Int. Congr. Pharmacology, San Francisco. Basel: Karger.

example, the reporting of uterine weight increase as the indicator of estrogenic effect for certain levels of DES may not be the most sensitive indicator. To a certain extent, the person who is trying to disprove the existence of linear dose response is seemingly faced with trying to prove a negative because of the recurring criticism of the "need for a better assay."

BIOASSAYS AND THE NO-EFFECT LEVEL

An example of the search for more and more sensitive assays that will serve as biologic indicators of pollutant toxicity can be seen in the case of inorganic lead. It is generally accepted that adults who are exposed to lead such that they exhibit a blood lead (Pb-B) level of equal to or greater than 70 μg lead per 100 g blood have an increased risk of developing neurological or nephrotic damage (Baloh, 1976; Repko, 1976; Joselow, 1976). However, a broad range of biochemical tests for lead toxicity have emerged that are also employed as indicators of lead exposure [e.g., aminolevulinic acid levels in urine (ALA-U) and ALA dehydrase levels in blood (ALAD-B)]. These tests are often used to complement other diagnoses of lead exposure such as Pb-B and Pb-U. It is interesting to note that although lead affects several of the precursors of heme biosynthesis, blood lead levels must usually reach greater than 110 μg lead per 100 g blood before anemia is noted (Stopps, 1974). Consequently, the neurological effect is seen at a lower dose than is the anemic effect. However, do these metabolic changes (e.g. decrease in ALAD activity) in heme biosynthesis markedly affect the health of the individual? In testing this hypothesis, Stopps (1974) reported that when ALAD levels were inhibited to a level of only 1–2% of normal by lead exposure, no adverse effect (as measured by recovery from severe hemorrhage) was noticed in dogs. This undoubtedly results because there is a hugh excess of ALAD available for use in the heme biosynthesis process. The activity of ALAD is clearly not the rate-limiting step in the synthesis of heme. However, ALAD is a very sensitive indicator of lead exposure. Even slight exposure to lead is known to reduce ALAD values (see chapter 8).

The use of ALAD as a bioindicator for childhood lead exposure has been widely discussed. However, these questions arise: What is the significance of a reduced ALAD level in and of itself? Is a reduction of the ALAD level really an adverse health effect? One could agree with the linear threshold position. Yet Dinman's concept of effect would certainly be used as a rebuttal by the opposing camp.

LATENCY AND DOSE IN CANCER DEVELOPMENT

An important consideration in the development of any health assessment of a pollutant's toxicity is the time dimension over which a particular range of doses effects given responses. If one is concerned with cumulative noncarcinogens (possibly Pb) that may act according to a threshold response, toxicity will manifest itself only after the body burden of lead has exceeded the individual's threshold limit. If the level of excretion is in equilibrium with total environmental exposures below the threshold limit, toxicity will not manifest itself. Noncarcinogens that have rather long biologic half-lifes (i.e., slow excretion rates), such as cadmium, lead, and mercury, may exhibit their toxicity considerably after initial exposure. This time interval between initial exposure and subsequent toxicity should not be considered a latency period in the traditional sense of the word. A latency period is usually meant to be that time interval from the initial exposure of an individual to the time at which a tumor resulting from such an exposure is recognized. Therefore, latency periods are usually considered with respect to carcinogens. The dose-response curve for carcinogens must be represented in a three-dimensional way because each dose must be considered with respect to time and effect. For example, during dose-response studies for carcinogens, one may consider the number of animals effected, the number of tumors, and the time at which the tumors manifest themselves. Bingham (1971) developed a two-dimensional, but multifaceted dose-response curve that addresses the complexity of carcinogenic effects. Figure 5-3 represents the dose-response curve for a hypothetical carcinogen with respect to the percentage of affected animals with tumors and the percentage of life free from tumor. The latter consideration is a functional way of relating latency to dose and response.

Jones and Grendon (1975) refined the traditional concept of latency by attempting to integrate the "pattern of accumulation of the dose," because they rightly recognized the uncertainty of whether the first exposure or the first multiple series of exposures initiates the carcinogenic response. They have developed what is called a "weighted latent period" which designates a certain weight to each interval in the degree to the dose absorbed at the start of that interval. In practical terms,

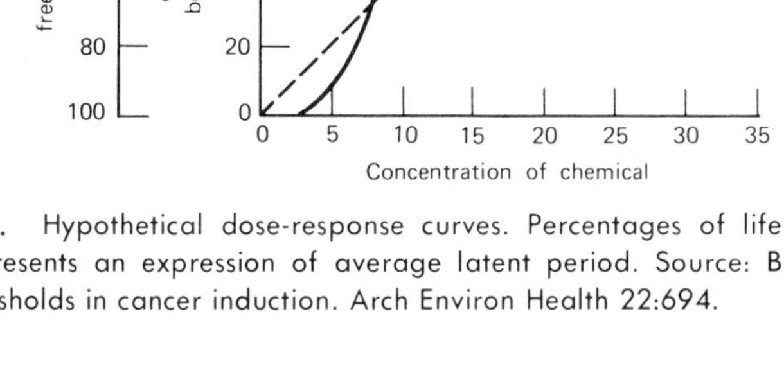

Figure 5-3. Hypothetical dose-response curves. Percentages of life free from tumors represents an expression of average latent period. Source: Bingham, E. (1971) Thresholds in cancer induction. Arch Environ Health 22:694.

when dealing with radiation-associated tumors, Jones and Grendon calculated the latent period by dividing the cumulative rad years (CRY) "which is the total of the products of the increments of dose and the prospective elapsed time intervals since each increment was received by the cumulative dose (CR) which is the cumulative dose in rads averaged over the entire skeleton, from initial exposure to the time of the survey."

Bingham (1971) described a schematic representation of the process that leads to the appearance of experimentally induced tumors. Figure 5-4 shows a model for the development of a squamous carcinoma following topical treatment of a polycyclic aromatic hydrocarbon like benzo(a)pyrene. The relationship of initiation (or primary event), latent period, and malignancy recognition are demonstrated. Natural defense mechanisms, including DNA repair and immunologic defenses are also included.

The latency period is an important consideration in standard setting because it provides an extra-dimensional assessment of carcinogenic activity; substances with short latency periods would certainly be given first consideration in the development of priority schemes. In contrast, if a latency period was found to occupy a rather large percentage of the individual's natural life span, it would most likely assume a low priority. Table 5-2 lists the proposed latency periods for a variety of human carcinogens. The variation between carcinogens with respect to latency is considerable. Such information should assume a critical role in the derivation of standards for carcinogens, since it is expected to modify cost/benefit ratios that may be developed during the process.

The relationship of dose to latency is a relatively new concept that had its origin with Druckrey (1967). Jones and Grendon (1975) succinctly summarized the way this concept developed. They indicated that when laboratory animals were given doses of the nitrosamines, dialkylnitrosamines, dialkylaminoazobenzenes, and dialkylaminostilbenes close to the acutely lethal level, they developed malignant tumors in a uniform fashion. In addition to the capability of causing cancer, these agents have been therapeutically used to destroy cancer cells. The incidence of tumors resulting from this application at high dosage in rodents is so high that only the time delay in tumor appearance can distinguish one potential therapeutic agent from another—thereby determining an effective measure of activity. Thus, by trying to distinguish the relative carcinogenic potency of these compounds, Druckrey developed the dose-latency relationship. This research lead Druckrey to conclude that the latent period was an inverse functional power of the dosage of the carcinogen. According to Jones and Grendon, it was not

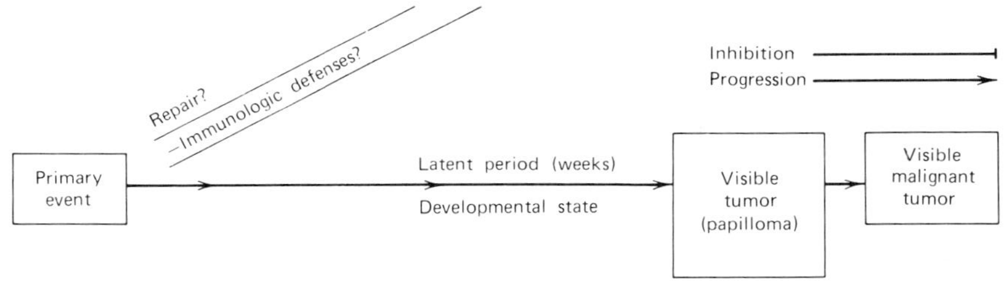

What is the primary event?	What occurs during developmental stage?	What is a malignant tumor?
Experimental procedure: Exposure to carcinogen. Speculations concerning mechanism: Cellular change—inheritable Binding to DNA—carcinogen or its metabolite Activation of virus Alteration of cellular respiration Enzyme deletion, etc.	Experimental procedure: Exposure to carcinogen may continue Noncarcinogen may be applied No further treatment Speculations concerning mechanism: Hyperplasia Inflammatory response Release of lysosomal enzymes Immunologic defenses overwhelmed, etc.	Criteria for diagnosis as a cancer: Morphologic features, degree of invasiveness, metastases, and transplantability are considered in making judgments Biochemistry of tumors: Enzymatic characterization Respiration Immunologic status of host

Figure 5-4. Events leading to the appearance of experimentally induced skin cancer in mice. Source: Bingham, E. (1971) Thresholds in cancer induction: If they exist, do they shift? ***Arch Environ Health*** 22:694. Copyright 1971, American Medical Association.

Table 5-2. Classification of Occupational Carcinogens

Agents	Affected Organ(s)	Incubation Period (yr)	Risk Ratio	Occupation
A. Organic Agents				
1. Aromatic hydrocarbons				
Coal soot Coal tar Other products of coal combustion	Lung, larynx, skin, scrotum, urinary bladder	9–23	2–6	Gashouse workers, stokers, and producers; asphalt, coal tar, and pitch workers; coke-oven workers; miners; still cleaners; chimney sweeps
Petroleum Petroleum coke Wax Creosote Anthracene Paraffin Shale Mineral oils	Nasal cavity, larynx, lung, skin, scrotum	12–30	2–4	Contact with lubricating, cooling, paraffin, or wax fuel oils or coke; rubber fillers; retortmen; textile weavers; diesel jet testers
Benzene	Bone marrow (leukemia)	6–14	2–3	Explosives, benzene, or rubber cement workers; distillers; dye users; painters; shoemakers
Auramine Benzidine α-naphthylamine β-naphthylamine Magenta 4-aminodiphenyl 4-nitrodiphenyl	Urinary bladder	13–30	2–90	Dyestuffs manufacturers and users, rubber workers (pressmen, filtermen, laborers), textile dyers, paint manufacturers

2. Alkylating agents				
Mustard gas	Larynx, lung, trachea, bronchi	10–25	2–36	Mustard gas workers
Isopropyl oil	Nasal cavity	10+	21	Producers
Vinyl chloride	Liver (angiosarcoma), brain	20–30	200(liver) 4(brain)	Plastic workers
Bis(chloromethyl) ether Chloromethyl methyl ether	Lung (oat cell carcinoma)	5+	7–45	Chemical workers
B. Inorganic agents				
1. Metals				
Arsenic	Skin, lung, liver	10+	3–8	Miners, smelters, insecticide makers and sprayers, tanners, chemical workers, oil refiners, vintners
Chromium	Nasal cavity and sinuses	15–25	3–40	Producers, processors, and users; acetylene and aniline workers; bleachers; glass, pottery, and linoleum workers; battery makers
Iron oxide	Lung, larynx	—	2–5	Iron ore (hematite) miners, metal grinders and polishers, silver finishers, iron foundry workers
Nickel	Nasal sinuses, lung	3–30	5–10(lung) 100+(nasal sinuses)	Nickel smelters, mixers, and roasters; electrolysis workers

Table 5-2. *(Continued)*

Agents	Affected Organ(s)	Incubation Period (yr)	Risk Ratio	Occupation
2. Fibers				
Asbestos	Lung, pleural, and peritoneal mesothelioma	4–50	1.5–12	Miners; millers; textile, insulation, and shipyard workers
3. Dusts				
Wood	Nasal cavity and sinuses	30–40	—	Woodworkers
Leather	Nasal cavity and sinuses, urinary bladder	40–50	50(nasal sinuses) 2.5(bladder)	Leather and shoe workers
C. Physical agents				
1. Nonionizing radiation				
Ultraviolet rays	Skin	varies with skin pigment and texture	—	Farmers, sailors
2. Ionizing radiation				
X-rays	Skin, bone marrow (leukemia)	10–25	3–9	Radiologists, medical personnel
Uranium Radon Radium Mesothorium	Skin, lung, bone, bone marrow (leukemia)	10–15	3–10	Radiologists, miners, radium dial painters, radium chemists
3. Other				
Hypoxia	Bone	—	—	Caisson workers

Source: Cole, P. and Goldman, M.B. (1975) Occupation. In *Persons At High Risk of Cancer: An Approach To Cancer Etiology and Control*, pp. 172–176. New York: Academic.

until the use of substances such as the nitrosamine, which demonstrates a high cancer potency and a low acute toxicity, that researchers were able to obtain precise information over a broad range of dosages to establish quantitative relationships between dose and latency. Previous research centered on the occurrence of a few tumors in the low-dose range, whereas precise determination of the occurrence of the tumor was ignored as a result of the considerably smaller degree of its variation throughout this narrow range of doses.

With respect to nitrosamines, Jones and Grendon reported that the logarithm of the dose versus the time of the occurrence of the tumor is generally recognized to be linear over a thousandfold range of dosage. This dose-latency response is similar for other carcinogens, including radiation and several polycyclic aromatic hydrocarbons. The latency period seems to vary according to the inverse cube root of the dose. Thus, if the dose decreases by a factor of 1000, the latency period increases by a factor of 10, and the reverse is true.

Albert and Altshuler (1973 and 1976) established a novel approach for the extrapolation of laboratory studies to the derivation of standards by incorporating the concept of age at the time of the appearance of an adverse health effect as well as its incidence. The traditional technique of evaluating carcinogenic hazards such as determining the dose to incidence relationship may be very misleading and can provide an inappropriate degree of importance in the decision-making process. This traditional approach does not consider the age at which new instances of cancer occur or the possibility that additional carcinogenic exposure could affect people who would have developed cancers from different causes. Albert and Altshuler offer a more fundamental approach to the task of quantifying carcinogenic risks because the adverse effects are characterized not only with respect to excess cancer incidence, but also in terms of life shortening. According to the NAS (1975), commenting on the work of Albert and Altshuler (1973), it may be possible to establish standards for carcinogens that would limit the possibility of environmentally induced tumors only to those very advanced in age, with approximately no more than a 10% increase in the chance of cancer at 95 years of age. This concept is very much like the proposal of Jones and Grendon that the latent period may be so long at low doses that it may extend beyond the normal life span of the individual.

This dose-latency mathematic relationship is highly striking because it provides the possibility of offering regulatory agencies a way in which to effectively deal with carcinogens in the standard derivation process. If this hypothesis is validated, it offers a situation whereby one can philosophically accept the existence of a linear dose-response relationship, while not necessarily accepting the necessity of a zero exposure level to prevent human cancers. This is possible because, under certain conditions, the latency period may be such that it exceeds the normal life span of the organism. Thus, according to Druckrey (1959), it is impossible to determine a tolerable dose by extrapolating from higher to lower doses of carcinogens. However, the reason why very small doses of known carcinogens produce no cancer is not that there is a threshold response, but that the latent period of some carcinogens exceeds the total life span of the organism.

The role of dose as a critical factor in the development of environmentally induced tumors has also been the focus of biomedical research in the Soviet Union. This notion has been studied by Yanysheva and Antomonov (1976), who reported on the dose-time-effect relationship with respect to the carcinogen benzpyrene. They found that the number of animals with tumors decreases as exposure to the benzpyrene decreases. Furthermore, and in agreement with Druckrey (1967), the latent period varies inversely with the dose. Based on their results, Yanysheva and Antomonov developed a dose-time-effect relationship, as seen in Table 5-3. Based on this information, they recommended a

Table 5-3. Calculated Time For Appearance Of The First Lung Tumor Following Administration Of Various Total Benzpyrene Doses In Ten Portions Intratracheally

Benzpyrene Dose (mg)	Time of Tumor Occurrence (mo)
0.1	27.0
0.05	38.0
0.02	67.9
0.01	118.9
0.005	221.0
0.002	527.3

From Table 3 of Yanysheva, N.Ya. and Antomonov, Yu.G. (1976) Predicting the risk of tumor occurrence under the effect of small doses of carcinogens. *Environ Health Perspect* 13:99.

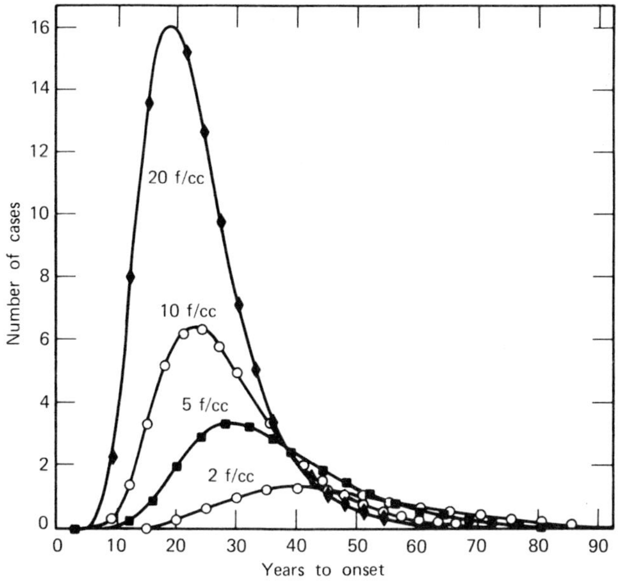

Figure 5-5. Respiratory cancer at four intensity levels for a linear dose-response relationship. Source: Enterline, P.E. (1976) Pitfalls in epidemiological research. ***J Occup Med*** 18(3):151.

permissible dose of benzpyrene of 0.02 mg that could be effectively used in the calculation of a safe air standard. They concluded that any carcinogenic effect would occur considerably after the normal life span of the individual had been exceeded.

The use of this dose-time-effect relationship applies to other carcinogens. Enterline (1976), in a review of 11 major epidemiologic studies of asbestos exposure and respiratory cancer, attempted to apply the inverse cube root relationship to the asbestos-cancer response. Figure 5-5 shows the respiratory cancer incidence at multiple exposure levels for a linear dose-response relationship. Consequently, as Enterline pointed out, this figure shows that workers who have been exposed to low levels of asbestos (i.e., 2 fibers/cc) for 20–30 years very rarely exhibit enhanced respiratory cancer responses. Conversely, groups with high exposure to asbestos may exhibit considerably reduced latent periods with respect to respiratory cancer. Further evidence supporting the inverse dose relationship with respect to diethylnitrosamine in rats was noted by Ariens et al. (1976) and Suss et al. (1973). Figure 5-6 represents data corroborating the previously noted conclusions of Druckrey (1967).

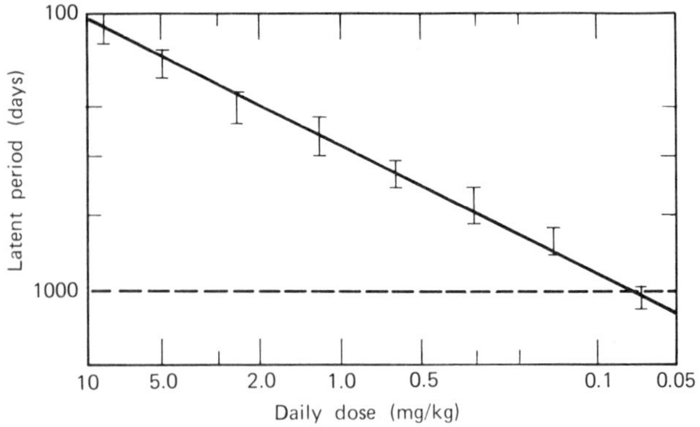

Figure 5-6. The relationship between the latent period for tumor (hepatocellular carcinoma) development and the dose of the carcinogen (diethylnitrosamine) in rats. Since plotting on a double logarithmic scale results in a straight line, the conclusion must be that dosages too low to make the carinogenic action manifest within the life span of the rats (± 1000 days) are not free from carcinogenic action. From Süss, R., V. Kinzel, and J. D. Scribner (1973) ***Cancer, Experiments and Concepts***, p. 50. Berlin and New York: Springer.

The inverse cube root concept of carcinogen dose-time relationships is based on limited data that have been summarized by Jones and Grendon (1975). It is highly appealing because of its (1) simplicity, (2) potentially widespread application to a multitude of environmental carcinogens, (3) consistency with the linear dose-response relationships, and (4) direct application to the derivation of health standards for environmental and occupational carcinogens. In fact, Jones and Grendon attempted to use this relationship to guesstimate safe doses for ionizing radiation, nitrosamines in foods such as cured meats, and DES as an additive to animal feed. However, before there is widespread application of this concept, it would be wise to recall that the latent period is affected by many factors in addition to dose. It is generally acknowledged by those studying environmental carcinogenesis that numerous biologic factors may affect the carcinogenic response of experimental animals. These factors include age, metabolic activity, including hormonal status, sex, diet, genetic factors (species susceptibility), and strain differences (inbred vs. random bred), and viruses (Bingham, 1971). Furthermore, Epstein (1973) contended that it is impossible to effectively predict carcinogenic responses because interactions of certain carcinogens with other substances may drastically affect their threshold for carcinogenic action. Sinhuber et al. (1969) supported this view with data that indicate that the production of hepatomas in trout by aflatoxin B_1 is markedly enhanced by the addition of noncarcinogenic oils to the diet. A similar experiment supporting the previous Epstein contention may be taken from the research of Bingham and Falk (1969), which showed that the threshold for the benzo(a)pyrene carcinogenic response can be decreased by a factor of 1000 by the use of the noncarcinogen n-dodecane as a solvent. In the same article, Bingham and Falk noted that the average latent period for benzo(a)pyrene-induced tumor development was also markedly reduced by the use of the application vehicle 1-phenyldodecane. On the other hand, the effective dosage of a carcinogen may be increased if anticarcinogens are present (Falk et al., 1964).

SAFE DOSES FOR CARCINOGENS

There are a number of scientific problems that beset regulatory agencies such as the EPA, OSHA, and the FDA during the standard derivation process. Perhaps the most difficult general type of problem involves the process of extrapolation. Extrapolation is the capacity to predict activity in one area based on activity in another. There are several different types of extrapolation concurrently in operation during the assessment of health risk with respect to pollutant toxicity. For example, there is the common type of extrapolation that involves predicting human responses based on animal studies. Other types of extrapolation involve (1) the prediction of adverse health effects on individuals who are considered at high risk with respect to the pollutant under consideration when only healthy subjects are used and (2) the prediction of potential adverse health effects on organisms exposed to a low dosage of toxicant when only data from relatively acute exposures are available. The recognition of these multiple, concurrent extrapolations should illustrate the complexity of the problem of trying to assess safe or no-effect doses for toxic substances including carcinogens. Chapters 2 and 3 considered the nature of animal predictive models and high-risk group extrapolation procedures, respectively. This section considers the nature of extrapolation from high to low doses and its limits in the standard derivation process.

Toxicity and carcinogenesis testing have been traditionally conducted using animal models exposed to levels of toxicants usually far in excess of those to which an average human would normally be exposed. The controversy about the carcinogenic potential of saccharin, the artificial sweetener, illustrates this point. One of the major studies supporting the carcinogenicity in rats that the FDA has publicly cited involved exposing the rats to such a high level of saccharin that an equivalent human dose would be 800 cans of dietary soft drink each day for one's entire life. Therein lies the major problem regarding extrapolation from high to low doses. Is it reasonable to assume that what occurs at extremely elevated levels of exposure predicts with sufficient reliability what will occur at concentrations orders of magnitude lower? When humans will be exposed to typically low or even trace amounts of potential toxicants in their food, water, or air, why do researchers continually conduct their experiments at such high and unrealistic levels of exposure? The answers to such critical questions have been attempted by a governmental study group and published in "Reports of the Advisory Panels on Carcinogenicity, Mutagenicity and Teratogenicity" (HEW, 1969). The report indicated that "for carcinogenicity, teratogenicity and mutagenicity pollutants must be tested at higher levels than those of general human exposure . . . [since it] is essential to the attempt to reduce the gross insensitivity imposed on animal tests by the small size of samples routinely tested, such as 50 or so rats or mice per dose level per chemical, compared with the millions of humans at presumptive risk." Epstein (1973) and Epstein et al. (1969) illustrated the logic of the Panel's decision with the following reasoning and example. They conjectured these assumptions: (1) man is equal in sensitivity to the rat or mouse model to a certain carcinogen or teratogen, (2) the chemical agent will cause cancer or teratogenic effects in 1 in 10,000 people who become exposed. With these assumptions in mind, it becomes obvious that the chances of noticing any adverse health effect in groups of 50 rats or mice when tested at the projected level of human exposure would be miniscule. They go on to state that the testing of 10,000 rats or mice would be required to detect one cancer or teratogenic condition; thus, to be able to claim that this event could not be reasonably explained by chance, at least 30,000 rodents may be required for study.

Stokinger (1977) sharply criticized the assumed validity of extrapolating from high experimental doses to low levels of exposure. The major problem with this downward extrapolation is that high-dose exposures to pollutants may totally overwhelm many of the natural endogenous antagonists to carcinogenic responses that are functional at low-level exposures. Stokinger cited the example of vinyl chloride in which elevated exposure of greater than 150 ppm completely overwhelmed the liver's resources of glutathione, a detoxifying agent. However, at 50 ppm or less, the vinyl-free radical was completely detoxified and was thus incapable of producing its carcinogenic response. This type of information lead Stokinger (1977) to provide not only a strong criticism of traditional toxicologic extrapolation techniques, but also the rationale on which a case can be made for the existence of thresholds for carcinogenic responses. Table 5-4 presents evidence for the existence of thresholds for the carcinogenic action of a variety of well-known carcinogens. In each case, high doses induced tumor development, whereas low doses were unable to do so. According to Stokinger (1977), this is presumably because the normal detoxifying responses have been exposed to such a huge and unrealistic dose (i.e., many orders of magnitude greater than an expected human exposure) that they have not been able to handle the load. However, when the exposures are lowered to more reasonable doses, the situation is easily handled by the adaptive (i.e., detoxifying) responses of the body.

A 1971 Food and Drug Administration Advisory Panel on Carcinogenesis Testing for food additives and pesticides also addressed the question of predicting safe doses of carcinogens. In effect, their answer is really one of a statistical nature similar to that of Epstein (1973) and Epstein et al. (1969). The panel concluded that a substance can never be proven to be noncarcinogenic. All that can be determined is the upper limit of statistical reliability of an agent's possible carcinogenicity. Obviously, the larger the number of animals studied with negative findings, the closer the upper limit approaches zero probability. However, there is no finite test size that causes the possibility to equal zero. The panel further pointed out that "even with as many as 1000 test animals and using only 90% confidence limits, the upper limit yielded by a negative experiment is 2.3 cancers per 1000 test animals. No one could wish to introduce an agent into a human population for which no more could be said than that it would probably produce no more than 2 tumors per 1000. To reduce the upper limit of risk to 2 tumors per one million with confidence coefficient 0.999 would require a negative result in somewhat more than three million test animals."

Realizing that extrapolating from high to low doses with only a small sample size (e.g., 50 specimens per treatment level) has built-in extrapolative uncertainties and that many thousands of animals per treatment would be needed to substantially improve statistical reliability, it has been suggested that megamouse animal studies (extremely large numbers of experimental animals) be used. The purpose of such studies is to overcome the problems of extrapolating from high to low doses, since there would be sufficient animals to test large numbers per treatment group over a broad range of concentrations.

In his article entitled "The Delany Amendment," Dr. Samuel Epstein strongly disagreed with this megamouse concept, both on economic and scientific grounds. Epstein (1973) quoted Dr. U. Saffiotti,

Table 5-4. Evidence for Thresholds in Carcinogenesis

Test Substance	Route	Species	Dose Levels Eliciting Tumors	Dose Levels Not Eliciting Tumors	Duration
Bis-chloromethyl	Inhalation	Rat	100 μg/liter	10 μg/liter 1 μg/liter	6 months daily
1,4-dioxane	Oral	Rat	1% in H_2O	0.1% in H_2O 0.01% in H_2O	2 years
	Inhalation	Rat	>1000 mg/liter	111 mg/liter	2 years daily
Coal tar	Topical	Mouse	6400 mg 640 mg 64 mg	<0.64 mg	64 weeks twice weekly
Beta-naphthylamine	Inhalation Topical	Man	>5% beta in alpha form	<0.5% beta in alpha form	22 years
Hexamethyl phosphoramide	Inhalation	Rat	4000 μg/liter 400 μg/liter	50 μg/liter	8 months
Vinyl chloride	Inhalation	Rat	2500 mg/liter 200 mg/liter 50 mg/liter	<50 mg/liter >10 mg/liter	7 months
Vinylidene chloride	Inhalation	Man	>200 mg/liter	1950–1955, 160 mg/liter average; 30–170 mg/liter range 1960, <50 mg/liter decreasing to 10 mg/liter	25 years

Source: Reprinted from Stokinger, H.E. (July 1977) Toxicology and drinking water. *J Am Water Works Assoc* p. 400 by permission of the Association, Inc.

the former Director of the National Cancer Institute, at a Congressional Hearing concerning the megamouse proposal. Saffiotti contended

> that certain approaches to the problem of identifying a "safe threshold" for carcinogens are scientifically and economically unsound. I have in mind some proposals to test graded doses of one carcinogen down to extremely low levels, such as those to which a human population may be exposed through, say, residues in food. In order to detect possible low incidences of tumors, such a study would use large numbers of mice, of the order of magnitude of 100,000 mice per experiment. This approach seems to assume that such a study would reveal that there is a threshold dose below which the carcinogen is no longer effective, and, therefore, that a "safe dose" can be identified in this manner. Now, there is presently no scientific basis for assuming that such a threshold would appear. Chances are that such a "megamouse experiment" would actually confirm that no threshold can be determined. But let us assume that the results showed a lack of measurable tumor response below a certain dose level in the selected set of experimental conditions and for the single carcinogen under test. In order to base any generalization for safety extrapolations on such a hypothetical finding, one would have to confirm it and extend it to include other carcinogens and other experimental conditions such as variations in diet, in the vehicles used, in the age of the animals, their sex, etc. Each of these tests would then imply other "megamouse experiments". The task would be formidable: suffice it to say that an experiment on 100,000 mice would cost about 15 million dollars; if one did 20 such experiments, it would cost 300 million dollars. All this to try and estimate the possible shape of a dose response curve which would still leave most of our problems in the evaluation of carcinogenesis hazards unsolved. This effort would also block the nation's resources for long-term bioassays for years to come and actually, prevent the use of such resources for the detection of potent carcinogenic hazards from yet untested environmental chemicals. If two million mice are made available as resources, they can be used effectively to test 4000 new compounds, each on 500 mice, thereby detecting among them those that are highly carcinogenic in the test conditions.

If the megamouse approach is fundamentally an unfeasible venture, we are back to estimating safe doses by downward extrapolation. The fundamental problem is that downward extrapolation is founded on the assumption, difficult to verify, that the nature of the dose-response relationship for elevated doses is similar to that at low-level exposures (FDA Panel, 1971). Cornfield et al. (1956) noted that the lack of reliability of extrapolations outside the observable experimental range proved to be the fundamental reason for the striking failure of the initial safety evaluation program for the Salk vaccine, even though the observed curve connecting log titer and inactivation time had some theoretical physical chemical basis. Furthermore, Bliss (1935) noticed a straight-line relationship for the effects of carbon disulfide on *Tribolium confusum* between the log dose and response down to the level of 33% response; however, at lower doses there was a significant change in the response curve. Bliss concluded that the effect at low doses was qualitatively different from that at high concentrations. In addition to commenting on the paper by Cornfield et al., the FDA panel noted that even a relationship with as sound a molecular foundation as Boyle's law becomes very unreliable at extremes of pressure and temperature. They concluded in an ominous fashion by stating that it would be unwise to place excessive faith in downward extrapolation to predict safe concentrations for carcinogens when dose-response relationships are primarily empirical and lack a theoretical, physical, or chemical basis.

To predict the type of response a dose outside of the observable range will produce is the essence of the downward extrapolation process. Several potential dose-response curves are possible. Figure 5-7 illustrates three distinctly different curves that have exceptionally separate health implications. Curve A represents the traditional threshold (nonlinear) dose-response relationship. It indicates that effects at the low dose end of the spectrum are negligible despite the fact that elevated doses produce adverse health effects. Using curve A, it is possible to establish a reasonably safe level of exposure at approximately dose 3. The important point is that the shape of this curve would not have been predicted based on the observed data points. In contrast, curve B represents the linear (nonthreshold) dose-response relationship. It indicates that there is an adverse health effect at each level of pollutant exposure. Curve C represents a dose-response relationship that is linear at high and moderate doses, but at low doses it veers upward. Relatively speaking, this curve implies that the substance in question is more serious at low doses than the linear dose response would have predicted. In fact, the carcinogenic risk per unit of high linear energy transfer (LET) radiation (e.g. alpha particles) appears to increase with decreasing dose and dose rates for which the lower limits are not as yet determined (Riddiough et al, 1977). An example of Curve C may be seen with respect to the health effects of low levels of radium-226 in drinking water on the occurrence of bone neoplasms. This conclusion was based on the data derived from an epidemiologic study by Peterson et al. (1966), which considered mortality by bone neoplasm in Illinois and Iowa communities with elevated levels of naturally occuring radium-226 in the drinking water. Based on earlier studies of high-dose radiation exposure and utilizing a linear dose-response relationship (curve B), significantly lower numbers of bone neoplastic deaths were predicted than actually occurred—thereby establishing data points to support the existence of curve C for high LET radiation.

Riddiough et al. (1977) presented an assessment of four theoretical approaches (EPA, 1975; Gofman and Tamplin, 1970; Evans, 1967; and Peterson et al., 1966) that predicted the number of cancers per 10^6 persons per year at various levels of radium-226 in drinking water. All except Peterson et al. predicted carcinogenic risk from extrapolation from high to low doses. In the case of Peterson et al. the original data were derived from low-level exposures occurring in natural drinking water sources. It is interesting to note that the EPA and Gofman and Tamplin accepted the linear dose-response model (Figure 5-7, curve B), whereas Evans felt that the data were more consistent with a threshold

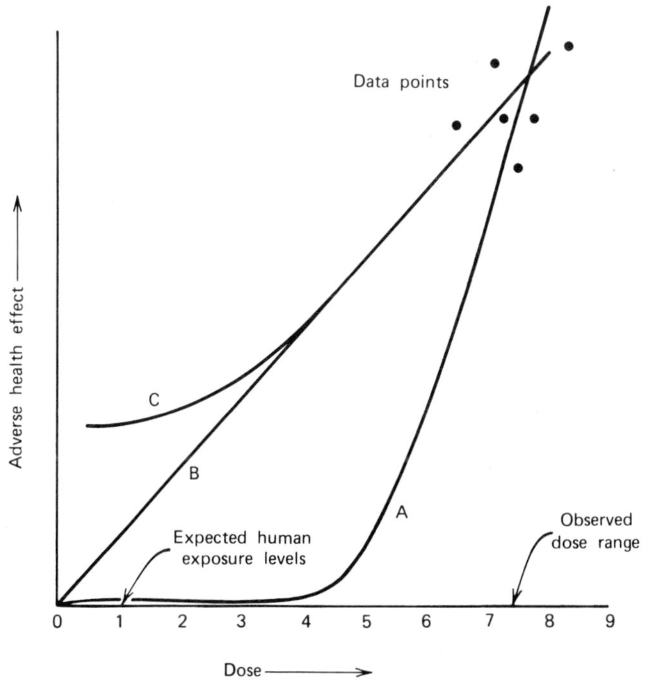

Figure 5-7. Possible dose-response relationships at low level exposures.

model (Figure 5-8, curve A). Consequently, the EPA and Gofman and Tamplin predicted adverse effects at all levels of exposure, whereas Evans contended that the threshold will not be reached even at 5 pCi/liter (Table 5-5). Although the EPA and Gofman and Tamplin are in fundamental agreement, their values differ because Gofman and Tamplin included a special consideration of high-risk groups (the very young). The predictions based on the data of Peterson et al. were also in very close agreement with the linear dose-response predictive model of the EPA and Gofman and Tamplin (Table 5-5). Riddiough et al. concluded by noting that there is exceptional agreement between the linear extrapolation and epidemiologic predictive models. Furthermore, when two highly divergent approaches to the problem of predicting safe doses generally agree, it tends to amount to an independent verification.

The critical issue in the sense of radium-226 in drinking water is that the EPA actually increased the PHS (1962) standard of 3 pCi/liter to 5 pCi/liter. Riddiough et al. pointed out that only the Gofman and Tamplin model specifically addressed the nature of high-risk groups in the derivation process. Although Gofman and Tamplin are in basic agreement with the linear approach of the EPA, the numbers of predicted cancers between the two methodologies at 5 pCi/liter differ by $22/10^6$ at a worst case estimate. Consequently, it was concluded that the action by EPA to raise the radium-226 drinking water standard was premature.

A somewhat different approach to deriving safe levels of carcinogen exposures has been proposed by Weil (1972). He proposed that a "tentatively" safe level of exposure to a carcinogen should be arbitrarily

Table 5-5. Risk Estimates For Cancer

^{226}Ra Concentration in Drinking Water (pCi/liter)	Number of Cancers due to ^{226}Ra per 10^6 Persons per Year			
	EPA	Gofman and Tamplin	Evans	Peterson
1	1.14–0.6	0.5–5	0	1.26
2	0.28–1.2	0.9–10	0	1.96
3	0.45–1.8	1.4–15	0	2.66
4	0.56–2.4	1.9–20	0	3.36
5	0.70–3.0	2.41–25	0	4.01

Source: From Riddiough, C.R., Musselman, R., and Calabrese, E.J. (1977) Is EPA's ^{226}Ra drinking water standard justified? *Med Hypoth* 3(5): 171–174.

defined as a dose 5000 times lower than the minimal cancer-producing dosage. Weil explained that one should allow for the typical 10-fold degree of difference for both intra- and interspecies differences. However, an additional factor of 10 should be incorporated because of the irreversible nature of carcinogenic effects and a factor of 5 to take into account variations in experimental designs.* Thus we have $10 \times 10 \times 10 \times 5 =$ the 5000 safety factor. In support of this proposal, Weil commented that such a safety factor, even though it does not specifically consider chemical interactions (synergism), would have provided safety for the enhanced carcinogenicity (by a factor of 1000) reported by Bingham and Falk (1969) of cocarcinogens.

The proposal of Weil is more conservative than Druckrey's (1967) suggestion of "1% of the lowest dosage, which given daily over the whole life span in susceptible experimental animals, producing cancer only at the end of the life span can be considered as the maximum tolerable dose for human beings." This is especially interesting since Druckrey's proposal is consistent with the linear dose-response perspective. However, Druckrey's view also includes the inverse relationship of dose to latency period. Despite their divergence by a factor of 50, both Weil and Druckrey are in fundamental agreement that total banning of essential materials, even though carcinogenic, is not consistent with "common-toxicological-sense and is not necessary to ensure safety for man."

SAFETY FACTORS FOR NONCARCINOGENS

The term safety factor is intimately associated with the extrapolation process. In fact, if it were not for the uncertainty inherent in the various types of extrapolation processes, there would be little need for safety factors. Safety factors are mathematical expressions of uncertainty that are used to protect individuals from situations they probably do not completely understand and certainly cannot control. Safety factors are established most often by the FDA with respect to food additives. The EPA has used the safety factor concept in the derivation of their drinking water standards for noncarcinogenic hydrocarbon insecticides and herbicides. Finklea et al. (1977) have even evaluated the current EPA national ambient air quality standards in light of safety factors. Earlier, Stokinger and Woodward (1958) considered industrial TLVs with regard to the safety factor concept for drinking water contaminants.

* "The minimum measured cancer-producing dose-level in a proper experimental design should not be more than 5 times that of the no effect level" (Weil, 1972).

Consequently, the concept of safety factors has been widely applied to conditions in which human safety has been threatened by potential toxicants. However, it should be noted that when permitted doses of carcinogens are the practical equivalent of zero, safety factors are not especially relevant. In fact, the EPA has safety factors, as indicated above, for all hydrocarbon insecticides and herbicides that were considered noncarcinogenic. At this time, the EPA has not instituted any maximum contaminant limit (MCLs) for carcinogens in drinking water.* The current thinking on carcinogens, although clouded with a high degree of uncertainty, is clearly conservative and supporting the linear dose-response relationship; that is, for each dose there is expected some frequency of the carcinogenic response. Thus there really is no safety factor in the traditional sense of the word. In the case of carcinogens, the term safety factor should be replaced by the term acceptable risk.

According to Barnes and Denz (1954), one of the reasons for the existence of safety factors evolved from studies such as that of Bliss (1935), since they clearly indicated the uncertainty of toxicologic predictions. Barnes and Denz noted that the uncertainty that Bliss found with lower animals may also occur when trying to predict the toxicologic responses of higher animals. They suggested that a practical way of dealing with this problem is the use of the 100-fold margin of safety. By margin of safety they meant that substances to be used in food should show no adverse health effect in animals when fed at a dose at least 100 times greater than the likely human daily dose. They noted that the margin of safety concept is a reasonable approach to the matter, but that its acceptance should not fool researchers and/or the public into believing that there is any experimental or theoretical basis for its existence.

In contrast to the comments of Barnes and Denz, Bigwood (1973) attempted to offer a systematic rationale for the safety factor of 100. He noted that in experimental studies there is a variation of approximately 1 to 10 for different individuals of a species, and the species to species variation may also be upwards to a factor of 10. When taken together, therefore, the overall variation between an individual in one species as compared to an individual in another species is on the order of 100. Thus, although the figure of 100 may indeed be called arbitrary, it is also a rational attempt at approximation. Bigwood tried to analyze the different types of variations between species more specifically by

* EPA has recently recommended that communities with greater than 75,000 people adopt activated carbon treatment of their drinking water to ensure that levels of potentially carcinogenic trihalomethane compounds do not exceed 100 ppb (Environ. Rep., 1977).

assigning a certain portion of the 100-fold safety factor to them. Thus he identified five sources of variation:*

1. Body size (using either weight or surface area) of the experimental animals as compared to man
2. Food needs, depending on factors such as age, sex, activity periods, and environmental conditions
3. Water balance
4. Hormonal functions and how they modify food intake
5. Different inherent susceptibility to a toxic effect of a given substance.

After "arbitrarily" assigning approximate values to each of the first four categories, Bigwood concluded that a safety factor of slightly greater than 60 was determined. The first four categories were analyzed because considerably more data on which to derive approximations exist as compared to category number 5. However, Bigwood (1973) speculated that, by extending the safety factor from 60 to 100, all five factors taken together should be reasonably treated.

Consequently, he concluded that there is a relatively high degree of validity to the safey margin of 100, with variations below or above (i.e., 50 to 500) very questionable. He concluded by stating that only in unusual situations should changes in the 100 safety factor be made.

It should be noted that Bigwood felt that standards for carcinogens could be adopted in light of their potency. He noted that for highly carcinogenic substances, such as aflatoxin or certain nitrosamines, which are active at the 1 ppm or less level, it would not be possible to determine a "no-effect level." However, for slightly carcinogenic substances that must be ingested at the rate of grams per kilogram of body weight per day, "safe" standards are possible.

According to Oser (1971), the reliability of safety factors derived from animal experimentation depends on the slope of the response; that is, the steeper the slope, the more confidence in the calculated safety factor. The reasoning behind this concept is that at the low dose region, the effects associated with a gradual curve may still be evident. Figure 5-8 compares these different dose-response relationships. As the curves indicate, the steeper curve (A) offers greater safety at low dose levels. Oser also noted that the degree of safety included in any standard depends on the type of potential toxic effect as well as on the slope. Thus the more toxic the potential response, the greater the safety factor

* The basis of animal extrapolation is presented in detail in Chapter 2.

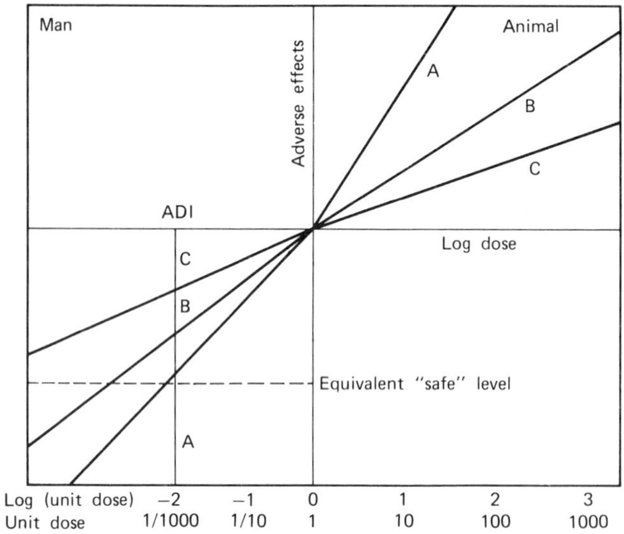

Figure 5-8. Relation of factors used in extrapolation of safe intake levels. Ordinate represents degrees of severity, or incidence, of the effect. Source: Oser, B. L. (1971) Food additives: the no effect level. ***Arch Environ Health*** 22:696–698. Copyright 1971, American Medical Association.

needed. Oser, in contrast to Bigwood, concluded that "on the basis of the slope and severity of the toxic effect observed in the experimental animal and its relevance to man under the conditions of expected use of exposure, the safety factor can be varied to give the closer degree of assurance of safety, instead of projecting down from a constant ¹⁄₁₀₀ of the no-effect dose to varying degrees."

Selected Cases of Safety Factors in Standard Setting for Drinking Water and Ambient Air Pollutants (Table 5-6)

Drinking Water Standards—See Chapter 6 for a detailed discussion

Cadmium

Based on animal toxicity studies (rats), it was determined that at approximately four times greater than projected human consumption of cadmium for a lifetime, kidney toxicity would result. Thus a safety factor of 4 has been assumed.

safety factor

Table 5-6. Safety Factors Contained in Primary Ambient Air Quality Standards

Pollutant	Lowest Best Judgment Estimate for an Effects Threshold	Adverse Effect	Standard	Safety Margin for Lowest Best Judgment Estimate (%)[a]
Sulfur dioxide	300–400 μg/cu m (short term); 91 μg/cu m (long term)	Mortality harvest; increased frequency of acute respiratory disease	365 μg/cu m (24 hour); 80 μg/cu m (yearly)	None; 14
Acid aerosols	8 μg/cu m (short term); 15 μg/cu m (long term)	Increased asthmatic attack; increased infections in children	None; none	None; none
Total suspended particulates	70–250 μg/cu m (short term); 100 μg/cu m (long term)	Aggravation of respiratory disease; increased prevalence of chronic bronchitis	260 μg/cu m (24-hour); 75 μg/cu m (yearly)	None; 33
Nitrogen dioxide	141 μg/cu m (long term)	Increased severity of acute respiratory illness	100 μg/cu m (yearly)	41
Carbon monoxide	23 (8-hour) mg/cu m; 73 (1-hour) mg/cu m	Diminished tolerance in heart patients	10 mg/cu m (8-hour); 40 mg/cu m (1-hour)	130; 82
Photochemical oxidants	200 (short term)	Increased susceptibility to infection	160 (1-hour)	25

[a] Safety margin—effects threshold minus standard divided by standard × 100.

Source: Finklea et al. (1977) The role of environmental health assessment in the control of air pollution. In J.N. Pitts, Jr. and R.L. Metcalf, eds. *Advances in Environmental Sciences and Technology*, Vol. 7, p. 342. New York: Wiley-Interscience.

Fluoride

Since the occurrence of mottling (white specks) of teeth has been reported in a very low percentage of the population even with fluoride at permissible levels, it is concluded that there is a very low safety factor of about 2–3 or less.

Nitrates

The occurrence of elevated levels of MetHb in infants at close to the present MCL supports the notion that there is a negligible safety factor for infants.

Barium

Although the methodologic basis for the derivation of the barium drinking water standard is highly questionable, the standard does have a safety factor of 2 added to it.

Lead

There is less than a factor of 2 between what the average American is exposed to (via all sources) and the amount that starts to cause an increased body burden. Since about ¼ to ⅓ of the total lead absorbed is derived from water, the standard was established to prevent any greater exposure. Examples of safety factors for ambient air pollutants are provided in Table 5-6.

Ambient Air Standards

It can be seen that there is no margin of safety for both sulfur dioxide and total suspended particulates with regard to short-term standards. In contrast, long-term standards for these two substances do have minimal built-in safety margins. Carbon monoxide has the largest safety margins of [130% (8-hour standard) and 82% (1-hour standard) greater than their respective standards] all the pollutants listed in Table 5-6.

NIOSH APPROACH TO STANDARDS FOR CARCINOGENS

From an historical perspective, approaches for regulating the exposure of workers to carcinogens involved the use of engineering controls and personnel protection apparatus. The impetus from this perspective was derived from the American Conference of Governmental Industrial Hygienists (ACGIH) in 1962 as part of their yearly suggested threshold limit values (TLVs). However, by 1972 the ACGIH unified their approach for dealing with separate carcinogens by placing each carcinogen in one of two categories: (1) a known human carcinogen or (2) experimental (when the knowledge was derived exclusively from animal studies in controlled toxicologic studies) (Rose, 1976).

The first actual attempt at banning an industrial carcinogen in the United States occurred in 1961, when the state of Pennsylvania banned the use of beta-napthylamine. However, not all states followed the example of Pennsylvania, thereby permitting exposure of beta-naphthylamine to their own workers. Such a piecemeal approach to regulating health and safety issues in the workplace among the different states was ultimately a critical factor in the development of the subsequent federal legislation (OSHAct, 1970) governing safety and health issues in the workplace. Six years later, the Pennsylvania Department of Health became empowered to regulate a variety of pollutant exposures by the use of a permit system that regulated the exposures to a variety of toxic/carcinogenic substances. Concurrently, in England the control of various carcinogenic substances was approached by a number of means: (1) banning their use, manufacture, or importation, (2) the adoption of personnel protection, and (3) medical examinations (Rose, 1976).

The approach taken by the United States government for the regulations of carcinogens in the workplace has been explained in a personal opinion by Rose (1976), then the director of the Office of Research and Standards Development, NIOSH. He noted that the possible approaches toward carcinogen control considered by OSHA and NIOSH involve environmental limits, nondetectable levels and reduction to ambient concentrations, use permit system, best available technology, substitution, and prohibition.

The major problem that OSHA faces with respect to setting acceptable standards for carcinogens in the workplace is that American society has a overriding desire to "eliminate cancer in our lifetime." With such a perspective permeating society, it is difficult to generate much support for positions that permit the continued presence of carcinogens in the air. The concept of environmental limits as discussed by Rose clearly states that, historically, in the United States "permissible levels of carcinogens which were thought to protect against the initiation of cancer have been derived. Asbestos may be used as an example of this type of environmental standard. Here it is assumed, based on available data, that a dose-response relationship exists and that below a certain concentration no adverse health effect will occur (equal to or less than one case in 100,000) or that the latency period may extend beyond the normal lifetime."

The critical problem in setting any standard, especially for carcinogens, is that adequate data necessary for establishing accurate and predictive dose-response relationships are often lacking. Consequently, it is not unexpected when future studies detect adverse health effects occurring at levels below the standard. These new data, of course, lead to the development of newer, more accurate, and more (although not completely) protective standards.

According to Rose, it is not in the best interests of worker health to establish TLVs for carcinogens. He based this position on the lack of appropriate human health effects data, the seemingly insurmountable problems in extrapolating from animal studies to man with a high degree of precision, and the difficulty in obtaining any type of consensus with respect to that highly personal term "acceptable risk."

A variation of the environmental limit standard is the "non-detectable exposure concept." According to Rose, the nondetectable exposure concept is quite contrary to the zero exposure concept, since the implementation of a zero exposure could require the banning of substances or redundant controls once nondetectable limits are achieved. He further stated that the nondetectable approach implies that workers could be exposed to any detectable levels while employing practical monitoring strategies during usual operation. Furthermore, it suggests that, as new and improved analytic techniques are developed, the standard—as defined by the lowest detectable limit—could be lowered. The use of the nondetectable limit has been applied in several instances, for example, vinyl chloride and inorganic arsenic. In the case of the 1974 permanent standard on vinyl chloride, OSHA required that employers limit worker exposure to 1 ppm as a TWA and 5 ppm as a ceiling.

OSHA defined 1 ppm as the "no detectable level." If feasible engineering controls and worker practices are not capable of reducing levels to 1 ppm, exposures must be reduced to the "lowest practical level," and respirators must be made available to each worker. No specific deadline for achieving the 1-ppm level via engineering controls was given because the technology has been generally inadequate to achieve the standard. The legality of the nondetectable level approach was challenged in court by the Society of the Plastics Industry. However, the United States Court of Appeals upheld the use of the nondetectable limit concept in a critical decision on the vinyl chloride standard (Rose).

The second major approach OSHA considered in regulating worker exposure to carcinogens was the use permit system. The use permit system would force employers to receive permission from OSHA before exposing the employees to carcinogens. According to Rose (1976), the critical components in the use permit system would involve (1) the availability of appropriate substitute industrial chemicals with less toxic characteristics, (2) administrative and/or on-site evaluation of the industry's application for the control procedures, and (3) proposed procedures for medical and environmental monitoring. Thus every permit, in effect, becomes a standard in and of itself.

Whether to adopt the use permit system became a legal controversy during the hearings concerning OSHA's proposed standards for 14 carcinogens. OSHA finally decided to drop the use permit system as a viable approach to carcinogen regulation. OSHA based its decision on the lack of feasibility in administrating such a complex and large undertaking (Ashford, 1976). Each permit would have had to be adapted to the particular workplace for which it was requested. When one considers the thousands of workplaces in which carcinogens are present, the task would not have been cost effective. Consequently, OSHA decided that employers would be required to notify OSHA of the extent and procedure of operation for how carcinogens would be used along with reports of the release of carcinogens into the worker atmosphere (Ashford, 1976).

The philosophy of OSHA for the control of carcinogens in the workplace is to strive for the adoption of isolating the suspected or known carcinogen in a closed system. Furthermore, OSHA requires companies to provide total body coverage of workers by personal protective apparatus when it is necessary. Also employers must be educated on the nature of carcinogenic hazards. Finally, annual medical surveillance is required before and after exposure to a carcinogen is thought to have occured (Ashford, 1976). All the requirements by OSHA for carcinogen control and worker health certainly establish a positive framework on which the health of hundreds of thousands of workers will rest. The procedure adopted by OSHA has been criticized on the basis that it requires a performance standard of no detectable limit (Ashford, 1976) (this is the case for vinyl chloride and inorganic arsenic). Also, the value of a medical examination immediately after exposure to a carcinogen that may have a latency period of 30 years seems rather meaningless. The annual medical examination should provide a good opportunity to institutionalize preventive medicine in industry. However, the most important regulation is the adoption of closed vessels, for therein lies the means by which elevated worker exposure to carcinogens will be curtailed.

Another approach to the control of carcinogens in the workplace

Table 5-7. Cancer Policy: Classifying Substances

Category	Scientific Evidence	Model Standard	Exposure
I, confirmed carcinogen	Carcinogenic in humans, in two mammalian species, or in repeated tests in the same species	Emergency temporary standard; permanent standard in 6 months	Lowest feasible level; banned if suitable substitute is available
II, suspect carcinogen	Evidence from only one animal species or if evidence is inconclusive	Permanent standard	Low enough to prevent acute or chronic toxic effects
III	Insufficient evidence to classify it in a higher category	No standard, but would be listed as needing more data	
IV	Substances that could fall into the three higher categories, but not found in United States work-places	No standard, listed to alert to potential danger	

Source: Federal Register, October 4, 1977.

includes the substitution of a less toxic for a more toxic substance. Rose favorably considered this approach, but pointed out that it is not a foolproof system. He noted the example of the replacement of methyl butyl ketone for methyl isobutyl ketone, which actually led to the occurrence of peripheral neuropathy in exposed workers. Finally, Rose suggested that carcinogens could be curtailed by banning them completely, but he doubted whether OSHA had the authority to ban the use, manufacture, or handling of the chemicals. In light of the passage of the Toxic Substances Control Act of 1976, it seems that the EPA Administrator may actually possess the authority to ban the use of potentially toxic substances. Court challenges will most likely decide this issue in the near future.

Recently OSHA has proposed a novel approach for regulating carcinogens in the workplace. Their new policy is designed to minimize regulatory procedures, thereby making the standard-setting process more efficient. The new policy requires that substances submitted to OSHA for rule making would be classified in one of four categories, according to the degree of scientific evidence concerning carcinogenicity (Table 5-7).

Besides establishing the biomedical basis for defining and classifying substances, the new protocol establishes three standardized formats for use in deriving standards for category I (i.e. confirmed carcinogen) and category II (i.e. suspect carcinogen) substances (Table 5-7). This policy is not expected to become operational for several years in light of anticipated delays due to litigation.

REFERENCES

Albert, R. E. and Altshuler, B. (1973) Considerations relating to the formulation of limits for unavoidable population exposures to environmental carcinogens. In C. L. Sander, R. H. Busch, J. E. Ballou, and D. D. Mahlum, eds., *Radionuclide Carcinogenesis,* pp. 233–253. Proceedings of the Twelfth Annual Hanford Biology Symposium. AEC Symp. Ser. No. 29, Conf.—720505. NTIS, Springfield, Va.

Albert, R. E. and Altshuler, B. (1976) Assessment of environmental carcinogen risks in terms of life shortening. *Environ Health Perspect* 13:91–94.

Ariens, E. J., Simonis, A. M., and Offermeier, J. (1976) *Introduction to General Toxicology,* p. 170. New York: Academic.

Ashford, N. (1976) *Crisis in the Workplace: Occupational Disease and Injury.* Cambridge; MIT Press.

Baloh, R. (1976) Neurological and behavioral toxicology of increased lead absorption. In *Health Effects of Occupational Lead and Arsenic Exposure: A Symposium,* pp. 51–58. U.S. HEW NIOSH.

Barnes, J. M. and Denz, F. A. (1954) Experimental methods used in determining chronic toxicity. *Pharmacol Rev* 6:191–242.

Bigwood, E. J. (1973) The acceptable daily intake of food additives. *CRC Crit Rev Toxicol* June, 1973, pp. 41–93.

Bingham, E. (1971) Thresholds in cancer induction. *Arch Environ Health* 22:692–695.

Bingham, E. and Falk. H. L. (1969) Environmental carcinogens: The modifying effect of carcinogens on the threshold response. *Arch Environ Health* 19:779–783.

Bliss, C. I. (1935) The calculation of the dosage-mortality curves. *Ann Appl Biol* 22:134–167 and 307–333.

Cleaver, J. E. (1968) Defective repair replication of DNA in xeroderma pigmentosum. *Nature* 218:652–656.

Cole, P. and Goldman, M. B. (1975) Occupation: In *Persons At High Risk of Cancer: An Approach to Cancer Etiology and Control,* pp. 172–176. New York: Academic.

Cornfield, J., Halperin, M., and Moore, F. (1956) Some statistical aspects of safety testing the salk poliomyelitis vaccine. *Public Health Rep* 71(10):1045–1056.

Dinman, B. D. (1972) "Non-concept" of "No-threshold" chemicals in the environment. *Science* 175:495–497.

Druckrey, H. (1959) Pharmacological approach to carcinogenesis. In Weistenholme, G. and O'Connor, M., eds., *CIBA Foundation Symposium on Carcinogenesis—Mechanisms of Action,* pp. 110–130. New York: Little-Brown.

Druckrey, H. (1967) Quantitative aspects in chemical carcinogenesis. In Truhaut, R., ed., *Potential Carcinogenic Hazards from Drugs,* vol. 7, p. 60. *Evaluation of Risks.* UICC Monograph Series. Berlin: Springer.

Enterline, P. E. (1976) Pitfalls in epidemiological research. *J Occup Med* 18(3):150–156.

Enviroment Reporter (December 30, 1977). EPA plans treatment standard to reduce trihalomethanes. p. 1334.

EPA (August 15, 1975). Statement on basis and purpose for the proposed National Interim Primary Drinking Water Regulations: radioactivity.

Epstein, S. S. (1973) The Delaney amendment. *Ecologist* 3(11):424–430.

Epstein, S. S., Hollaender, A., Lederberg, J., Legator, M., Richardson, H., and Wolff, A. (1969) Wisdom of cyclamate ban. *Science* 166:1575.

Evans, R. D. (1967) The radium standard for boneseeker-evaluation of data on radium patients and dial painters. *Health Phys* 13:267.

Falk, H. L., Kotin, P., and Thompson, S. (1964) Inhibition of carcinogenesis: the effect of polycyclic hydrocarbons and related compounds. *Arch Environ Health* 9:169–179.

Finklea, J., Shy, C. M., Moran, J. B., Nelson, W. C., Larsen, R. I., and Akland, G. G. (1977) The role of environmental health assessment in the control of air pollution. In J. N. Pitts and R. L. Metcalf, eds., *Advances in Environmental Sciences and Technology,* pp. 315–389. New York: Wiley-Interscience.

Food and Drug Administration Advisory Committee on Protocols for Satety Evaluation: Panel on Carcinogenesis Report on Cancer Testing in the Safety Evaluation of Food Additives and Pesticides (1971). *Toxicol Appl Pharmacol.* 20:419–438.

Friedman, L. (1973) Problems of evaluating the health significance of the chemicals present in foods. In *Pharmacology and the Future of Man,* Vol. 2, pp. 30–41. Proc. 5th Int. Congr. Pharmacology. Basel: Karger.

Gofman, J. W. and Tamplin, A. R. (1970) In LeCam, J., Neyman, J., and Scott, E. L., eds., *Proceedings of the Sixth Berkeley Symposium on Mathematical Statistics and Probability,* p. 235. Berkeley: University of California Press.

Hatch, T. F. (1971) Thresholds: Do they exist? *Arch Environ Health* 22:687–689.

Hatch, T. F. (1972) Permissible levels of exposure to hazardous agents in industry. *J Occup Med* 14:134–137.

Herman, E. R. (1971) Thresholds in biophysical systems. *Arch Environ Health* 22:699–706.

HEW, (1969) Reports of the Advisory Panels on Carcinogenicity, Mutagenicity, Teratogenicity. Report of the Secretary's Commission on Pesticides and Their Relationship to Environmental Health.

Hutchinson, G. E. (1964) The influence of the environment. *Proc. Nat Acad Sci USA* 54:930–934.

Izmerov, N. F. (1973) *Control of Air Pollution in the U.S.S.R.,* pp. 42–60. Geneva: WHO.

Jones, H. B. and Grendon, A. (1975) Environmental factors in the origin of cancer and estimation of the possible hazard to man. *Food Cosmet Toxicol* 13:251–268.

Joselow, M. (1976) Biological monitoring—problems of blood lead levels. In B. W. Carnow, ed., *Health Effects of Occupational Lead and Arsenic Exposure: A Symposium,* pp. 27–38. U.S. Dept. of HEW PHS NIOSH.

Krisko, G. C. I. and Bolander, K. (1974) Chemical carcinogens in the environment and in the human diet: Can a threshold be established. Fd. Cosmet. Toxicol. 12:737–746.

NAS (1975) *Principles for Evaluating Chemicals in the Environment.* Washington, D.C.: National Academy of Sciences.

Oser, B. L. (1971) Food additives: the no-effect level. *Arch Environ Health* 22:696–698.

OSHAct of 1970, Public Law 91–596.

Parke, D. V. (1968) *The Biochemistry of Foreign Compounds.* New York: Pergamon Press.

Peterson, N. J., Sammuels, L. D., Lucas, H. F., and Abraham, S. P. (1966) An epidemiologic approach to low-level radium-226 exposure. *Public Health Rep* 81:805.

PHS (1962) *Drinking Water Standards.* Washington, D.C.: Public Health Service.

Repko, J. (1976) Behavioral toxicology in inorganic lead exposure. In *Health Effects of Occupational Lead and Arsenic Exposure: A Symposium,* pp. 59–73. U.S. HEW NIOSH.

Riddiough, C. R., Musselman, R., and Calabrese, E. J. (1977) Is EPA's ^{226}Ra drinking water standard justified? *Med Hypoth* 3(5):171–174.

Rose, V. E. (1976) Standards for the control of carcinogens in the workplace. *J Occup Med* 18(2):81–84.

Saffiotti, U. (1971) Statement before the Subcommittee on Executive Reorganization and Government Research, Senate Committee on Government Operations, April, 1971.

Sinhuber, R. O., Wales, J. H., Ayers, J. L., Engerbrecht, R. H., and Armend, D. L. (1969) Dietary factors and hepatoma in rainbow trout (Salmo gairdneri). Aflatoxin in vegetable protein feedstuffs. *J Natl Cancer Inst* 41:711–718.

Stopps, G. J. (1974) Is there a safe level of lead exposure? *J Wash Acad Sci* 61(2):103–120.

Stokinger, H. E. (1972) Concepts of thresholds in standards setting. *Arch Environ Health* 25:153–157.

Stokinger, H. E. (1977) Toxicology and drinking water contaminants. *J Am Water Works Assoc* July pp. 399–402.

Stokinger, H. E. and Woodward, R. L. (April, 1958) Toxicologic methods for establishing drinking water standards. *J Am Water Works Assoc* 50:515–529.

Suss, R., Kinzel, V., and Scribner, J. D. (1973) *Cancer: Experiments and Concepts,* p. 50. Berlin: Springer.

Upton, A. C. (1961) The dose-response relationship in radiation-induced cancer. *Cancer Res* 21:717–729.

Weil, C. S. (1972) Statistics vs. safety factors and scientific judgment in the evaluation of safety for man. *Toxicol Appl Pharmacol* 21:454–463.

Yanysheva, N. Ya. and Antomonov, Yu. G. (1976) Predicting the risk of tumor occurrence under the effect of small doses of carcinogens. *Environ Health Perspect* 13:95–99.

6 Drinking Water Standards: Their Origin And Rationale

HISTORICAL PERSPECTIVE

In contrast to federal standards for ambient air quality and industrial hygiene, drinking water standards have existed for a comparatively long time, that is, since 1914. At that time, the United States Public Health Service (PHS) first adopted drinking water standards to protect the health of the traveling public. The original PHS drinking water standards were designed to regulate interstate water carriers so that people traveling on boats, buses or trains would be assured a safe water source. Intrastate or community drinking water sources were not under federal control.

In the years since 1914, the PHS has modified its standards on several occasions, including 1925, 1942, 1946, 1956, and 1962, in light of new knowledge generated from technical and health-related research. During the initial decades subsequent to the development of the first drinking water standards, the main goal was to prevent the occurrence of diseases such as typhoid and cholera. However, in more recent years, that is, since World War II, there has been an increase in the awareness of the toxicity of chemical pollutants in drinking water and the need for regulation. Thus the evaluation of the drinking water standards represented an initial concern with the control of bacterial contamination, followed by the gradual recognition of toxic heavy metals, radiation, and now carcinogenic pesticides and other organics.

In addition to a growing recognition of new hazards in drinking water, it was also realized that regulations for safety standards should be equally applied to intrastate and community drinking water. Particularly important in encouraging this perspective was the membership of the American Water Works Association (AWWA), who recommended as early as 1946 that the PHS standards on interstate water be adopted for community drinking water (PHS Standards, 1946). Despite this encouragement by the AWWA to develop federally regulated drinking water standards, it is significant to note that even the 1962 amendments to the PHS standards were only applicable for potable water used by interstate carriers.

Even though the federal government was not directly involved with the regulations of intrastate drinking water quality, this does not mean that no controls existed. Many individual states did adopt their own programs to ensure drinking water safety. However, there are often problems with state programs that are not regulated by the federal government. For example, there are differences in uniformity of standards (Table 6-1), enforcement, and implementation when the different states are compared (Stokinger and Woodward, 1958). The existence of completely separate state programs does not ensure that all people are uniformly protected. The similar lack of uniformity among state industrial health standards also acted as a stimulus for the development of national federal occupational regulations (Ashford, 1976).

The first major federal legislation that included regulation of inter- and intrastate drinking water was the Safe Drinking Water Act of 1974. One of the major goals of the 1974 act was the establishment of National Drinking Water Regulations. Several specific factors strongly influenced the passage of this act. For example, in 1969 approximately 1000 representative public water supply systems were surveyed by the Bureau of Water Hygiene (of the U.S. Environmental Health Service, now part of the EPA). The survey reviewed eight large urban areas and the entire state of Vermont. It was found that people who lived in cities with a population of greater than 100,000 usually had drinking water of good and safe quality. However, people in smaller communities were often found to have drinking water with less than satisfactory taste, odor, and appearance (National Water Commission Report, 1973). Although these data did not strongly support the need for federal regu-

Table 6-1. Comparison of Early Drinking Water Standards For the States of Colorado, North Daokta, and Ohio[a]

	Typical Drinking Water Standards (ppm)		
Substance	Colorado	North Dakota	Ohio
Arsenic	0.05	—	—
Cadmium	—	0.4	—
Chloride	—	—	250.0
Chromium	0.05	2.0	2.0
Copper	3.0	0.4	0.4
Cyanide	0.5	0.15	0.15
Iron	—	—	5.0
Lead	0.1	0.35	0.35
Manganese	—	—	5.0
Nickel	—	0.5	—
Nitrate	—	10.0	10.0
Selenium	0.05	—	—
Zinc	—	1.0	15.0
Oils	15.0	—	—
Phenol	0.02	—	0.03

[a] It is of interest to note that for all substances for which these three states had standards, there was not one instance of complete uniformity.

Source: H. E. Stokinger and R. L. Woodward. (1958) Toxicological methods for establishing drinking water standards. *J Am Water Works Assoc* 50:516.

lation of drinking water, the EPA noted that state and local water quality control programs were not giving sufficient regulation of the quality of local water supply systems as a result of deficiencies in planning, training, and enforcement programs. As a result of these and other factors, the National Water Commission Report of 1973 noted that there was a "need for a comprehensive restatement of policy to govern the role of the federal agencies meeting the nation's needs for municipal and industrial water supplies." Another important factor assisting in the passage of the 1974 act was the widespread and highly emotional publicity in the fall of 1974 to an EPA water survey that revealed the presence of potential human carcinogens in many of the nation's drinking water systems. It is commonly recognized that the timing of the release of these results coincided with its most appropriate legislative potential. In any case, the Safe Drinking Water Act was signed into law on December 14, 1974 by President Ford. Ninety days later, on March

14, 1975, the EPA published proposed national regulations in the Federal Register. (On August 14, 1975, the EPA proposed national regulations for radioactivity.) During the interim between the publication of the proposed regulations and the issuance of Interim Regulations on December 24, 1975, the EPA held public hearings in Boston, Chicago, San Francisco, and Washington, D.C. In addition, nearly 500 written comments from the general public were received by the EPA. As a result of such comments from interested individuals and groups, as well as a further review of available data, the EPA made a number of modifications in the proposed standards, and these were reflected in the Interim Regulations. A similar process with respect to regulations for radionuclides was conducted. Thus written responses from the public were requested and a public hearing in Washington was held on September 10, 1975. In similar fashion to the earlier set of proposals, various changes were made in the proposed regulations and incorporated into the Interim Regulations for radionuclides issued on July 9, 1976. Since the Interim Regulations were not finalized until 18 months after issuance, June 24, 1977 was the date of the promulgation of the final National Drinking Water Standards. Even though regulations for radioactivity have proceeded at a somewhat slower pace than the bacterial and inorganic chemical regulations, final regulations were also promulgated on June 24, 1977.

DRINKING WATER STANDARDS

As a result of the adoption of the newly issued federal drinking water quality regulations, it is highly appropriate to take a critical look at the toxicologic and epidemiologic basis on which several of the standards are based. Specific contaminants are reviewed, including representatives of both inorganic and organic constituents. Table 6-2 offers a summarization of the historical framework and biologic basis of all (not only representative examples) the drinking water standards for chemicals. Bacterial water quality is not discussed.

Barium

The first specific numerical drinking water standard for barium was published in the 1962 amended regulations of the USPHS. At that time, the literature contained specific references to various limited toxicologic studies of barium activity (Lorente de No and Feng, 1946; Fite, 1955). However, there was a total lack of evidence about the levels of barium that may be safe in drinking water. Despite such limitations of appropriate experimental data, Stokinger and Woodward (1958) presented a theoretical attempt at deriving a drinking water standard for barium based entirely on the TLV of 0.5 mg Ba/m^3 air as determined by the American Conference of Governmental Industrial Hygienists (1958). According to Stokinger and Woodward, TLVs for industrial chemicals in air can be reasonably adapted to water standards if certain assumptions are used. These assumptions state that the TLV values offer safety to individuals who are exposed to the toxic substance for 8 hours per day, 40 hours per week over a working lifetime. Also, it must be acknowledged that the TLV is not designed to ensure the protection of all individuals, that is, some of the hypersusceptible segments of the population may be adversely affected. In contrast, community drinking water standards are designed to protect people 24 hours per day, 7 days per week for their entire life including the 25 or so years prior to entering the work force and for those years of retirement.

It is equally important that community standards are required to ensure the protection of all the citizens, including the segments of the population at high risk, to the extent that it is technologically and economically feasible. With these differences between TLVs and community drinking water standards in mind, let us consider how the barium standard was derived.

The TLV for barium and its compounds in air is 0.5 mg/m^3, based on the working hypothesis that during an 8-hour day one inhales 10 m^3 of air and that 75% of the barium inhaled is absorbed from the lungs into the blood. Consequently, the quantity of barium permitted to be absorbed into the body per day via industrial exposure amounts to 3.75 mg (10 m^3 $\times$ 0.5 mg $\times$ 0.75 = 3.75 mg) and is considered to be a noninjurious daily exposure. To apply this calculation to drinking water, the following assumptions and determinations are necessary. People are assumed to consume 2 liters/day of water with a gastrointestinal tract absorption factor of 90%. Thus (2 liters/day)(concentration of barium mg/liter (0.90, absorption factor) = 3.75 mg = the calculated noninjurious dose per day.

Solving for the unknown concentration yields 4.17 mg barium per 2 liters of drinking water, or nearly 2 ppm. According to Stokinger and Woodward (1958), this numerical derivation of a possible barium standard is an *approximate* limiting concentration for a healthy adult population. The authors firmly stated that this represents "only a first approximation in development of tentative drinking water standards." To consider the health of potentially hypersusceptible segments of our

Table 6-2. Historical Development of Drinking Water Standards for Inorganic and Organic Chemicals[a]

Substance	USPHS 1925	USPHS 1942	USPHS 1946	USPHS 1962	Recommended 1969[b]	Proposed 1975[c]	Adopted 1977[d]	Comments
					Inorganics			
Arsenic	—	M 0.05	M 0.05	R 0.01 M 0.05	M 0.1	M_p 0.05	M_p 0.05	No mathematically derived rationale for a standard except that arsenic is a carcinogen or cocarcinogen and should be kept as low as possible; the frequency of skin cancer as reported by Neubauer (1947) was quite high in areas of England where levels of arsenic in drinking water were 12 mg/liter
Barium	—	No standard; not to be added	No standard; not to be added	M 1.0	M 1.0	M_p 1.0	M_p 1.0	No specific human studies; based on an industrial air standard; USEPA sponsored an epidemiologic study of long-term exposure, results due in 1978
Cadmium	—	—	—	M 0.01	M 0.01	M_p 0.01	M_p 0.01	It would take 352 μg of Cd per day for 50 years to reach a critical level in the kidney (200 ppm) based on animal studies; at the

								proposed standard of 0.01 mg/liter, 20 μg/day would be additionally contributed to an average of 75 μg/day from food; thus a safety factor of 4 is assumed
Chloride	R 250.0	R 250.0	R 250	R 250	R 250	R_s 250	R_s 250	Based on taste
Chromium	—	No standard; not to be added	M 0.05	M 0.05	M 0.05	M_p 0.05	M_p 0.05	1946 standard was based on the lowest amount able to be measured; later studies showed a family of 4 known to have drunk water for 3 years at a level of 1 mg/liter with no harmful effects as determined by a single medical examination; only hexavalent chromium known to be toxic, but total chromium will be used as limit since atomic absorption cannot distinguish between different valences
Copper	R 0.2	R 3.0	R 3.0	R 1.0 M 3.0	M 1.0	R_s 1.0	R_s 1.0	Based on taste
Cyanide	—	—	—	R 0.01 M 0.2	M 0.2	M_p 0.2	—	50–60 mg, single dose fatal; 19 mg/liter in water is safe dose as derived from industrial air standard; 10

Table 6-2. *(Continued)*

Substance	USPHS 1925	USPHS 1942	USPHS 1946	USPHS 1962	Recom-mended 1969[b]	Proposed 1975[c]	Adopted 1977[d]	Comments
								mg, single dose-noninjurious; 2.9–4 mg/liter noninjurious; NOT considered in interim regulations because level in most drinking water sources is 1/10 of the proposed standard which is itself 1/100 of a toxic dose; also chlorination reduces toxicity of cyanide
Fluoride	—	M 1.0	M 1.5	R L 0.6–0.9 O 0.7–1.2 U 0.8–1.7 depending on temp. (D.T.) M 1.4–2.4 D.T.	M 1.1–1.8 D.T.	R L 0.6–1.1 O 0.7–1.2 U 0.8–1.3 M_p 1.4–2.4 D.T.	M_p 1.4–2.4 D.T.	Level that prevents dental caries yet low enough to prevent dental fluorosis
Foaming agents as methylene blue active substances	—	—	—	R 0.5	M 0.5	R_s 0.5	R_2 0.5	Prevents undesirable tast and foaming; indicative of undesirable level of sewerage pollution

Heavy metal glucoside	—	—	Should not be added	—	—	—	—	
Hydrogen sulfide	—	—	—	—	—	R_s 0.05	R_s 0.05	
Iron	R 0.3	A limit of 0.3 for Fe and Mn combined	A limit of 0.3 for Fe and Mn combined	R 0.3 just for Fe	M 0.3	R_s 0.3	R_s 0.3	High levels affect color of laundry; affects taste of beverages
Lead	M 0.1	M 0.1	M 0.1	M 0.05	M 0.05	M_p 0.05	M_p 0.05	Since Pb from other sources was not closely regulated and since the total daily intake of Pb that results in progressive retention is less than twice the average normal intake, this standard was adopted
Magnesium	R 100	R 125	R 125	—	—	—	—	
Manganese	—	See Fe	See Fe	R 0.05	M 0.05	R_s 0.05	R_s 0.05	To prevent esthetic and economic damage—brown laundry and taste of beverages; to avoid truly excessive levels of intake
Mercury	—	—	—	—	M 0.005	M_p 0.002	M_p 0.002	No precise calculation; total intake from all sources must be equal to or less than 30 μg/day which is 1/10 of a known toxic dose; total Hg levels used in regulations

Table 6-2. *(Continued)*

Substance	USPHS 1925	USPHS 1942	USPHS 1946	USPHS 1962	Recommended 1969[b]	Proposed 1975[c]	Adopted 1977[d]	Comments
Nitrate	—	—	—	R 10 as N	M 10 as N	M_p 10 as N	M_p 10 as N	To prevent methemoglobinemia in infants and young children; adults are less susceptible
Phenol	—	R 0.01	R 0.001	M 0.001	—	—	—	1946—based on undesirable taste resulting from reaction with chlorine following chlorination
Radioactivity Ra—226 Sr—90	—	—	—	M 3 pCi Ra/liter; 10 pCi Sr/liter; beta: up to 1000 pCi/liter permitted when alpha emitters & Sr-90 are absent	M 1 pCi Ra/liter; 10 pCi/liter for gross alpha; 10 pCi/liter for Sr-90	M 5 pCi Ra/liter; 15 pCi/liter gross alpha; beta: 50 pCi/liter and if levels of H^3 are <20,000 pCi/liter and Sr-90 <8 pCi/liter	M 5 pCi Ra/liter; 15 pCi/liter gross alpha; beta: 50 pCi/liter and if levels of H^3 are <20,000 pCi/liter and Sr-90 <8 pCi/liter	Standard is based on the carcinogenic potential of the different radionuclides
Selenium	—	M 0.05	M 0.05	M 0.01	M 0.01	M_p 0.01	M_p 0.01	Standard was lowered in 1962 because of the potential carcinogenic effects of selenium

Silver	—	—	—	M 0.05	M 0.05	M_p 0.05	M_p 0.05	Causes blue-gray discoloring of skin, eyes, mucous membranes; 50 μg/liter from water —if all of it reaches the site of action-it would take 27 years to reach intolerable level
Sodium	—	—	—	—	270	—	—	American Heart Association recommends a standard of 20 mg/liter; not enough information on which to base a standard
Sulfates	R 250	R 250	R 250	R 250	R 250	R 250	R_s 250	Elevated levels may be a laxative in transients who are not accustomed to such levels
Total solids	R 1000	R 500 M 1000	R 500 M 1000	R 500	—	—	—	Based on taste
Zinc	R 5.0	R 15	R 15	R 5	R 5	—	R_s 5	11–27 mg/liter in human studies was not harmful; 30 mg/liter caused nausea; 30 mg/liter caused water to be milky
Organics								
Chlorinated hydrocarbons								
Aldrin	—	—	—	—	M 0.01	—	—	Approval limits for non-
Chlordane	—	—	—	—	M 0.01	M_p 0.03	—	carcinogenic chlori-

Table 6-2. *(Continued)*

Substance	USPHS 1925	USPHS 1942	USPHS 1946	USPHS 1962	Recommended 1969[b]	Proposed 1975[c]	Adopted 1977[d]	Comments
DDT	—	—	—	—	M 0.1	—	—	nated hydrocarbons in drinking water have been calculated primarily on the basis of the extrapolating from observed minimal effects in animals to predicted human responses; aldrin, dieldrin, and DDT will be considered for a MCL only if they are found to be widely present in drinking water. They are also suspended for most pesticide uses; chlordane, heptachlor, and heptachlor epoxide were deleted because the EPA has suspended their usage
Dieldrin	—	—	—	—	M 0.01	—	—	
Endrin	—	—	—	—	M 0.003	M_p 0.0002	M_p 0.0002	
Heptachlor	—	—	—	—	M 0.02	M_p 0.0001	—	
Heptachlor epoxide (HE)	—	—	—	—	M 0.02	M_p 0.0001	—	
Heptachlor and HE	—	—	—	—	M 0.02	—	—	
Lindane	—	—	—	—	M 0.1	M_p 0.004	M_p 0.004	
Toxaphene	—	—	—	—	M 0.1	M_p 0.005	M_p 0.005	
Methoxychlor	—	—	—	—	M 0.5	M_p 0.1	M_p 0.1	
Organophosphate[b] and carbamate insecticides	—	—	—	—	M 0.1	—	—	No evidence that organophosphate and carbamate insecticides reach the tap; this is the reason the EPA did not issue limit; accidental spills can be dealt with in other administrative ways

Herbicides								
2,4-D	—	—	—	—	M 1	M_p 0.1	M_p 0.1	2,4,5-T was deleted in 1975; the EPA's ban on the use of 2,4,5-T for aquatic uses made a drinking water limit unnecessary
2,4,5-T	—	—	—	—	M 0.005	—	—	
2,4,5-TP	—	—	—	—	M 0.2	M_p 0.01	M_p 0.01	
Organic chemicals determined by the chloroform extract method								CCE test was originally used as a measure of undesirable tastes and odors; recently consideration was given to using it as a health standard in light of the growing concern over organic chemicals; however, it is very nonspecific for the different organic chemicals and is not considered a good choice as a health standard.
(CCE)	—	—	—	R 0.2	M 0.3	M_p 0.7	—	
(CAE)	—	—	—	—	M 1.5	—	—	

[a] All values are given in milligrams per liter. R = recommended limit. M = maximum limit. R_s = recommended limit, secondary standard. M_p = maximum limit concentration, primary standard. L = lower limit. O = optimum limit. U = upper limit. [b] Recommended by Advisory Board, 1969. [c] Proposed Regulations by EPA, March, 1975. [d] Adopted Interim Regulations by EPA, December 1975; finally promulgated on June 24, 1977. [e] Expressed in terms of parathion equivalent cholinesterase inhibition.

heterogeneous human population, the calculated level of 2 ppm was lowered to 50% of its value, that is, 1 ppm.

This procedure of utilizing the industrial TLV for the development of the barium drinking water standard was not only officially accepted in the PHS standards of 1962, but is essentially repeated by the EPA as their justification for the present national standard. The major drawback to the standard has been the lack of human epidemiologic evidence to support the validity of the derivation. However, a major epidemiologic study supported by EPA for $160,000 over 2 years is now being conducted in Illinois under the direction of the University of Illinois School of Public Health on residents of a community with long-term exposure to elevated levels of barium in their drinking water (5.0 mg/liter). This study is designed to test the hypothesis that elevated levels of barium in drinking water may adversely affect cardiovascular disease and blood pressure distribution levels. It is expected that results may be forthcoming within a year*.

Finally, it should be pointed out that the barium standard did not incorporate exposure from multiple sources. This is a process that is typical, especially for the heavy metals. A brief survey of ambient barium exposure revealed an average urban air level of 0.025 μg Ba per cubic meter (PHS, 1962). Daily exposure would have resulted in only 0.385 μg/day, assuming that 20 m^3 of air are inhaled (with a 0.75 absorption rate). Exposure from food would be significant only if there were a high content of Brazil nuts in the diet (Seaber, 1933). Thus exposure to barium from ambient air and food is so insignificant that it would not effect even 5% of the total exposure if levels in the water were 1.0 mg/liter. However, individuals who are working in industrial settings in which the levels of barium are elevated (i.e., approaching 0.5 mg/m^3) and whose communities have drinking water with a barium level of approximately 1.0 mg/liter are exposed to potentially excessive levels. Thus it would seem that in such regions of the country, special consideration should be made to ensure the health of these workers. Stricter industrial standards, additional hygiene equipment, stricter water standards, or any type of collective compromise of all these suggestions should be seriously entertained.

Cadmium

The USPHS did not adopt a cadmium drinking water standard until 1962, when a maximum limit of 0.01 mg/liter was issued. Since that

* Refer to page 210 for a summary of recently completed study results.

time, the 0.01 mg/liter standard has been reaffirmed by an EPA Advisory Committee in 1969, by the proposed standards of March 14, 1975, and by the final standards of June 24, 1977. The 1962 standard for cadmium was derived, in part, from both animal and human studies. For example, rat studies of dietary cadmium exposure demonstrated the accumulation of cadmium in soft tissues (kidney and liver) at concentrations in drinking water as low as 0.1 mg/liter. Also, extremely small quantities ($5 \times 10^{-6} M$) of cadmium in liver mitochondria of rats can interrupt oxidative phosphorylation. This information, in conjunction with the work of Schroeder (1965) that indicated that extremely small quantities of cadmium in the kidney may cause potentially harmful renal arterial changes in man, led to the conclusion that cadmium in concentrations greater than 0.01 mg/liter should not be permitted. In the documentation for the 1962 standard, it was mentioned that the cadmium levels in both food and tobacco were unknown and thus could not be considered in the basis for this standard derivation. However, the levels of cadmium in the urban air were known to average 0.005 $\mu g/m^3$ and ranged from 0.000 to 0.599 $\mu g/m^3$ (PHS, 1962). Considering only the average Cd level in ambient urban air, it would seem that the amount contributed from air would be insignificant. However, at concentrations of 0.599 $\mu g/m^3$, the contribution may be substantial and would add an exposure of approximately 0.012 mg/day, not including an assumed absorption factor of 0.25 (Stokinger and Woodward, 1958). It would be more helpful if the PHS (now the EPA) had included population weighted averages in addition to the listing of the average and range of concentrations. This information is especially relevant in deciding whether the contribution from air is significant enough to be specifically included in the standard derivation process.

By 1975 considerably more scientific data had emerged concerning the toxicology of cadmium as well as its concentrations in tobacco smoke and different food substances. For example, cadmium has been found to be an important component of cigarette smoke and may contribute up to about 1.5 μg/day to a person's body burden (EHRC, 1972). Dietary intake is not well defined, but it has been reasonably viewed as approximately 75 μg/day (EPA, 1975) with a 25% absorption factor (EHRC, 1972). The average ambient air concentration of cadmium in urban areas was reported in 1960 to be 0.025 $\mu g/m^3$. In 1968 the Chicago-Hammond area had a mean value for all samples of 0.019 $\mu g/m^3$, with a maximum of 0.08 $\mu g/m^3$ and a minimum of less than 0.005 $\mu g/m^3$ (EHRC, 1972). Although human studies are needed to establish more firmly the response to long-term exposure, more recent rat studies have revealed that when the level of cadmium reaches 200 ppm in the kidney, renal damage is initiated. The EPA calculated that for humans to reach a level of 200 ppm in their kidneys, it would take 352 μg of cadmium per day for 50 years. Consequently, in the derivation of the cadmium standard, the EPA assumed a daily dietary (i.e., via food) exposure of 75 μg and 20 μg/day from water. This 20 μg/day would result from a standard of 0.01 mg/liter. The total daily cadmium exposure amounts to approximately 95 μg/day and thus a safety factor of 4. Of course, this safety factor is extrapolated directly from rat data and may not be particularly relevant to the human condition. During the proposal of the cadmium standard in the Federal Register, the EPA requested comments from the public concerning whether the standard should incorporate additional protection for smokers. Of 52 comments received on this question, only three suggested that the cadmium standard should be revised to protect cigarette smokers, whereas the remaining 49 were strongly opposed to the concept of having the nonsmoker bear the financial burden for lowering the standard to benefit smokers. The EPA administrators agreed with the majority.

One point that may have been overlooked by those 49 opposing comments, however, was the recently published information by Menden et al. (1973) that cadmium is present in substantially greater quantities in sidestream smoke than in mainstream smoke (three to four times greater) and may present an even greater exposure to nonsmokers. At present, published articles on the effects of smoking on the nonsmoker with respect to cadmium are lacking. If such experiments demonstrate that nonsmokers are at considerable risk to cadmium exposure via passive smoking, the cadmium standard may have to be revised.

Fluoride

Of all the substances for which the PHS and EPA have proposed drinking water standards, the derivation of the fluoride regulations is based on the most human exposure data. Since fluorides are naturally present in the drinking water of many areas of the country, numerous epidemiologic studies have been performed to elucidate the effects of fluorides on health. In addition, the adoption of artificial fluoridation in numerous American communites and foreign countries to prevent dental caries has provided the systematic experimental epidemiologic studies that complement the observation of humans in naturally fluoridated areas.

A summarization of the large volume of studies on human health in relation to fluoride exposure has revealed that (1) fluoride in drinking water reduces dental caries, (2) excessive levels of fluoride in drinking water cause dental fluorosis, (3) dental fluorosis appears to be the only harmful effect caused by fluoride levels of less than 8 mg. fluoride per liter, and (4) at levels between 8 and 20 mg./liter, bone changes may occur, and at levels greater than 20 mg/liter for 20 or more years, crippling fluorosis may occur (PHS, 1962).

As early as 1946, the USPHS had a maximum standard for fluoride of 1.5 mg/liter. However, by 1962 the USPHS instituted a new concept in standard setting for fluoride. The new approach was to have a standard that was lowered or raised depending on the average ambient temperature range. Thus, in warm regions of the country where people tend to drink more water than those in colder areas, the maximum level of fluoride in the drinking water would be diminished, and vice versa. The reason for having this drinking water standard as a function of the average ambient temperature was that this standard does not have a large built-in safety factor. The concept of temperature regulation of fluoride exposure has been maintained and is incorporated into the final 1977 standards. According to Dean (1942), the safety factor present when drinking water contains 1.8 ppm fluoride is actually less than 2. Consequently, exposure should be closely regulated.

It should be recognized that the establishment of a maximum limit concentration for fluoride is not intended to encourage programs geared to having communities adopt artificial fluoridation.

In 1953 the United States National Research Council suggested that the lowest toxic daily intake for young children was 2 mg of fluoride. Krepkogorsky (1963) recommended that the uppermost total daily intake for adults should not exceed 3.2 mg. The FDA has often stated that 2–3 mg of dietary fluoride should, in most conditions, constitute a safe amount (Jerard and Patrick, 1973).

In light of these reports of "safe limits," it is important to note a report by Spencer et al. (1969) which indicated that certain Illinois residents in fluoridated areas ingest from 3.6 to 5.4 mg of fluoride per day. If people are exposed to only 1.0–1.5 mg/liter, how is it possible that they are exposed to levels as high as 5 mg/day? Marier and Rose (1966) indicated that one major consequence of widespread fluoridation of municipal water supplies is that food and beverage processing plants use the fluoridated water in their operations. Martin (1951), in studies conducted during the Evanston, Illinois dental caries study, reported that the fluoride content of vegetables significantly increases when the vegetables are cooked in water containing fluoride. For example, peas boiled in a saucepan containing water with 0.0 ppm of fluoride had 0.21 ppm, but when boiled in 1.0 ppm fluoridated water, the concentration increased to 1.22 ppm. Marier and Rose reported similar results, except that they extended the food types to such items as pork and beans, tomato soup, ginger ale, and beer. They reported total daily fluoride intakes that ranged between 2 to 5 mg in seven male members (age 30–50 years) of their research staff (average daily intake of 1–3 liters per person) during a 1-month winter period.

A more recent estimate of total daily fluoride intake in the United States for individuals whose drinking water contains low levels or essentially no fluoride and for whom there are no special fluoride intakes is between 0.5 to 1.5 mg. (Hodge and Smith, 1965). Marier and Rose, commenting on Krepkogorsky (1963), reported total daily fluoride intake data for people living in districts in which 1.0 ppm of fluoride is present in the drinking water. Their results indicate that such people in England take in 2.5 mg/day, in the Ukraine, 3.3 mg, and up to 2.1 mg in other regions of the Soviet Union. Furthermore, Elliot and Smith (1960) have shown that traditional dietary habits can markedly affect the amount of fluoride ingested. For example, the staple diet of Newfoundlanders can contribute up to 2.74 mg of fluoride per day in an area in which the drinking water was fluoride free.

It seems that the use of fluoridated water in food processing causes a significant increase in the fluoride content of food and beverages. Preliminary data indicate that healthy individuals may ingest up to or greater than 5.0 mg/day. It should also be pointed out that Marier and Rose, commenting on Krepkogorsky, reported that fluorosis can occur in areas in which the fluoride content is high in foods, but negligible in drinking water.

The results of 10 years of "optimal" fluoridation in Newburgh, New York indicated "mild" dental fluorosis (mottling) in about 2% of the children who had been born since fluoridation began (Ast et al., 1956). "Mild" dental fluorosis is present when 25 and 50% of the surfaces of two or more teeth are opaque, paper white. No brown staining from fluoride was reported. However, Dean (1942) found that some brown staining occurs in 1–2% of the children who use a water supply containing 1.8 or 1.9 ppm fluoride. Therefore, the "safety factor" for these effects on teeth is less than 2.

This information leads one to tentatively hypothesize that fluoride intake has risen substantially in recent years with the widespread adoption of artificial fluoridation. In fact, the evidence seems to suggest that the degree of fluoride intake, especially for children with developing secondary permanent teeth, may be approaching or exceeding the "safe" level of fluoride with respect to mottling. A reasonable approach to this situation could involve the following: The teeth of representative children in fluoridated areas be periodically surveyed for fluoride-induced brown stains. This information would then serve as an excellent guide in monitoring total fluoride exposure to ascertain whether children are being exposed to the "proper" amount of fluoride.

During the public comment period for the proposed drinking water standards, it was suggested that lower MCLs be considered, since there was an increased level of fluoride intake from other sources—probably based on data similar to those presented here. The EPA concluded that, even though the presented evidence of enhanced body burden may be true, the MCL for fluoride is based primarily on epidemiologic evidence derived from areas in which fluoride is a natural constituent of the water. Thus the EPA felt that it should be assumed that in those areas, much of the food was prepared in the community water; thus the amount of fluoride taken from this source is automatically considered.

Nitrate

The first federal drinking water standard for nitrate was established in 1962, when the USPHS issued a recommended limit of 45 mg of nitrate as nitrate per liter (e.g., 45 mg NO_3 per liter* (PHS, 1962). It is of interest that of the nine inorganic chemicals for which the EPA has instituted MCLs, only nitrates and silver were not considered for drinking water standards until 1962, when the USPHS formally promulgated such limits.**

The primary reason for this relatively recent recognition of elevated nitrate levels as a potentially harmful constituent in drinking water is quite simple: the first cases of human morbidity and mortality associated with high nitrate levels in drinking water were reported as late as 1945 by Comly. He noted 21 cases of methemoglobinemia in infants from Iowa who consumed well water with elevated levels of nitrates. Over the next 10 years, a number of other similar associations of nitrates and infant methemoglobinemia were reported in the United States (Bosch et al., 1950; Campbell, 1952; Lecks, 1950; Walton, 1951).

As a result of the recognition of the potential health problems of elevated nitrate levels in drinking water, the American Public Health Association (APHA) sent a survey questionnaire to each state department of public health, including those of Alaska and Hawaii, to obtain data concerning reported cases of methemoglobinemia associated with elevated levels of nitrate in drinking water. With one exception, all states cooperated in the study. The survey revealed that 14 states had recorded a cumulative total of approximately 300 cases of methemo-

* 45 mg NO_3 per liter is equal to 10 mg NO_3 as nitrogen per liter.

** When the standard was exceeded the public was to be warned not to use such water for infant feeding (PHS, 1962).

globinemia and 39 deaths associated with high nitrate levels in drinking water. Most cases of methemoglobinemia, as well as deaths, were noted in the north central states. Also, this survey did not find any of these cases of methemoglobinemia to be associated with municipal drinking water supplies (APHS Committee, 1949/1950). [In 1965, one such case of methemoglobinemia associated with a municipal water supply was noted (Vigil et al., 1965)]

A precise dose-response relationship for nitrate toxicity in humans was difficult to derive from these data. The primary reasons for this difficulty were (1) the presence of nitrite, which is drastically more dangerous than nitrate in the water along with the nitrates, (2) in numerous cases in which methemoglobinemia was reported, the water was not sampled for nitrate levels until several months after recognition of the illness; consequently, the levels may not have been identical with those which poisoned the infant, and (3) analytic techniques may have resulted in erroneously low nitrate values in the presence of high chloride levels or erroneously high nitrate values perhaps caused by the presence of high levels of organic matter in the water sample.

The possible sources of contamination of drinking water by nitrates are municipal and industrial effluents and run off from barnyards, feedlots, septic tank disposal, and farm soils with high levels of commercial fertilizers (EHRC, 1974). It should be noted that nitrogen usage in the United States has increased from 378,543 tons (of nitrogen) in 1940 to 6,912,700 tons in 1969 (EHRC, 1974). Thus it appears that the relatively low usage of nitrogen products, especially fertilizers, prior to the 1940s is a reason for the nonrecognition of the nitrate-methemoglobinemia relationship.

The major problem with excessive nitrates in drinking water is that once they are consumed by the infant, the nitrates may be converted to nitrites; furthermore, infants are particularly predisposed to the formation of methemoglobin (MetHb) in the presence of nitrite. Thus there are two issues (1) Why is the infant hypersusceptible to the toxic effects of nitrites? (2) How are nitrates converted to nitrites?

The formation of MetHb is dependent on the conversion of nitrate to nitrite. In the presence of nitrite, hemoglobin is oxidized to MetHb. In this oxidized state, the hemoglobin (that is, the MetHb) is not capable of reversibly combining with oxygen. When carried to the extreme, the end result of methemoglobinemia is acute oxygen deprivation and death (Lee, 1970). Levels of 1–2% MetHb are usually considered normal in the blood of adults (Committee on Nitrate Accumulation, 1972), and levels of 2–5% are not uncommon in infants (Gruener and Shuval, 1970). When concentrations are less than 5% MetHb, there are no obvious signs of toxicity. However, as the levels of MetHb increase from 5 to 10%, clinical symptoms of toxicity, such as cyanosis, may appear (Committee on Nitrate Accumulation, 1972; EHRC, 1974).

Factors that predispose infants to the development of MetHb formation include

1. The infant's total liquid intake per unit body weight is approximately three times that of an adult (Hansen and Bennett, 1964).
2. The infant has an incompletely developed ability to secrete gastric acid; this permits the gastric pH to become high enough (5–7) to allow nitrate-reducing bacteria to grow high in the gastrointestinal tract and thereby convert the nitrate to nitrite prior to absorption into the circulation (PHS, 1962).
3. Infants are born with hemoglobin F (fetal hemoglobin); this form of hemoglobin is more susceptible than adult hemoglobin to be oxidized to MetHb (Betke et al., 1956).
4. Infants have a somewhat reduced enzymatic capability to reduce MetHb to hemoglobin (Ross and Des Forges, 1959).

Prior to the issuance of the 1962 USPHS nitrate standard, prestigious research bodies, foreign countries, and several researchers proposed recommended standards. For example, the International Drinking Water Standards of 1958 by the World Health Organization indicated that the ingestion of water containing nitrate with greater than 50 mg as NO_3/liter may lead to infantile methemoglobinemia. In 1951 the Committee of the Division of Medical Science of the United States National Research Council regarded any well water with a nitrate concentration greater than 10 mg/liter as unfit for infants (EHRC, 1974). Several South American countries have recommended maximum permissible concentrations of from 0.5 to 228 mg as NO_3/liter (0.1–51 mg NO_3 as N per liter) (Caballero 1949–1950). Taylor (1958) in England and Bosch et al. (1950) in the United States suggested a limit of 20 mg NO_3 as N/liter and 10–20 mg NO_3 as N/liter, respectively, to protect infants from cyanosis.

In addition to nitrate exposure from drinking water, further exposures to nitrate (or nitrite) from use as a food preservative in various products often occurs. For example, a former limit of 200 ppm nitrite in certain meat products was based on the premise of a 1.4–5.7% conversion of hemoglobin to MetHb per 100 g of food product. High levels of nitrates have also been found in various vegetables, including spinach, beets, lettuce, and several other common vegetables. However, despite these other possible sources of nitrate exposure, the PHS (1962) did not consider the contribution of nitrates (or nitrites) from food during the standard derivation process.

Since the issuance of the 1962 standard, several studies have extended the previously reported research. For example, reports from foreign literature of 467 cases of methemoglobinemia in infants where drinking water nitrate levels were taken, revealed that a small percentage (3%) of the cases were associated with nitrate levels of 9.2 mg NO_3 as N per liter or less (Sattlemacher, 1962). These results are in excellent agreement with 745 cases of methemoglobinemia reported in Germany between 1955 and 1964 (Simon et al., 1964; EHRC, 1974). (Figure 6-1 shows a dose-percentage population response for the development of methemoglobinemia.) It seems evident that as the level of nitrate increases in the drinking water, so does the frequency of cases of methemoglobinemia. Of particular significance are the data that indicate the occurrence of a low but consistent prevalence of such cases at nitrate levels lower than the former USPHS and present EPA national standard. Precisely why certain infants were more susceptible was not determined. However, low levels of dietary vitamin C (EHRC, 1974) along with other sources of nitrate exposure are factors that may partially account for the variable responses in the population.

Although most of the previous research emphasized infants as the principal subgroup of the population at high risk with respect to nitrates, several Soviet studies suggested that elevated nitrates may

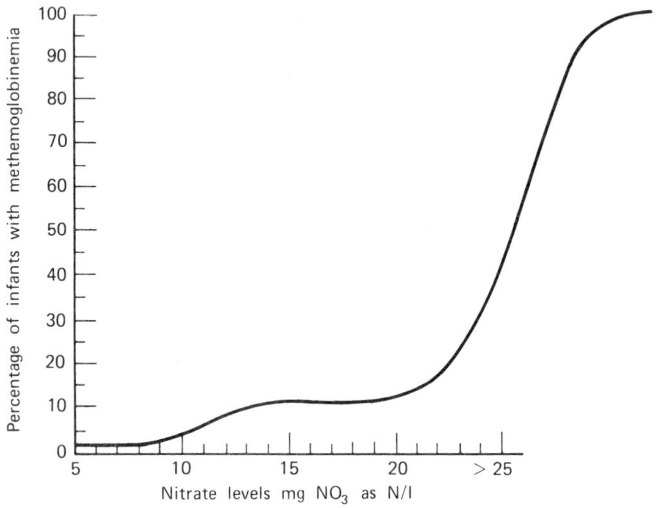

Figure 6-1. Nitrate Levels mg NO_3 as N/l. (Based on results of Sattlemacher (1962) and Simon et al (1964).

have deliterious effects in older children (up to at least 14 years of age in one study). For example, children in day nurseries and kindergartens who consumed water with 20–40 mg NO_3 as N per liter frequently had MetHb levels above 5% (Subbotin, 1961). Another Soviet study reported diminished responses to both auditory and visual stimuli in 21 children (aged 12–14 years) exposed to 26 mg NO_3 as N per liter as compared to a control group exposed to 2 mg NO_3 as N per liter. Methemoglobin levels in the elevated nitrate group were 5.3%, whereas the control group had 0.75% (Petukhov and Ivanov, 1970).

As a result of these and other supportive studies (Diskalenko, 1968; Shearer et al., 1971; Gelperin et al., 1971) concerning the effects of nitrates on the occurrence of methemoglobinemia in humans, especially infants, the EPA retained the earlier USPHS standard of 10 mg NO_3 as N per liter. It should be emphasized that unlike most of the standards (excluding fluorides), there is little if any safety margin for the most susceptible segments of the population. In marked contrast to the initial toxicity from fluorides (e.g., mottling of developing teeth), the development of methemoglobinemia is extremely dangerous and potentially fatal.

The EPA's national standard does recognize the enhanced potential toxicity of elevated nitrate levels in drinking water as compared to the other eight inorganic substances for which there are MCLs in the following manner. Normally, if a pollutant level exceeds the MCL, the supplier of the water must report the occurrence of this elevated level to the state within 7 days and then take three additional water samples within the next 30 days. An average value of the concentration of the pollutant in question is made; if this average value exceeds the MCL, the supplier is required to notify the state and the public. In contrast to this methodology for the other eight inorganic chemicals with MCLs, when a level greater than the maximum contaminant level for nitrate is found, a subsequent analysis is initiated not within 30 days but within 24 hours. If the mean of two analyses is greater than the MCL, the supplier must notify the state and the general public.

Lead

In contrast to the rather recent recognition of health hazards for certain inorganic elements (e.g., barium, cadmium, mercury, nitrates) in drinking water, there has been a drinking water standard for lead since 1925 (Public Health Reports, 1925). In the past 50 years an enormous amount of research has been directed toward trying to elucidate the health effects of lead on the human body. Considerable attention has been directed to industrial exposure in the United States and foreign countries (NIOSH, 1973). The development of exposure limits to lead has been a continuous concern of such groups as the American Conference of Governmental Industrial Hygienists (ACGIH) for at least the past 35 years (ACGIH, 1976). More recently, lead has been recognized as a community health problem with respect to ambient air (EPA, 1972; NAS, 1972; EHRC, 1972). However, despite this recognition, lead was not included in the original group of national ambient air standards promulgated as a result of the Clean Air Act Amendments of 1970. More recent attempts to regulate the emission of lead into the ambient environment (e.g., catalytic converter and low lead fuels) (EHRC, 1975) have been attempted by the EPA (Environment Reporter-Cases (ER-C), 1975).

In light of the growing awareness of lead as a serious environmental pollutant, it is of interest to consider how the rationale for the lead standard has evolved and how it has been influenced by data from both the industrial and community air environments. One of the first steps in the process of standard derivation is the establishment of an accurate dose-response relationship. In the case of lead, the PHS cited work by Kehoe (1947) which indicated that at daily lead exposures of 0.5–0.6 mg for periods of 1 year or more, there were no indications of adverse health effects in healthy adult humans. However, there was a small increase in the body burden for lead. These results were consistent with data derived from studies of long exposure to lead in occupational settings. Further evidence presented by the PHS (1962) indicated that significant increases in lead exposure above 0.6 mg/day would lead to proportional increases in the body burden. The PHS finally suggested that exposures greater than 0.6 mg/day for greater than a 5-year period may lead to the accumulation of a potentially harmful quantity of lead in the body.

Having established what they considered a reasonable dose-response relationship for lead, the PHS focused next on the various sources of lead exposure from the environment. Thus exposures to lead from food, air, and cigarette smoke were considered in the derivation of the drinking water standard. With regard to exposure from food, it was suggested that the intake of lead via food sources was probably approaching an irreducible minimum. Lead is present in food as a result of its natural occurrence in the earth's crust, its former use as an insecticide, as a contaminant in food processing and packaging, and from deposits on soil from air pollutants that have settled out. Specific food types seriously and unavoidably contaminated by lead were required by law not to exceed specific limits of lead. Consequently, lead levels in food were regulated; however, this applied only to a limited and actually a minor part of the typical diet. It was suggested that the typical daily diet of the average American adult resulted in a maximum exposure of 2.0 mg, although the mean amount was approximately 0.3 mg.

In 1962 air exposure to lead was viewed as a potentially serious problem, based on the observation that ambient concentrations had been steadily increasing over the past decade. In fact, from the early 1950s to the early 1960s, there was an increase by at least a factor of 10 in major urban areas. By 1962 the national average for lead concentrations in urban ambient air had reached 1.4 $\mu g/m^3$. Based on an inhalation of 20 m^3 air per day and a 10% absorption factor, the total daily exposure to lead from air alone in an urban area would be approximately 3.0 μg/day (Tabor, 1957; PHS, 1962). Exposure to lead as a result of smoking cigarettes was indicated to be as high as 0.3 μg per puff. Thus a heavy smoker could be assumed to have an additional daily exposure to lead of at least several micrograms (Cogbill and Hobbs, 1957; PHS, 1962).

Based on (1) these multiple sources of lead exposure, (2) the limited governmental regulation of lead exposure from these sources, and (3) the fact that typical daily lead intake is greater than half the amount needed before progressive retention of lead occurs (i.e., increasing body burdens), it was decided to adopt a 0.05 mg/liter standard for lead in drinking water (PHS, 1962).

Evaluating this standard, the following observations can be made: (1) The dose-response relationship was based on adult exposure, yet young children are known to be more efficient in the absorption of lead from the digestive tract than adults (Alexander et al., 1973). (2) Young children also are known to take in more lead on a milligram of lead per kilogram of body weight basis (Alexander et al., 1973). (3) Evidence indicates that young children are known to be more susceptible to lead toxicity than adults (Millar et al., 1970; NAS, 1972). Furthermore, other potentially high-risk groups were not considered, especially those with dietary deficiencies of calcium and/or iron (Mahaffey-Six and Goyer, 1972; Mahaffey, 1974).

In fairness to those who derived the standard, this information has become more convincing during the past 10 years, and in most cases has been incorporated into the rationale for the 1975 standard. Also, whether smokers should be specifically protected by a standard is a debated point (see section on cadmium; in the case of the 1962 standard for lead, the PHS specifically considered this issue.

From 1962 to 1975 (when EPA proposed the new national lead standard which incidently is identical to that issued by the PHS in 1962), considerable information was published concerning lead toxicity.

As a result of this new information, the EPA derived the 1975 standard. A discussion of the new rationale for the drinking water standard for lead follows.

There were some obvious differences between the 1962 and the 1975 approaches to the derivation of the respective lead standards. Although the EPA Advisory Board (1975) did mention that several previous studies had indicated that people who smoke cigarettes had higher blood lead levels than nonsmokers (see Cogbill and Hobbs, 1957; Hofreuter et al., 1961; Survey, 1965), this factor does not seem to be emphasized as it was in the 1962 standard derivation. Although not mentioned in their justification, several studies available to the EPA indicated that the contribution of smoking to blood lead levels has considerably diminished as a result of the removal of lead arsenate as an insecticide used by tobacco growers (Patterson, 1965; McLaughlin and Stopps, 1973). Despite their lack of citation of these references, the diminished concern with smoking as an important contributing factor is clearly consistent with these data.

The originally proposed dose-response relationship of Kehoe (1947) used by the PHS in the 1962 standard, has been challenged in the interim by a World Health Organization Committee (1972), which indicated that 3 mg of lead per week (or 428 μg/day) was the maximum lead exposure an average person is able to take in without increased body burden. This is in contrast to Kehoe's value of 600 μg/day.

According to the EPA Advisory Board (1975), if the average diet contained 100–300 μg lead per day, and the average urban air had 1–3 μg lead per m^3, the average urban resident would absorb 16–48 μg of lead per day, assuming that 10% of the ingested amount is absorbed via the digestive tract and 30% is absorbed via the lungs, which have inhaled 20 m^3 of air per day. (Note: the 1962 standard assumed that 10% of the lead was absorbed via inhalation, whereas in the 1975 EPA Advisory Board's statements of justification, a 30% absorption rate from the lungs was assumed. Neither assumption was specifically supported by literature references despite the importance of this factor in determining susceptibility to toxicity.) Consequently, if one accepted the WHO Committee recommendations, certain urban residents would tend to be exposed to quantities of lead that approach the WHO standard.

Although no mention was made of high-risk groups such as children in the 1962 standard justification, strong consideration is directed toward this group in the present EPA justification. For example, recognition was made of the greater exposure from air and food per unit of body size and enhanced absorption of lead via the digestive tract in children as compared to adults (Alexander et al., 1973). It was also suggested that children may have greater susceptibility to lead than adults with respect to neurological, hematologic, and immunologic parameters (Millar et al., 1970; Betts et al., 1973; Selye et al., 1966; Gainer, 1973; NAS, 1972; EPA Advisory Board, 1975).

Regardless of the specific differences in the data base between 1962 and 1975, there existed sufficient commonality such that the identical standard (as indicated above) with a similar reasoning process occurred in the derivation of the standard; that is, there is a factor of less than 2 between what the average American is exposed to in his/her daily life and the amount of lead causing an increase in the body burden (especially in children). It was pointed out that at the standard of 0.05 mg lead per liter, 25–33% of the 1–3-year-old child's daily lead exposure would be derived from the drinking water as compared to food. (This assumes an intake of 1.0 liter of water per day and a dietary lead exposure of between 150 and 200 μg/day.) A similar comparison (at 0.05 mg lead per liter) for adults revealed that lead exposure from drinking water would be 33% of that from food.

It is difficult to understand why the proposed (and now promulgated) standard of the EPA was not lowered from that issued in 1962. The evidence was clearly much stronger with regard to the dangers of lead toxicity, especially with respect to children. Furthermore, the unresolved controversy between the WHO Committee (1972) and Kehoe (1947) concerning a precise dose-response relationship should have suggested a more conservative approach. Finally, since exposure to lead from water was the only source that had lent itself to widespread regulation and it was capable of contributing up to 33% of the lead exposure contributed from food, it would seem to be more reasonable to have a stricter water standard.

It should be noted that the EPA has made vigorous yet highly controversial efforts to reduce ambient lead exposure. For example, the adoption of the catalytic converter in new automobiles since 1975 was considered an indirect approach to reducing ambient lead exposure (EHRC—catalytic converter, 1975; ER-C, 1975). Although the catalytic converter was designed to reduce the levels of hydrocarbons and carbon monoxide, the catalyst would not function in the presence of lead. Thus low-lead gasoline has been used in automobiles with the catalyst. In theory, therefore, after 5–10 years of catalyst usage (with low-lead gasoline employed), lead levels in the ambient air would be expected to decrease. This reasoning, however, was not used in the adoption of the drinking water standard for lead. It was wise that it was not used, for the converter is a highly controversial mechanism in light of its enhanced emissions of sulphuric acid mist and certain automobile models can achieve hydrocarbon and carbon monoxide emission regulation without the use of the catalyst (EHRC-catalytic converter, 1975).

Sodium

Since the mid 1950s, there has been a profound increase in the use of salt (principally NaCl) to melt ice and snow from Massachusetts state highways (Huling and Hollocher, 1972; Mass. DPW, 1974). In fact, the rate of application of salt to state highways was 22 metric tons of total salt ($NaCl + CaCl_2$) per lane mile in 1972. One of the environmental consequences of the increased application of salt to Massachusetts roadways has been the unintentional contamination of various community water supplies. For example, 15 communities in eastern Massachusetts have reported greater than (>) 100 mg of chloride per liter of drinking water, and one of these communities (Burlington) had one well with >200 mg Cl per liter. Subsequently, the salting of ice- and snow-covered roads was suspended in Burlington (Huling and Hollocher, 1972). Furthermore, in 1970 the Massachusetts Department of Public Health (1969/1970) warned that 62 communities in the state had levels of sodium in their drinking water exceeding the American Heart Association's (AHA) (1957) recommended standard of 20 mg Na per liter. By 1975, 87 communities (Massachusetts Department of Public Health, 1975) exceeded the AHA standard, and in 1976 the number of communities exceeding the AHA standard had increased to 117 (Springfield Daily News, 1976).

In light of the widespread occurrence of elevated levels of sodium in the drinking water of Massachusetts, the State Office of Environmental Affairs (1977) recently proposed that the state adopt a sodium drinking water standard of 20 mg/liter. Although in direct agreement with the proposed sodium standard of the AHA, the newly proposed Massachusetts standard is in marked contrast to the Advisory Board's proposed sodium standard of 270 mg/liter (EPA, 1971) and the more recent recommendation of NAS (1977) of 100 mg/liter. It should be mentioned that the EPA's present drinking water standards (which were finalized on June 24, 1977) do not contain a standard for sodium, because the EPA felt that sufficient studies are lacking on which to base a proper judgment (Federal Register, Dec. 24, 1975). In light of the controversy surrounding the issue of a sodium drinking water standard, this section critically reviews the data on which a sodium drinking water standard may be based.

Health Effects of Elevated Sodium Exposure

Considerable research on the effects of sodium on the health of animal models and humans has been conducted since 1900. Much of the health-related research has focused on the association of sodium intake and the development of hypertension.

Animal Research. Of importance in the elucidation of the effects of sodium on hypertension is the extensive research with rat, chicken, rabbit, dog, and monkey models (Sapirstein et al., 1950; Toussaint et al., 1953; Meneely et al., 1953, 1957, 1961; Dahl, 1961; Fukuda, 1951; Lenel et al., 1948; Selye, 1943; Langston et al., 1963; Koletsky, 1958, 1961; Dahl et al., 1968; Haddy and Overbeck, 1976). The summarized results indicate that

1. The greater the ingestion of salt, the more severe the hypertension.
2. The younger the animal at the time it is fed a high salt diet, the more sensitive it is to developing rapid hypertension.
3. Even a brief exposure (2–6 weeks) to high salt intake in early life may influence the development of a permanently elevated blood pressure.
4. Genetic factors influence the individual's response to NaCl.

Human Studies. Several of the most important earlier human studies revealed that salt intake may significantly affect blood pressure levels (Ambard and Beaujard, 1904; Allen and Sherill, 1922). More specifically, Allen and Sherill (1922) reported that a reduction of salt in the diet could reduce hypertension. However, it was not until the 1940s that large-scale human studies showed that low sodium diets were successful in reducing the frequency of hypertensive disease (Kempner, 1944, 1948). Observations of humans from the 1940s to the present in both epidemiologic and controlled settings have tended to confirm the relationship of sodium exposure as a causative factor in the development of hypertension (Dahl and Love, 1954; Jossens et al., 1971; Watson and Langford, 1967; Murphy, 1950; Watkin et al., 1950; Parijs et al., 1973; Sasaki, 1964; Weinsier, 1976). For example, Murphy, Watkin et al. and Parijs et al. were able to demonstrate that blood pressure levels can be modified up or down depending on the amount of salt in the diet. Furthermore, observations have revealed that the high prevalence of hypertension (30–40%) in the fifth decade of life in northeastern Japan is associated with a 25–30-g/day Na intake (Sasaki, 1964).

In marked contrast to "civilized" populations, hypertension is not usually present in unacculturated societies, nor does it increase as a function of age. Numerous anthropological-medical studies of unacculturated societies in New Guinea (Maddocks, 1967), the highlands of Malaysia (Burns-Cox and MacLean, 1970), the Easter Islands (Cruz-Coke et al., 1967), rural Uganda (Shaper, 1972), and the Kalahari Desert of Africa (Kaminer and Lutz, 1960) demonstrated this difference to "civilized" societies. However, when unacculturated societies that are generally free from hypertension take on "modern" life styles, blood pressure increases and hypertension becomes evident. Part of the modern life style assumed by the natives is a substantial increase in salt intake. For example, urban Zulu develop hypertension, whereas the nonurban Zulu do not (Scotch, 1960); semiacculturated Cook Islanders have more hypertension than more primitive neighbors (Prior et al., 1968); nomadic Ugandan tribes without hypertension are those with a low salt intake, whereas tribes with hypertension have significantly higher salt intakes (Shaper, 1972). In studies of six Solomon Island societies, investigators reported a rise in blood pressure with age in the three most acculturated societies and no rise in the three most unassimilated peoples (Page et al., 1974). In almost every modern epidemiologic study of unacculturated peoples, the significance of salt has been observed as the most important factor for determining the presence or absence of hypertension (Lowenstein, 1961; Kean, 1944; Prior et al., 1968; Page et al., 1974; Sinnet and Whyte, 1973; Floyer and Richardson, 1963). Other factors that have been suggested as explanations for the lack of hypertension in unacculturated people, such as disease factors, noncompetitive life styles, and lack of obesity, do not offer as consistent an explanation of the occurrence of variations in hypertension within and between societies as does salt intake, although their influence in particular cases may be important (Freis, 1976). However, it should be emphasized that although the level of salt intake may play an important role in the development of hypertension, it is only one of the factors that affects hypertensive vascular disease, and the effects of salt load or salt restriction should be interpreted in this light.

Mode of Action

Dietary salt and extracellular fluid. Dietary salt is distributed predominantly to the extracellular space. Excess quantities of salt or water are excreted primarily through the kidneys. Freis (1976) presented evidence that in acculturated peoples the extracellular fluid volume may be as much as 15% greater than that of unacculturated peoples who ingest much less salt and exhibit much less hypertension. Numerous other studies have indicated a relationship between extracellular fluid volume (ECF) and the development of arterial blood pressure (Murphy, 1950; Watkin et al., 1950; Ledingham, 1953; Borst and Borst, 1963; Guyton et al., 1974; Tobian, 1972). These animal and human studies have shown that a rise in extracellular volume is accompanied by an increase in cardiac output and a rise in blood pressure. The elevated extracellular volume then decreases while total peripheral resistance increases to maintain the hypertension. Therefore, the chronic state of hypertension is characterized by a normal cardiac output and an increased peripheral resistance. The precise manner by which the increased peripheral resistance is maintained is controversial and may involve both autoregulation (Freis, 1960) and/or humoral agents (Haddy and Overbeck, 1976).

An enhanced urinary output along with a decrease in ECF accompanies the increase in blood pressure. The common factor required for the development of any chronic elevation of blood pressure is an increase in urine volume and sodium excretion by the kidney to prevent a chronically enlarged ECF (Guyton et al., 1974). The increased level of blood pressure needed to effect the enhanced diuresis is directly affected by the capacity of the particular kidney to excrete an excess of sodium. The more efficient the functional capacity of the kidney, the more limited the rise in blood pressure. The reverse is also true (Guyton et al., 1974). Observations that the ECF of hypertensive patients is similar to that of normotensive individuals is consistent with this volume-load hypothesis, since the kidney maintains ECF because of the increased blood pressure (Ledingham, 1953).

Selectively inbred rat strains with either a predisposition or resistance to hypertension have been developed. The transplantation of kidneys from the hypertension-prone rats into the hypertension-resistant rats affects the development of hypertension in the host. The reverse experiment, with the hypertension-prone animal serving as the host, revealed that the presence of kidneys from a hypertension-resistant rat would lower hypertension (Haddy and Overbeck, 1976; Dahl, 1972; Bianchi et al., 1974; Dahl et al., 1974).

Renin-Angiotensin-Aldosterone (RAA) and Homeostatic Control of the ECF. Despite the fact that the RAA system plays a role in homeostatic control of ECF, its capacity to regulate a continued increase in ECF is limited. Guyton et al. (1974) indicated that the RAA system lacks the infinite gain inherent in the arterial blood pressure–

urine volume relationship. Thus they concluded that even though the RAA system assists the kidney in control of ECF volume, it is not able to efficiently regulate the homeostasis of the ECF for long periods in a situation of high salt intake and diminished functional capacity of the kidney. Freis (1976) discussed the inconsistencies of the RAA system as the critical factor in the pathogenesis of hypertension.

Dietary Management of Blood Pressure

Murphy (1950) reported that when hypertensive patients were treated with the Kempner rice diet ($\leq$8 mEq (or 0.5 g) per day of Na), there was no change in serum sodium levels, but plasma volume decreased by 10% and ECF fell approximately 12% over a 3-week period. Blood pressure was also significantly diminished. Similar results were demonstrated by Watkin et al. (1950) with hypertensive patients on the Kempner diet. By adding 3.0 g of salt daily to the diet of the hypertensive patients, the original plasma volume levels were achieved, along with an increase in blood pressure similar to that of the control group. However, an addition of 1.0 g of salt effected only a slight increase in plasma volume and did not increase blood pressure. Such results prompted Watkin et al. to conclude, "The critical level of sodium intake with respect to hypertension appears to be extremely low; at least in many patients with advanced stages of hypertension, a sodium intake above the critical level causes a more or less prompt return of hypertensive manifestations." In a similar fashion, continued reduction in volume is needed to sustain the antihypertensive action of diuretic drugs (Freis, 1976).

Parijs et al. (1973) reported the only well-controlled study that associated a moderate restriction in sodium intake with a definite reduction in blood pressure. In their study, patients with mild hypertension were provided with a sodium-restricted diet that lowered the 24-hour excretion of sodium from 191 to 93 mEq/day. The authors concluded that if the salt intake was diminished from approximately 10 to 5 g each day, blood pressure values could be diminished by 10/5 mm Hg.

The dose-response relationship of the effects of sodium on blood pressure remains to be more precisely determined. However, it is apparent, in a general sense, that as the salt dosage increases, the incidence of hypertension also increases.

The typical daily salt intake in the United States and other Western countries ranges from a low of about 3 g to a high of 30 g, with most people ingesting 5–15 g (Weinsier, 1976; Freis, 1976). It should be noted that a lunch of biscuits and canned soup may contain approximately 3.5 g of salt, and a TV dinner plus salad and dressing contains about 4 g. Thus a high salt intake can be easily achieved without the additional use of the salt shaker. In contrast, Eskimos, certain Chinese, American Indians, and African tribes customarily employ no additional salt in their diets. Their estimated daily intake of salt ranges from 1 to 2 g, which is more than sufficient for proper growth, development, and maintenance. In certain "no salt" cultures (e.g., Yanomamo Indians of Brazil), a chronic elevation (compared to Western levels) of plasma renin activity has been reported (Oliver et al., 1975). This appears to be a physiologic adaptation to assist the body in retaining sufficient sodium for proper physiologic functions.

Evidence Relating Sodium Ingestion Via Drinking Water to Human Health Problems

The chronic excessive consumption of salt has long been recognized as a risk factor in the development of hypertension in animals and humans. Since sodium intake is often substantially greater from food than water,* nearly all the previous studies of the relationship between the frequency of hypertension and sodium intake considered only the contribution of sodium from food and not from drinking water. However, Furstenberg et al. (1941) reported that high levels of sodium in the drinking water resulted in the initial failure of a low salt diet to effectively treat Ménière's disease. According to Elliot and Alexander (1961), the health of patients on sodium-restricted diets may be adversely affected by large amounts of sodium in the drinking water. They reported several cases of recurrent episodes of heart failure at home that ceased after substitution of a low-sodium drinking water source for a high-sodium source. Russell (1969) indicated that in patients following a sodium-restricted diet (500 mg Na per day) and consuming 2.5 liters of water per day containing 125 mg sodium/liter, 64% of the daily sodium allowance would come from the drinking water. It can therefore be seen that high levels of sodium in drinking water may be an important contributing factor in the development of a variety of heart disorders in high-risk patients. An interesting epidemiologic study in Romania revealed that the great prevalence of arterial hypertension in people in the town of Jurilovca is highly correlated with

* Water contributes from 0.5 to 9.0% of the total sodium consumed, with the important exceptions of those on a salt-restricted diet (Schroeder, 1974).

the very high concentration of electrolytes (especially sodium) in the drinking water (Steinbach et al., 1974).

Calabrese and Tuthill (1977) recently reported the first epidemiologic study in the United States of the effects of elevated levels of sodium in community drinking water on blood pressure distribution patterns. They studied 16-year-old high school students (300 students from each school) in two closely matched communities with sodium levels in drinking water of 8 and 107 mg/liter, respectively. Their results demonstrated the occurrence of statistically significant elevated systolic and diastolic blood pressures in both males (0.05 and 0.01 levels, respectively) and females (0.001 level) in the high sodium community. The difference averaged approximately 3–5 mm Hg for each specific comparison category. Thus the high sodium student group appears to display a blood pressure distribution characteristic of persons several years older.

Elevated levels of sodium in drinking water have been associated with the occurrence of a variety of heart-related diseases. Wolf and Esmond (1974) clearly demonstrated a concomitant variation in the increase and decrease in the NaCl concentration of the public water supply in Dallas and a similar pattern of deaths from both arteriosclerotic heart disease and hypertensive disease during the 1950s drought when water with high sodium levels was imported. Of particular interest is that the increase in heart-related diseases was first observed when the sodium levels began to exceed 125 mg/liter approximately 18 months after the inclusion of the imported water into the drinking water system. These results clearly implied the possibility of a threshold for adverse health associated with elevated levels of sodium in the drinking water. Finally, Tuthill (1976) assessed the relationship of 18 water constituents and 13 social variables to 12 categories (including eight for cardiovascular disease) of sex-specific, age-adjusted mortality ratios in 215 Massachusetts communities. The results indicated a low but positive correlation of ischemic heart disease for females and overall death rate for females under 65 with sodium levels.

Conclusion

Based on the toxicologic and epidemiologic studies reviewed here, it is evident that elevated levels of sodium in the diet are a factor in the development of hypertension in a large number of individuals. Recently several studies were published that indicate that levels of sodium in drinking water of 100–125 mg/liter may adversely affect the health of adolescents as well as adults. Although more definitive human epidemiologic studies remain to be conducted to elucidate a more precise dose-response relationship, the evidence clearly supports the need for a federal standard considerably below 100 mg/liter. Since United States residents usually receive considerably more than the daily requirement of sodium in their diets, the implication that any amount of sodium in water is excessive has begun to emerge. If it is assumed that sodium acts according to a nonthreshold response for a highly heterogenous population, all excessive exposure would entail some degree of risk. It has therefore become necessary for the American society, through regulatory procedures, to determine the level of risk they choose to tolerate, or conversely, the degree of safety for which they are willing to pay.

Chlorinated Hydrocarbon Insecticides

Perhaps the most striking change from the 1962 PHS drinking water standards to the newly developed EPA regulations is the widespread adoption of MCLs for a broad variety of synthetic organic insecticides (Table 6-3). Since 1945, synthetic organic insecticides have, for the most part, taken the place of the botanical and inorganic insecticides. The three general types of insecticides are chlorinated hydrocarbons, organophosphates, and carbamates. Specific examples of each of these three general types of insecticides include: chlorinated hydrocarbons—DDT, chlordane, lindane, methoxychlor, heptachlor, toxaphene, dieldrin, and aldrin; organophosphates—malathion, diazinon, metasystox, and nalad; carbamates—sevin and vapam (Kennedy and Hessel, 1971).

In attempting to deal with the establishment of proper regulations for safe limits of insecticides in drinking water, the EPA has considered chlorinated hydrocarbons separately from organophosphates and carbamates. The organophosphates and carbamates were collectively grouped since their "nonspecific" presence is detected by decreased levels of cholinesterase activity in both serum and red blood cells (EPA Advisory Board, 1975). Despite this recommendation, the EPA did not establish a MCL for such compounds since there is a lack of supporting evidence that organophosphates are actually present at the consumer's tap. It was suggested that the best way to deal with organophosphates in drinking water would be through an alternative administrative procedure that regulates major spills. Because of the speedy degradation of organophosphates, it was felt that only accidental spills would present a possible threat to drinking water sources (Fed. Reg., March 14, 1975). However, with the chlorinated hydrocarbons,

Table 6-3. Derivation of Approval Limits (ALs) for Noncarcinogenic Chlorinated Hydrocarbon Insecticides

	Lowest Long-Term Levels With Minimal or No Effects			Calculated Maximum Safe Levels			Intake from Diet		Water	
Compound	Species	ppm in Diet	mg/kg Body Weight/ day[a]	Safety Factor (X)	mg/kg/ day	mg/man/ day[b]	mg/man/ day	Percentage of Safe Level	Percentage of Safe Level	Recommended MAL (mg/ liter)[c]
	Rat	2.5	0.42	1/500	0.00084	0.0588[d]				
* Chlordane	Dog	N.A.	N.A.	—	—	—	T	T	6	0.003[e]
	Man	N.A.	N.A.	—	—	—				
	Rat	5.0	0.83	1/500	0.00166	0.1162				
Endrin	Dog	1.0	0.02	1/500	0.00004	0.0026[d]	0.00035	4.1	20	0.0002
	Man	N.A.	N.A.	—	—	—				
	Rat	0.5	0.083	1/500	0.000166	0.01162				
* Heptachlor	Dog	4.0	0.03	1/500	0.00016	0.0112[d]	0.00007	0.6	2	0.0001[f]
	Man	N.A.	N.A.	—	—	—				

* Heptachlor	Rat	0.5	0.083	1/500	0.000166	0.01162				
Epoxide	Dog	0.5	0.01	1/500	0.00002	0.0014[d]	0.0021	150.0	5	0.0001
	Man	N.A.	N.A.	—	—	—				
	Rat	50.0	8.3	1/500	0.0166	1.162				
Lindane	Dog	15.0	0.3	1/500	0.0006	0.042[d]	0.0035	8.3	20	0.004
	Man	N.A.	N.A.	—	—	—				
	Rat	100.0	17.0	1/100	0.17	11.9				
Methoxychlor	Dog	4000.0	30.0	1/100	0.8	56.0	T	T	20	0.1
	Man	—	2.0	1/100	0.02	1.4[d]				
	Rat	10.0	1.7	1/500	0.0034	0.238[d]				
Toxaphene	Dog	400.0	8.0	1/500	0.016	1.12	T	T	2	0.005[e]
	Man	N.A.	N.A.	—	—	—				

[a] Assume weight of rat = 0.3 kg and of dog = 10 kg; assume average daily food consumption of rat = 0.05 kg. and of dog = 0.2 kg. [b] Assume average weight of human adult = 70 kg. [c] Assume average daily intake of water for man = 2 liters. [d] Chosen as basis on which to derive maximum approval limit (MAL). [e] Adjusted for organoleptic effects. [f] Adjusted for interconversion to H. epoxide. N.A., no data available. T, Infrequent occurrence in trace quantities.

* These insecticides have subsequently been classified as potential human carcinogens.

Source: EPA (1973) Advisory Document on Drinking Water Standards, Washington, D.C.

each specific insecticide has its own MCL. The only exception to this methodology are those chlorinated hydrocarbons such as DDT, aldrin, and dieldrin, which are considered to be carcinogens.

With regard to carcinogens, the EPA has directly faced the unresolved question of whether any level of exposure, no matter how slight, will be associated with increased risk to carcinogenesis. The EPA's response was to "err on the side of safety" and assume a nonthreshold response. Consequently, it was assumed in dealing with carcinogens that there is a risk at every dose and that the risk is directly related to the dose. Of all the insecticides considered by the EPA for standard setting, three (DDT, aldrin, and dieldrin) were initially considered potential human carcinogens (now there are a total of six with the addition of chlordane, heptachlor, and heptachlor epoxide) based on studies with rats and mice. The EPA's specific methodologic approach for establishing limits for these potential "carcinogens will be derived by estimating the health risk associated with various concentrations and comparing these concentrations with ambient levels to assess the attainability of the proposed limits with presently known means of technology" (EPA Advisory Board, 1975.).

At present, there are extremely limited data with respect to the concentrations of aldrin, dieldrin, and DDT in drinking water supplies. Consequently, as an interim decision the EPA has chosen to delay the selection of MCLs for these substances until a survey of selected water supplies, intended to evaluate the level of these insecticides in the United States, is completed.

Drinking Water Standards for Noncarcinogenic Chlorinated Hydrocarbon Insecticides

The fundamental approach used by the EPA in the derivation of "safe" drinking water standards is similar to the classic methodology used by pharmacologists in the development of safe and acceptable medical drugs. This approach involves the development of safety criteria based on chronic toxicity in animal species. More specifically, the MCLs for the noncarcinogenic chlorinated hydrocarbons in drinking water have been derived from the extrapolated level of human exposure that equals that amount resulting in minimal toxic effects in several mammalian species (rats and dogs). Table 6-3, taken from the EPA Advisory Board (1975), presents the methodology used in the derivation of approval limits (ALs—these are equivalent to MCLs) for chlorinated hydrocarbons. As indicated above, all insecticides were tested in two animal species, the rat and dog. The chronic toxicity testing revealed the lowest level of pollutant (on a milligram of dose per kilogram of body weight basis) that the animal could take in with either minimal or no toxic effects. Once this had been determined, the policy makers selected that species which was the most sensitive to the substance in question (the dog in the case of endrin, heptachlor, heptachlor epoxide, and lindane; the rat in the case of chlordane, methoxychlor, and toxaphene). If there were no supporting human exposure data to consider (as was the case in all except methoxychlor), a safety factor of 500 was applied to the minimally toxic dose in the most sensitive of the two species of animals (i.e., the minimal toxic dose was divided by 500). This number was taken to be the total amount of insecticide to which a human could be exposed each day for an unspecified period of time without suffering any adverse health effects. Next, amounts of the chlorinated hydrocarbons normally consumed via the diet were derived from market basket surveys. If this amount was substantially less than $^{1}/_{500}$ of a known toxic dose in the most sensitive animal species tested (that is the assumed safe level of intake for humans), a drinking water limit was established that would permit 20% of the safe limit (see endrin and lindane as examples) to be consumed via water. There were several exceptions to this scheme. For example, heptachlor epoxide levels were adjusted from 20 to 5% because market basket surveys indicated that humans already consume amounts exceeding $^{1}/_{500}$ of a known minimally toxic dose in the most sensitive animal species and because heptachlor is often converted to heptachlor epoxide. This last reason is the justification of the low MCL for heptachlor (2% of the "safe" level). The other major methodologic point to note is that a safety factor of only 100 was utilized in the derivation of the standard for methoxychlor, because a limited amount of human data were available to help corroborate the animal extrapolation information. It should be noted that Table 6-3 contains the standards the EPA justification document recommended. As previously implied, chlordane, heptachlor, and heptachlor epoxide were deleted from this list of standards by the EPA.

Since the methodologic approach to standard setting for chlorinated hydrocarbons is so strikingly different from the previously discussed standards for inorganic substances, several comments are in order. The primary reason that the EPA adopted chronic toxicity studies is the lack of human data with respect to these pollutants. In the absence of chronic, long-term exposure studies on humans from either occupational and/or community environments (as in the case of chlorinated hydrocarbon insecticides) reliance on animal models became necessary.

Consequently, the question is not why were animal models used as the basis for deriving the human safety standards, but (1) what is the validity of utilizing only two animal species; (2) why were the rat and dog models chosen instead of any other potential models; (3) how reliable is a safety factor of 500; (4) why was it selected; and (5) should 20% of a "safe" exposure have been chosen as the maximum exposure to which chlorinated hydrocarbon insecticides would be permitted?

As discussed in the chapter on animal extrapolation, the use of multiple species in toxicity testing in animals enhanced the reliability of qualitative and quantitative predications for potential human health effects (Golberg, 1963; Freireich et al., 1966; Dixon, 1976). Precisely where to draw the line on how many species to use is difficult to decide and may often be determined by a cost-feasibility study. It is interesting to refer to the major paper by Freireich et al. (1966) which reported the toxicologic predictive potential of six different species with 18 different anticancer drugs. The principal conclusion was that each animal species could generally predict human toxicity; however, the monkey and Swiss mouse clearly offered the best two predictive models. It is of interest to note that the predictive potential offered by the monkey model was only slightly improved on when all six models were collectively considered. Thus, although the use of two species may seem insufficient for making effective extrapolations to humans, it is not obvious that significant improvements in prediction will be consistently made with two or three additional species.

The use of the rat and dog does not seem to offer specific predictive benefits compared to two different animal species from the following grouping: mouse, hamster, guinea pig, cat, and monkey. Rats are often used because of their durability, health, size, life span, and cost. In fact, the principal reason for selecting rats usually does not consider its predictive potential but its experimental efficacy (Barnes and Denz, 1954). With respect to the dog, it has been in vogue ever since it was found to be the first animal model to develop bladder cancers similar to humans following long-term exposure to carcinogenic dyes (Hueper et al., 1938). Yet their continued use, even with respect to uniquely predicting bladder cancer, has been strongly challenged (Deichmann and Radomski, 1969; Bonser, 1969). A priori, there does not seem to be a specific "predictive" reason why rats and dogs were the only two species selected. This does not imply that any other combination of two species would have offered greater predictive value with chlorinated hydrocarbons.

The use of a safety factor of 500 (for noncarcinogens) seems sufficiently protective, especially when the safety factor was derived directly from the most sensitive species of the two tested. However, the use of the number 500 is still quite arbitrary and subjective and does not offer any total guarantee of safety to humans.

Finally, the reason for adopting 20% of a safe dose as the maximum drinking water exposure (i.e., standard) is once again highly subjective. Surely it offers what appears to be reasonable protection and would not allow any truly excessive exposures. However, the argument could be made that since exposure to such insecticides is most directly controlled by limiting the levels in the drinking water, a stricter standard could easily be employed; this argument is quite similar to the reason the PHS used in the 1962 lead standard (see lead section).

Chlorophenoxy Herbicides

The potential contamination of community drinking water sources by various herbicides was first noted by the 1969 Advisory Committee when they recommended the adoption of maximum limits for three herbicides [2,4-D (2,4-dichlorophenoxyacetic acid), 2,4,5-T (silvex)(2-(2,4,5-trichlorophenoxy) propionic acid, and 2,4,5-TP (2,4,5-trichlorophenoxyacetic acid)]. Since that time, 2,4,5-T has been banned from most aquatic uses. Therefore, when the EPA proposed their 1975 drinking water quality standards, only 2,4,-D and 2,4,5-TP were issued MCLs.

The primary reason for the banning of the principal uses of 2,4,5-T was that toxicologic investigations revealed teratogenic and/or embryotoxic effects in a variety of animal models, including mice, hamsters, and rats (Courtney et al., 1970; Courtney and Moore, 1971; Collins and Williams, 1971). The EPA has developed a policy of treating substances with carcinogenic, teratogenic, and embryotoxic properties with considerably more caution than substances that produce other toxic effects. Such a policy certainly seems in the best interests of community health when one considers the irreversible nature of such toxic responses as cancer and birth defects.

With respect to 2,4,5-TP, toxicologic investigations have focused on acute, subacute, and chronic toxicity studies in several species, including rats and dogs (EPA, 1975; Drill and Hiratzka, 1953; Mullison, 1966). As a result of such studies, the EPA derived an approval limit for 2,4,5-TP based on the identical protocol previously described for chlorinated hydrocarbon insecticides. Table 6-4 represents the actual derivation the EPA used in the determination of the currently enforced MCL for 2,4,5-TP. As can be seen in the table, the dog was selected as the most sensitive of the two species tested. Consequently, the lowest

Table 6-4. Derivation of Approval Limits (ALs) for Chlorophenoxy Herbicides

	Lowest Long-Term Levels with Minimal or No Effects		Calculated Maximum Safe Levels From All Sources of Exposure			Water	
Compound	Species	mg/kg/day[a]	Safety Factor (X)	mg/kg/day	mg/man/day[b]	Percentage of Safe Level	AL (mg/liter)[c]
2,4-D	Rat	50	1/500	0.1	7.0	20	0.1
	Dog	8.0	1/500	0.016	1.12[d]		
2,4,5-TP	Rat	2.6	1/500	0.005	0.35	20	0.01
	Dog	0.9	1/500	0.002	0.14[d]		

[a] Assume weight of rat = 0.3 kg and of dog = 10 kg; assume average daily food consumption of rat = 0.05 kg. and of dog = 0.2 kg. [b] Assume average weight of human adult = 70 kg. [c] Assume average daily intake of water for man = 2 liters. [d] Chosen as basis on which to derive AL.

Source: EPA (1973) Advisory Document on Drinking Water Standards, Washington, D.C.

dose in chronic studies causing either minimal or no effects was divided by 500 and declared to be a reasonably "safe" amount for daily human exposure. Thus the calculated "safe" level of 2,4,5-TP in the human diet is approximately 0.002 mg/kg/day, which is equal to the "reasonably safe" limit of 0.14 mg/man/day. Since little 2,4,5-TP is thought to be present in the diet, 20% of a safe limit (0.014 ppm rounded off to 0.01 ppm) was permitted.

The derivation of a drinking water standard for 2,4-D followed a similar methodologic procedure as 2,4,5-TP. Data exist on the responses of animal models (rats and dogs) to acute, subacute, and chronic exposures (Hill and Carlisle, 1947; Rowe and Hymas, 1954; Drill and Hiratzka, 1953). In addition, however, there was also some limited data on 2,4-D toxicity to humans. Kraus (1946) reported that 500 mg/day for 21 days caused no observable adverse effects in one volunteer. Further research by Seabury (1963) considered the usage of 2,4-D as a treatment for disseminated caccidioidomycosis. A patient was given 18 intravenous doses over 33 days, with the final 12 doses being 800 mg or more and the 17th dose being 2000 mg. A 19th dose of 3000 mg caused minor adverse symptoms. Besides these human data, the EPA (1975) emphasized that, in 1965, 63 million pounds of 2,4-D were produced with no confirmed cases of occupational poisoning. However, several cases of illness due to ingestion were noted (Hayes, 1963; Nielson et al., 1965; Berwick, 1970).

Table 6-4 also demonstrates the methodologic approach used by the EPA to derive the MCL for 2,4-D. This is identical to the manner in which they derived the MCL for 2,4,5-TP. It is interesting to note that the EPA did not incorporate the use of the human data, as previously discussed. This is in marked contrast to the incorporation of limited human data (see Stein et al., 1965) in the derivation of the methoxychlor standard (see Table 6-3). In this case, the EPA did not specifically use the human data to derive the actual number, but it did convince them to use a safety factor of 100 instead of 500. Although the EPA did not say why it was influenced by the one human study with respect to methoxychlor but were not so moved in the case of 2,4-D human tests, the reason is probably professional judgment. However, in matters in which professional judgments make the difference between a 100 or 500 safety factor, a detailed explanation would seem to be in order.

COMMENTS ON THE NAS REPORT

EPA "Safe" Levels of Noncarcinogens

There is considerable agreement between the methodologic approaches used to establish maximum tolerance levels for noncarcinogenic organics in drinking water by the EPA and the NAS. The following includes a comparison of the working assumptions of the EPA and the NAS on chronic toxicity of noncarcinogenic organics:

1. A safety factor was employed with its size determined by the extent of the available data (see Table 6-5). The EPA and the NAS adopted different safety factors when there was only very limited applicable data available. In this instance, the EPA used a safety factor of 500, whereas the NAS used 1000.
2. The MCLs of the EPA and the recommended limits of the NAS did not consider chemical interactions such as synergism, additivity, and antagonism.
3. The NAS and the EPA assumed a 70-kg person who consumed 2 liters of water each day in calculation of "safe" doses.
4. The NAS developed two possible acceptable daily intake (ADI) levels; one was based on the ingestion of 20% of the total ADI from water, with the remaining 80% coming from other sources; the other involved a ratio of 1:99. As previously noted, the EPA followed a similar procedure using the 20:80 ratio; however, special circumstances (as previously explained) made them adopt a more conservative approach (i.e., the 2:98 ratio). Table 6-5 lists a number of organic pesticides and other organic contaminants in drinking water as well as their suggested no-adverse-health-effect levels for which the NAS had sufficient data to derive predictions. Based on the frequent use of the 1000 safety factor unit, it is obvious that more data are needed before truly reliable estimates can be made.

"Safe" Levels of Carcinogens

The Safe Drinking Water Act of 1974 required the Administrator of the EPA to request the NAS to prepare an extensive review of the literature concerning the relationship between constituents of drinking water and the public health. The NAS prepared a list of substances that are known or suspected carcinogens. For each of the listed carcinogens, the NAS attempted to derive a reasonable statistical estimate of the human cancer risk as a result of a lifetime (70 years) of daily exposure (ingestion) of 1 liter of drinking water with a concentration of 1 μg of carcinogen per liter. The procedure used by the NAS involved the conversion of the assumed human exposure dose (1 μg/liter) to the comparable dose in the animal model. This conversion was based on a microgram per cubic meter of surface area ratio. [See Chapter 2 on animal extrapolation techniques for a discussion of the comparability of relative surface areas and body weight; also Freireich et al. (1966) is a valuable reference of this principle.] The second step necessitated the development of a risk model associating dose to effect. The mathematical expression developed to predict the dose-response relationship is

$$P(d) = 1 - e^{-}(\lambda_0 + \lambda_1 d + \lambda_2 d^2 + \cdot \cdot \cdot \lambda_k d^k)$$

$P(d)$ is the lifetime probability that the dose d (total daily intake) will produce cancer. K is the number of events in the carcinogenic process, and λ_0, λ_1, λ_2 . . . are nonnegative parameters.

Upper confidence limits from the estimates of low-dose risk were determined by employing the maximum likelihood theory. Table 6-6 presents the various categories of known or suspected organic chemical carcinogens found in drinking water. The table indicates the probability (with 95% confidence) that exposure to various carcinogens for 70 years in the dosage noted alone would result in a cancer. The extrapolative technique represents an interesting approach toward trying to develop quantitative estimates of human cancer risks. It should be realized that this model is based in part on the study by Freireich et al. (1966) that used anticancer drugs that were theoretically not affected by the liver microsomal enzymes system. Transposition of such studies from the clinical cancer problems to environmental carcinogenesis should be made only with the recognition of the limitations inherent in such a methodology.

TLVs AND DRINKING WATER STANDARDS*

It was seen that the present drinking water quality standard for barium of 1.0 ppm was derived by the application of industrial TLV information. Although considerable space (in the barium section of this chapter) was devoted to the limitations of such a methodology, it does appear to represent a reasonable first attempt at standard derivation

* This section was previously published as "Can Drinking Water Standards Be Reliably Derived from Industrial TLVs? E. J. Calabrese. In: Medical Hypotheses (1978)—in press—with permission of the publisher.

Table 6-5. Organic Pesticides and Other Organic Contaminants in Drinking Water—Concentration, Toxicity, ADI, and Suggested No-Adverse-Effect Levels

Compound	Maximum Concentration in H_2O (μg/liter)	Maximum Dose Producing No Observed Adverse Effect (mg/kg/day)	Uncertainty Factor[a]	ADI[b] (mg/kg/day)	Suggested No-Adverse-Effect Level from H_2O (μg/liter) Assumptions[c] 1	2
2,4-D	0.04	12.5	1000	0.0125	87.5	4.4
2,4,5-T		10.0	100	0.1	700	35.0
TCDD		10^{-5}	100	10^{-7}	7×10^{-4}	3.5×10^{-5}
2,4,5-TP	Detected[d]	0.75	1000	0.00075	5.25	0.26
MCPA		1.25	1000	0.00125	8.75	0.44
Amiben		250	1000	0.25	1750.0	87.5
Dicamba		1.25	1000	0.001125	8.75	0.44
Alachlor	2.9	100	1000	0.1	700.0	35.0
Butachlor	0.06	10	1000	0.01	70.0	3.5
Propachlor		100	1000	0.1	700.0	35.0
Propanil		20	1000	0.02	140.0	7.0
Aldicarb		0.1	100	0.001	7	0.35
Bromacil		12.5	1000	0.0125	87.5	4.4
Paraquat		8.5	1000	0.0085	59.5	2.98
Trifluralin (also for nitralin and benefin)	Detected	10	100	0.1	700.0	35.0
Methoxychlor		10	100	0.1	700.0	35.0
Toxaphene		1.25	1000	0.00125	8.75	0.44
Azinphosmethyl		0.125	10	0.0125	87.5	4.4
Diazinon		0.02	10	0.002	14.0	0.7
Phorate (also for disulfoton)		0.01	100	0.0001	0.7	0.035
Carbaryl		8.2	100	0.082	574	28.7

Ziram (and Ferbam)		12.5	1000	0.0125	87.5	
Captan		50	1000	0.05	350	17.5
Folpet		160	1000	0.16	1120	56.0
HCB	6.0	1	1000	0.001	7	0.35
PDB	1.0	13.4	1000	0.0134	93.8	4.7
Parathion (and methyl parathion)		0.043	10	0.0043	30	1.5
Malathion		0.2	10	0.02	140	7.0
Maneb (and zineb)		5.0	1000	0.005	35	1.75
Thiram		5.0	1000	0.005	35	1.75
Atrazine	5.0	21.5	1000	0.0215	150	7.5
Propazine	Detected	46.4	1000	0.0464	325	16.0
Simazine	Detected	215	1000	0.215	1505	75.25
di-n-butyl phthalate	5.0	110	1000	0.11	770	38.5
di (2-ethyl hexyl) phthalate	30.0	60	100	0.6	4200	210.0
Hexachlorophene	0.01	1	1000	0.001	7	0.35
Methyl methacrylate	1.0	100	1000	0.1	700	35.0
Pentachlorophenol	1.4	3	1000	0.003	21	1.05
Styrene	1.0	133	1000	0.133	931	46.5

[a] Uncertainty factor—the factor of 10 was used where good chronic human exposure data were available and supported by chronic oral toxicity data in other species; the factor of 100 was used where good chronic oral toxicity data were available in some animal species; the factor of 1000 was used with limited chronic toxicity data or when the only data available were from inhalation studies.

[b] Acceptable daily intake (ADI)—maximum dose producing no observed adverse effect divided by the uncertainty factor.

[c] Assumptions: Average weight of human adult = 70 kg. Average daily intake of water for man = 2 liters.

1. 20% of total ADI assigned to water, 80% from other sources.
2. 1% of total ADI assigned to water, 99% from other sources.

[d] Detected but not quantified.

Source: NAS (May, 1977) *Drinking Water and Health—Summary Report.* Washington, D.C.

Table 6-6. Categories of Known or Suspected Organic Chemical Carcinogens Found in Drinking Water

Compound	Highest Observed Concentrations in Finished Water (μg/liter)	Upper 95% Confidence Estimate of Lifetime Cancer Risk per μg/liter[a]
Human carcinogen		
Vinyl chloride	10	5.1×10^{-7}
Suspected human carcinogens		
Benzene	10	I.D.[b]
Benzo(a)pyrene	D.	I.D.
Animal carcinogens		
Dieldrin	8	2.6×10^{-4}
Kepone	N.D.	4.4×10^{-4}
Heptachlor	D.	4.4×10^{-5}
Chlordane	0.1	1.8×10^{-5}
DDT	D.	1.2×10^{-5}
Lindane (γ-BHC)	0.01	9.3×10^{-6}
α-BHC	D.	6.5×10^{-6}
β-BHC	D.	4.2×10^{-6}
PCB (Aroclor 1260)	3	3.1×10^{-6}
ETU	N.D.	2.2×10^{-6}
Chloroform	366	3.7×10^{-7}
Carbontetrachloride	5	1.5×10^{-7}
PCNB	N.D.	1.4×10^{-7}
Trichloroethylene	0.5	1.3×10^{-7}
Diphenylhydrazine	1	I.D.
Aldrin	D.	I.D.
Suspected animal carcinogens		
bis(2-chloroethyl)ether	0.42	1.2×10^{-6}
Endrin	0.08	I.D.
Heptachlor epoxide	D.	I.D.

[a] See text for details. [b] I.D., insufficient data to permit a statistical extrapolation of risk. N.D., not detected. D, detected but not quantified.
Source: NAS (May, 1977) *Drinking Water and Health—Summary Report.* Washington, D.C.

when sufficient data are lacking. As indicated earlier, the present epidemiologic study in Illinois concerning elevated levels of barium in drinking water will help to reevaluate the validity of the 1.0 ppm standard. At this point, it would be useful to compare the calculated drinking water standards of Stokinger and Woodward (1958) as derived from existing TLVs in 1958 with present national primary drinking water standards. Table 6-7, column 7, lists the possible standards for certain inorganic substances (arsenic, barium, cadmium, chromium, fluoride, and lead) in drinking water as derived from TLVs for air (Stokinger and Woodward, 1958). If one is consistent and utilizes the same methodology as in the derivation of the barium standard (see page 151 for specific procedures), it is necessary to reduce the value in column 5 by a factor of 2. This reduction factor was adopted in an effort to protect the more susceptible segments of the population, since the original value was designed to protect healthy adults. Although the adoption of a safety factor of 2 may be considered by many as inadequate in this situation, it is not of immediate concern, since the present intention is to determine how effectively TLV-derived water limits actually compare with our national primary drinking water standards. From the table it can be seen that there is considerable disagreement with respect to columns 7 and 8. In the cases of arsenic, cadmium, chromium, and lead, the present drinking water standards actually provide 6, 200, 60, and 10 times greater protection than the proposed standards as derived from TLVs (see column 9). In contrast, for only one substance, fluoride, does the TLV application process offer greater protection (a factor of only 0.46).* Thus, of the five primary drinking water standards that have been compared, only one (fluoride) was within a factor of approximately 2 with respect to the present standard. The remaining four substances had exceptionally large divergences from the present standards by factors ranging from 6 to 200, depending on the substance.

Although there is a wide discrepancy between the TLV-derived and the 1977 EPA standards, this difference is considerably reduced if one uses a safety factor of 10 instead of 2 as was originally selected (see Table 6-7, columns 10–12). Thus the factorial differences between the two standards for arsenic have been reduced from 6 to 1.2, for cadmium from 200 to 40, for chromium from 60 to 12, and for lead from 10 to 2. With regard to fluoride, the opposite occurs; the difference is increased from 0.46 to approximately 0.089. Based on traditional (albeit subjective) protocol, a safety factor of 10 is usually adopted to account for normal variance within a highly heterogeneous population such as American society.

Another significant factor that should be considered in the comparison of the 1958 TLV conversion approach with the 1977 standards is the difference in absorption factors. Stokinger and Woodward (1958) emphasized that the most important uncertainty factor in the calculation of drinking water limits was the selection of accurate absorption

Table 6-7. Limits for Inorganic Substances in Drinking Water Calculated from Threshold Limits for Air[a]

1	2	3	4	5	6	7	8	9	10	11	12
	Absorption Factor										
Substance	Inhalation	Ingestion	TLV for Air-1958 Values[b]	1958 Proposed TLV-Derived Standards for Drinking Water without Safety Factor (mg/liter)	Reduction Factor (safety factor of 2)	Proposed 1958 TLV-Derived Standards for Drinking Water with a Safety Factor of 2	Present National Drinking Water Standards	Ratio of Column 7 to 8	Reduction Factor (safety factor of 10)	TLV-Derived Standards for Drinking Water with a Safety Factor of 10 (mg/liter)	Ratio of Column 11 to 8
Arsenic	0.2	0.8	0.5	0.6	÷ 2 =	0.3	0.05	6	0.6 ÷ 10 =	0.06	1.2
Barium	0.75	0.90	0.5	2.0	÷ 2 =	1.0	1.0	0	2.0 ÷ 10 =	0.2	0.2
Cadmium	0.25	0.03	0.1	4.0	÷ 2 =	2.0	0.01	200	4.0 ÷ 10 =	0.4	40
Chromium	0.75	0.06	0.1	6.0	÷ 2 =	3.0	0.05	60	6.0 ÷ 10 =	0.6	12
Fluoride	0.1	1.0	2.5	1.25	÷ 2 =	0.625	1.4–2.4	0.46	1.25 ÷ 10 =	0.125	0.089
Lead	0.2	0.2	0.2	1.0	÷ 2 =	0.5	0.05	10	1.0 ÷ 10 =	0.1	2

[a] Process of derivation for a TLV-based drinking water standard

1. First, calculate the total amount of substance taken into the body from inhalation—assume exposure to be equal to the TLV value and the total amount of air inhaled 10 cu m (a typical 8-hour exposure).

   e.g., Lead—1958 value—

   $$\text{Absorption factor} \times \text{total amount of air inhaled} \times \text{TLV}$$
   $$(0.2) \times (10 \text{ cu m}) \times (0.2) = 0.4$$

2. Next, calculate the amount from water absorbed into the body.

   $$\text{Absorption factor} \times \text{quantity of water ingested} \times \text{water limit}$$
   $$(0.2) \times (2 \text{ liters}) \times (x) = 0.4x$$

3. Solve the equation: $0.4x = 0.4$

   $x = 1$ Thus, the water limit (column 5) equals 1.

[b] Current TLVs are arsenic—0.5 mg/cu m, barium—0.5 mg/cu m, cadmium—0.05 mg/cu m, fluoride—2.5 mg/cu m, and lead—0.1 mg/cu m. Adapted from Stokinger and Woodward, 1958.

factors for inhalation and ingestion. Slight differences in the actual absorption factor may result in highly different standards. An example of the significance of absorption factors may be seen for the lead standard. Stokinger and Woodward selected a 0.2 lead inhalation absorption factor in 1958 for their TLV conversion study, whereas the PHS adopted a 0.1 lead inhalation absorption factor for the 1962 standards, and by 1975 the EPA chose 0.3. If one used 0.1, 0.2, and 0.3, respectively, in deriving the lead drinking water standard from the lead TLV, the respective water limits would be 2.0, 1.0 and 0.66 ppm. If a safety factor of 10 is applied to the 0.66 ppm, a value of 0.066 ppm results. This compares quite favorably with the present standard of 0.05 ppm. (In addition to current absorption factors, the present TLVs should be used. For example, the present TLV for lead is 0.15 mg/m^3 compared to 0.2 mg/m^3 in 1958.) Thus, when comparing the water limits of the Stokinger and Woodward conversion process to the 1977 EPA standards, it is necessary to substitute current data. At least with respect to lead, the TLV conversion process is highly comparable.

The intention of this section on the TLV conversion process is not to endorse or condemn, but to place it in a reasonable perspective so that those involved in the regulatory process will not dismiss it at once, but at the same time will not be overly optimistic that their problems will be quickly solved by a convenient and highly adaptable formula. Also, very rarely do individual studies prove to be "definitive" with respect to establishing a proper standard for the general population. Consequently, evidence must be considered from diverse but complementary sources. Data from sources that are independent and in related areas add an external validity that is quite reassuring when health limits must be established. An analogous situation may be that of evolutionary theory in which one type of supporting evidence (e.g., comparative anatomy) does not offer convincing evidence for the theory. However, when one considers supporting evidence from geology, embryology, biochemistry, animal behavior, and so on, evolutionary theory has an overwhelming (although not absolutely certain) intellectual credibility. So too with standard derivation, supporting evidence from toxicology (in vitro, in vivo studies, behavioral studies, etc.) and epidemiologic studies of the general population and various high-risk groups all converging toward the same conclusion add a high degree of certainty for any proposed standard. Thus the use of the TLV conversion factor should be viewed as one of many lines of potential evidence that should be reviewed in the standard derivation process—but it should not, if at all possible, be considered alone, as in the case of barium.

U.S. DRINKING WATER STANDARDS AND THOSE OF OTHER COUNTRIES

In 1956 the World Health Organization (WHO) issued its first report on drinking water quality standards. The report was entitled "Report of the European Study Group on Standards of Drinking Water Quality and Methods of Examination Applicable to European Countries." Two years later, WHO published its first International Drinking Water Standards, which included both maximum limits of exposure and acceptable laboratory methods. The 1958 International Standards were revised in 1963 following the response to questionnaires. The three primary sources of references for the Second International WHO Drinking Water Standards were (1) the USPHS Drinking Water Standards, (2) Standard Methods for the Examination of Water and Waste Water, and (3) the Department of Health and Social Security, England and Wales Report on Bacteriological Examination of Water Supplies. In the third edition of International Standards, the standards of other countries, especially the Soviet Union, were also considered (Subrahmanyam, 1975).

A comparison of the European Standards and the first edition of the WHO International Standards reveals that they were the same with respect to maximum limits for toxic chemicals and radioactivity, but somewhat different in bacteriologic standards. In the second edition of the International Standards further limits for both toxic chemicals and radioactivity were made. According to Subrahmanyam (1975), the intention of having separate European and International Standards was to provide the incentive for the more economically and technologically advanced countries in Europe to implement stricter standards. It was felt that the "advanced" countries had special health problems due to widespread industrial and agriculture pollution. Consequently, stricter standards were necessary in the more technologically advanced countries. By 1970 the second edition of the European standards had added special new sections on viruses, polyaromatic hydrocarbons, and pesticides.

Subrahmanyam presented a conceptual scheme for the development of national and international standards that took into account not only the development of health criteria assessment, but also the social and economic factors unique to individual countries. Subrahmanyam said that the utilization of criteria in the establishment of official national standards necessitates the realization of the cost of implementation of such standards as well as the hazards inherent if such implementation is postponed. Each country perceives its own priorities in a unique

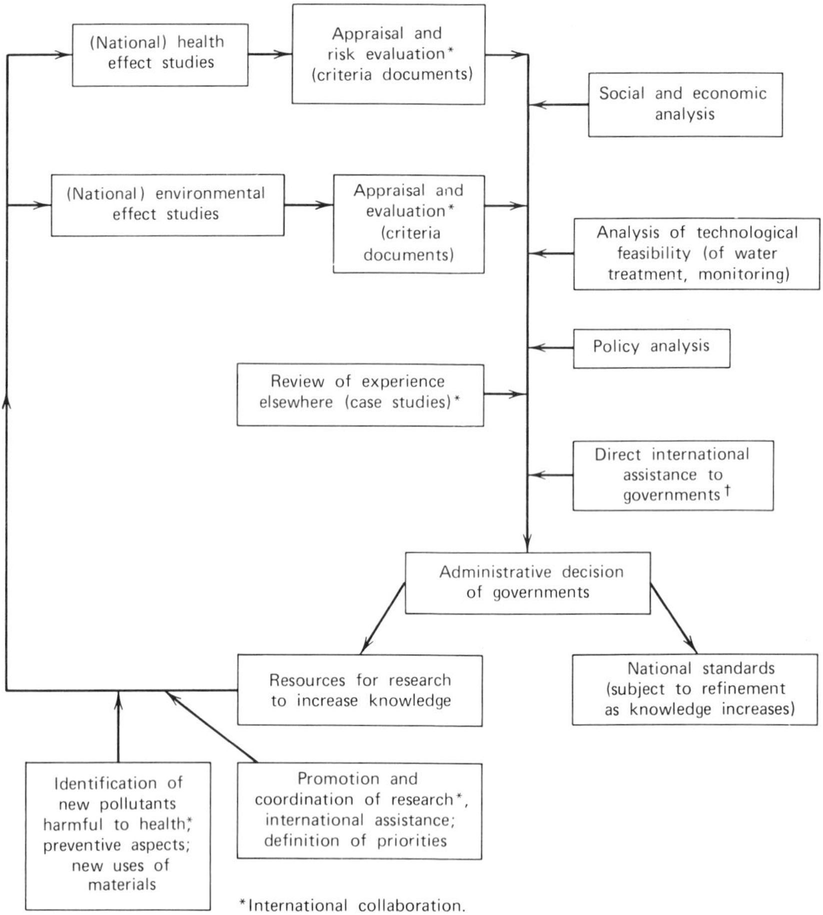

*International collaboration.

†Applicable to countries currently lacking adequate resources.

Figure 6-2. The standard-setting process in environmental health. (From WHO Doc. EP/73.1).

fashion; thus the cost/benefit ratio is likely to be quite different country by country. Figure 6-2 shows the WHO model for standard setting in matters of environmental health, be they air or water pollution problems. Although each country may vary somewhat in the actual standard selected, it follows the process as outlined in Figure 6-2. Furthermore, Subrahmanyam emphasized that all countries, regardless of their individual cultural, economic, and social outlooks, consider five fundamental factors in establishing their priorities.

a. severity of adverse effects on the population; irreversible or chronic effects, those which have adverse genetic implications, and those which are embryotoxic and teratogenic are considered to be of particular importance.

b. persistence of the agent in the environment, resistance to environmental degradation and accumulation in man or in the food chain;

c. metabolic degradation or synthesis in biological systems which may produce metabolites either more or less toxic than the parent compound;

d. ubiquity and abundance of the agents in man's environment particularly those occurring naturally or produced inadvertently;

Table 6-8. Comparison of WHO Limits For Inorganic Chemicals in Drinking Water with the U.S. and the U.S.S.R. (mg/liter)

	WHO Limits			
Substances	European	International	U.S. Values	U.S.S.R. Values
Arsenic	0.05	0.05	0.05	0.05
Barium	1.00	1.00	1.00	4.0
Cadmium	0.01	0.01	0.01	0.01
Chromium (Cr^6)	0.05	0.05	0.05	0.1
Cyanide	0.01	0.01	No standard—dropped in 1977	0.1
Fluoride	1.0–1.7	1.0–1.5	1.4–2.4	1.5
Lead	0.1	0.1	0.05	0.1
Mercury		0.001	0.002 (total Hg)	0.005 (inorganic compounds only)
Nitrate	50 as NO_3 recommended 100 as NO_3 acceptable	45 as NO_3	10 as N 45 as NO_3	10 as N
Selenium	0.01	0.01	0.01	0.001 (as SeO_3)
Silver			0.5	

e. size, type, and demographic characteristics of population exposed, the frequency and magnitude of exposure, selection of highly vulnerable groups of the population.

It is interesting to note that by the end of 1970, of 90 developing countries, 37 had either adopted in toto or adapted the WHO standards, five had National Standards even before the documentation of the WHO standards, and an additional eight countries had different standards.

Comparison with Soviet Union Drinking Water Standards

The first Soviet drinking water standards were published in 1944–1945. By 1970 the Soviet Union had established maximum permissible concentrations for 294 chemical substances in drinking water on the basis of toxicity, organoleptic properties, or potential impairment of the self-purification of water course (Stofen, 1973). It is interesting to note that Soviet drinking water standards are in reasonably close agreement with those proposed by the WHO and the United States. This contrasts with the rather large differences in occupational health air standards between the United States and the Soviet Union (Table 6–8) (See chapter 11).

Note: Results of barium epidemiological study: no significant ($p > 0.05$) differences in the incidence of hypertension, heart disease, stroke, kidney disease or distribution of blood pressure were found between individuals residing in two communities, one with low levels (< 1.0 ppm) and the other with higher levels (> 2.0 < 10.0 ppm) of barium. There was "no substantial cardiovascular health effects in such an exposed population." Brenniman, G. R. et al (May 31, 1978). Report of EPA: Health Effects of Human Exposure to Barium in Drinking Water. EPA Grant No. R803918.

REFERENCES

Advisory Committee (1969) Manual for evaluating public drinking water supplies. US EPA Office of Water Programs, Water Supply Program Division, Washington, D.C.

Alexander, F. W., Delves, T. H., and Clayton, B. E. (1973) The uptake and excretion by children of lead and other contaminants. *Proceedings of the International Symposium of Environmental Health Aspects of Lead.* Luxenbourg Commission of the European Communities, Amsterdam. October 2–6, pp. 319–331.

Allen, F. M. and Sherill, J. W. (1922) The treatment of arterial hypertension. *J Metab Res* 2:429.

Ambard, J. and Beaujard, E. (1904) Causes de l'hypertension arterielle. *Arch Gen Med* 1:520.

American Conference of Governmental Industrial Hygienists. (1958) Threshold limit values for 1958. *Arch Indus Hyg* 18:178–182.

American Conference of Governmental Industrial Hygienists. (1976) Threshold limit values for 1976. Cincinnati, Ohio.

American Heart Association. (1957) *Your 500 Milligram Diet.* New York.

APHS Committee (1949/1950) Water Supply: nitrate in potable waters and methemoglobinemia. *APHA Yearbook* 40:110.

Ashford, N. A. (1976) *Crisis in the Workplace: Occupational Disease and Injury.* Cambridge: MIT Press.

Ast, D. B., Smith, D. J., Wachs, B., and Cantwell, K. T. (1956) Newburgh-Kingston caries–fluorine study. IV. Combined clinical and roentgenographic dental findings after ten years of fluoride experience. *J Am Dent Assoc* 52:314.

Barnes, J. M. and Denz, F. A. (1954) Experimental methods used in determining chronic toxicity. *Pharmacol Rev* 6:191–242.

Berwick, P. (1970) 2,4-dichlorophenoxyacetic acid poisoning in man. *JAMA* 214(6):-114–117.

Betke, J., Kleihaver, E., and Lipps, M. (1956) Vergleichende untersuchugen uber die spontanoxydation von Nabelschnur and Erwachsenenhamoglobin. *Ztschr Kinderh* 77:549.

Betts, P. R., Astley, R., and Raine, D. N. (1973) Lead intoxication in children in Birmingham. *Br Med Jour* 1:402.

Bianchi, G., Fox, V., Difrancesco, G. F., Giovannetti, A. M., and Pagetti, D. (1974) Blood pressure changes produced by kidney cross-transplantation between spontaneously hypertensive rats (SHR) and normotensive rats (NR) *Clin Sci Mol Med* 47:435.

Bonser, G. M. (1969) How valuable the dog in the routine testing of suspected carcinogens? *J Natl Cancer Inst* 43(1):271–274.

Borst, J. G. G. and Borst, DeGa. (1963) Hypertension explained by Starling's theory of circulatory homeostasis. *Lancet* 1:677.

Bosch, H. M., Rosenfield, A. B., Huston, R., Shipman, H. R., and Woodward, R. L. (1950) Methemoglobinemia and Minnesota well supplies. *J Am Water Works Assoc* 42:161–170.

Burns-Cox, C. J. and MacLean, J. D. (1970) Splenomegaly and blood pressure in an Orang Asli community in West Malaysia. *Am Heart J* 80:718.

Caballero, P. J. (1949–1950) Discussion sobre las de normas de calidad para aqua potable. *Organo official de la Associacion Interamericana de Ingenieria Sanitaria* 3:-53–64.

Calabrese, E. J. and Tuthill, R. W. (1977) The effects of sodium levels in community drinking water on blood pressure distribution patterns. *Arch Environ Health* (Sept/Oct) 32:200.

Campbell, W. A. B. (1952) Methemoglobinemia due to nitrates in well water. *Br Med J* 2:371–373.

Cogbill, E. C. and Hobbs, M. E. (1957) Transfer of metallic constituents of cigarettes to the main-stream smoke. *Tobacco Sci* 144:68–73.

Collins, T. F. X. and Williams, C. H. (1971) Teratogenic studies with 2,4,5-T and 2,4-D in hamster. *Bull Environ Contam Toxicol* 6(6):559–567.

Comly, H. H. (1945) Cyanosis in infants caused by nitrates in well water. *JAMA* 129:112–116.

Committee on Nitrate Accumulation, Agricultural Board, Div. of Biology and Agriculture, National Research Council. (1972) Accumulation of nitrate. National Academy of Sciences, Washington, D.C.

Courtney, K. D., Gaylor, D. W., Hogan, M. D., and Falk, H. L. (1970) Teratogenic evaluation of 2,4,5-T. *Science* 168:864.

Courtney, K. D. and Moore, J. A. (1971) Teratology studies with 2,4,5-trichlorophenoxyacetic acid and 2,3,7,8-trichlorodibenzo-p-dioxin. *Toxicol Appl Pharmacol* 20:396.

Cruz-Coke, R., Etcheverry, R., and Nagel, R. (1967) Influence of migration on blood pressure of Easter Islanders. *Lancet* 1:697.

Dahl, L. K. (1961) Effects of chronic excess salt feeding. *J Exp Med* 114:231.

Dahl, L. K. (1972) Salt and hypertension. *Am J Clin Nutr* 25:231.

Dahl, L. K., Heine, M., and Thompson, I. (1974) Genetic influence of the kidneys on blood pressure. Evidence from chronic renal homografts in rats with opposite predispositions to hypertension. *Cir Res* 40:94.

Dahl, L. K., Knudsen, K. D., Heine, M. A., and Leitl, G. J. (1968) Effects of chronic excess salt ingestion. Modification of experimental hypertension in the rate of variations in the diet. *Cir Res* 22:11.

Dahl, L. K. and Love, R. A. (1954) Evidence for relationship between sodium (chloride) intake and human essential hypertension. *Arch Intern Med* 94:525.

Dean, H. T. (1942) In F. R. Moulton, ed., *Fluorine and Dental Health,* pp. 23–32. Washington, D.C.: Assoc. Adv. Amer. Science.

Deichmann, W. B. and Radomski, J. L. (1969) Carcinogenicity and metabolism of aromatic amines in the dog. *J Nat Cancer Inst* 43(1):263–269.

Diskalenko, A. P. (1968) Methemoglobinemia of water nitrate origin in the Moldavian, U.S.S.R. *Hyg Sanit* 33:32–37.

Dixon, R. L. (1976) Problems in extrapolating toxicity data for laboratory animals to man. *Environ Health Persp* 13:43–50.

Drill, V. A. and Hiratzka, T. (1953) Toxicity of 2,4-dichlorophenoxyacetic acid and 2,4,5-trichlorophenoxyacetic acid. A report of their acute and chronic toxicity in dogs. *Arch Indust Hyg Occup Med* 7:61–67.

Elliot, G. B. and Alexander, E. A. (1961) Sodium from drinking water as an unsuspected cause of cardiac decompensation. *Circulation* 23:562.

Elliot, G. B. and Smith, M. D. (1960) Dietary fluoride related to fluoride content of teeth. *J Dent Res* 39:93.

Environment Reporter—Cases (ER-C) (1975) Ethyl corporation vs. EPA. The Bureau of National Affairs, Inc. 7ERC1353-7ERC.

Environmental Health Resource Center (EHRC) (1972) Health effects and recommended standard for cadmium. Illinois Institute for Environmental Quality, Chicago, Ill.

Environmental Health Resource Center (EHRC) (1974) Advisory report on health effects of nitrates in water. Illinois Institute for Environmental Quality (IIEQ Document No. 74-5), Chicago, Ill.

Environmental Health Resource Center (EHRC) (1975) Health effects of the catalytic converter. Illinois Institute for Environmental Quality. Chicago, Ill.

Environmental Protection Agency (EPA) (1971) *Manual for Evaluating Public Drinking Water Supplies: A Manual for Practice,* p. 7. Previously published in 1969 as PHS Publication 1820.

Environmental Protection Agency (EPA) (Nov. 29, 1972) EPA's position on the health effects of airborne lead. Prepared by Health Effects Branch, Processes and Effects Division, Office of Research and Monitoring. USEPA. Washington, D.C.

Environmental Protection Agency (EPA) Advisory Board on Drinking Water Standards (March 14, 1975) Statement of basis and purpose for the proposed national interim primary drinking water standards. USEPA, Washington, D.C.

Federal Register (December 24, 1974) National interim primary drinking water regulations. *Federal Register* 40(248):59566–59574.

Federal Register (March 14, 1975) Interim primary drinking water standards. *Federal Register* 40(51):11990–11998.

Federal Register (August 14, 1975) Interim primary drinking water regulations—proposed maximum contaminant levels for radioactivity. *Federal Register* 40(158):-34324–34328.

Federal Register (Dec. 24, 1975). National interim primary drinking water regulations. Federal Register 40(248):59566–59474.

Federal Register (July 9, 1976) Drinking water regulations: radionuclides. *Federal Register* 41(133):28402–28405.

Fite, F. (1955) Granuloma of lung due to radiographic contrast medium. *Arch Pathol* 59:673–676.

Floyer, M. A. and Richardson, P. C. (1963) Mechanism of arterial hypertension. Role of capacity and resistance vessels. *Lancet* 1:253.

Freireich, E. J., Gehan, E. A., Rall, D. P., Schmidt, L. H., and Skipper, H. E. (1966) Quantitative comparison of toxicity of anti-cancer agents in mouse, rat, hamster, dog, monkey and man. Cancer Chemother Rep 50(4):219–244.

Freis, E. D. (1960) Hemodynamics of hypertension. *Physiol Rev* 40:27.

Freis, E. D. (1976) Salt, volume and the prevention of hypertension. *Circulation* 53(4):589.

Fukuda, T. R. (1951) L'hypertension par le sel chez les lapins et ses relations avec la glande surrenale. *Union Med Canada* 30:1276 (cited in Fries, 1976).

Furstenberg, A. C., Richard, G., and Lathrop, F. D. (1941) Ménière's disease: addenda to medical therapy. *Arch Otolaryng* 34:1083.

Gainer, J. H. (June 1973) Effects of metals on viral infections in mice. *Environ Health Perspect* 998–999.

Gelperin, A., Jacobs, E. E., and Kletke, L. S. (1971) The development of methemoglobin in mother and newborn infants from nitrate in water supplies. *Ill Med J* 82:40–44.

Golberg, L. (1963) The predictive value of animal toxicity studies carried out on new drugs. *J New Drugs* January–February, pp. 7–11.

Gruener, N. and Shuval, H. I. (1970) Health aspect of nitrates in drinking water. In H. I. Shuval, ed., *Developments in Water Quality Research,* pp. 89–106. Ann Arbor: Humphry Science.

Guyton, A. C., Coleman, T. G., Cawley, A. W., Manning, R. D., Jr., Norman, R. A., and Ferguson, J. D. (1974) A systems analysis approach to understanding long-range arterial blood pressure control and hypertension. *Cir Res* 35:159.

Haddy, F. J. and Overbeck, H. W. (1976) The role of humoral agents in volume expanded hypertension. *Life Sci* 19:935–948.

Hansen, H. E. and Bennett, M. J. (1964) In W. E. Nelson, ed., *Textbook of Pediatrics*, p. 109. Philadelphia: Saunders.

Hayes, W. J., Jr. (1963) *Clinical Handbook of Economic Poisons.* PHS Pub. No. 476, U.S. Government Printing Office, Washington, D.C.

Hill, E. C., and Carlisle, H. (1947) Toxicity of 2,4-Dichlorophenoxyacetic acid for experimental animals. *J Indust Hyg Toxicol* 29:85–95.

Hodge, H. C. and Smith, F. A. (1965) *Fluorine Chemistry*, vol. 4, pp. 155 and 171. New York: Academic.

Hofreuter, D. H., Catcott, E. J., Keenan, R. G., and Xinteras, C. (1961) The public health significance of atmospheric lead. *Arch Environ Health* 3:568–574.

Hueper, W. C., Wiley, F. H., and Wolfe, D. H. (1938) Experimental production of bladder tumors in dogs by administration of beta-naphthylamine. *J Indust Hyg Toxicol* 20:46–84.

Huling, E. E. and Hollocher, T. C. (1972) Groundwater contamination by road salt: steady-state concentrations in east central Massachusetts. *Science* 176:288.

Jerard, E. and Patrick, J. B. (1973) The summing of fluoride exposures. *Intern J Environ Studies* 4:141–155.

Jossens, J. V., Williams, J., Claessens, J., Claes, J., and Lissens, W. (1971) Sodium and hypertension. Fidanza, F., Keys, A., Ricci, G., and Somogyi, J. C., ed., *Nutrition and Cardiovascular Diseases,* p. 91. Morgagni Edizion: Scientifiche.

Kaminer, B. and Lutz, W. P. W. (1960) Blood pressure in Bushman of the Kalahari Dessert. *Circulation* 22:289.

Kean, B. H. (1944) The blood pressure of the Cuna Indians. *Am J Trop Med* 24:341.

Kehoe, R. A. (1947) Exposure to lead. *Occup Med* 3:156–171.

Kempner, W. (1944) Treatment of kidney disease and hypertensive vascular disease with rice diet. *N Carolina Med J* 5:125.

Kempner, W. (1948) Treatment of hypertensive vascular disease with rice diet. *Am J Med* 4:545.

Kennedy, D. and Hessel, J. (1971) The biology of pesticides. In J. P. Holden and P. R. Ehrlich, eds., p. 89. *Global Ecology.* New York: Harcourt-Brace-Jovanovich.

Koletsky, S. (1958) Hypertensive vascular disease produced by salt. Lab. Invest 7:377.

Koletsky, S. (1961) Pathogenesis of experimental hypertension induced by salt. *Am J Cardiol* 8:576.

Kraus, cited by Mitchess, J. W., Hogson, R. E. and Gaetjens, C. F. (1946) Tolerance of farm animals to feed containing 2,4-dichlorophenoxyacetic acid. *J Animal Sci* 5:-226–232.

Krepkogorsky, L. N. (1963) Fluorine in the traditional diet of the population of Vietnam. *Gigiena i Sanit* 8(1):30.

Langston, J. B., Guyton, A. C., Douglas, B. H., and Dorsett, P. E. (1963) Effect of changes in salt intake on arterial pressure and renal function in partially nephrectomized dogs. *Cir Res* 12:508.

Lecks, H. I. (1950) Methemoglobin in infancy. *Am J Dis Child* 79:117–123.

Ledingham, J. M. (1953) Distribution of water, sodium and potassium in heart and skeletal muscle in experimental renal hypertension in rats. *Clin Sci* 12:337.

Lee, D. H. K. (1970) Nitrates, nitrites, and methemoglobinemia. *Environ Res* 3:484–511.

Lenel, R., Katz, L. N., and Rodbard, S. (1948) Arterial hypertension in the chicken. *Am J Physiol* 152:557.

Lorente de No, R. and Feng, T. P.(1946) Analysis of effect of barium upon nerve with particular reference to rhythmic activity. *J Cell Comp Physiol* 28:397–464.

Lowenstein, F. W. (1961) Blood pressure in relation to age and sex in the tropics and subtropics. A review of the literature and an investigation in two tribes of Brazil Indians. *Lancet* 1:389.

Maddocks, I. (1967) Blood pressures in Malanesians. *Med J Aust* 1:1123. (cited in Freis, 1976).

Mahaffey, K. R. (1974) Nutritional factors and susceptibility to lead toxicity. *Environ Health Perspect* Exp. Issue No. 7, p. 107.

Mahaffey-Six, K. and Goyer, R. A. (1972) The influence of iron deficiency on tissue content and toxicity of ingested lead in the rat. *J Lab Clin Med* 79:128.

Marier, J. R. and Rose, D. (1966) The fluoride content of some foods and beverages. *J Food Sci* 31:941.

Martin, D. J. (1951) Fluorine content of vegetables cooked in fluorine-containing waters. *J Dent Res* 30:676.

Mass. Dept. Public Health (1969, 1970) Report of Routine Chemical and Physical Analyses of Public Water Supplies, Boston, Massachusetts.

Mass. Dept. Public Health (1975) Report of Routine Chemical and Physical Analyses of Public Water Supplies, Boston, Massachusetts.

Mass. Dept. Public Works (Nov. 1974) Environmental Impact Report. Snow and Ice Control Program 1973–74, 1974–75. Mass. Turnpike Authority and Mass. Bay Transit Authority.

McLaughlin, M. and Stopps, G. J. (1973) Smoking and lead. *Arch Environ Health* 23:131–136.

Menden, E. E., Elia, V. J., Michael, L. W., and Petering, H. G. (1973) Distribution of cadmium and nickel of tobacco during cigarette smoking. Environ. Sci. Technol. 6(9):830–832.

Meneely, G. R., Ball, C. O. T., and Youmans, J. B. (1957) Chronic sodium chloride toxicity: the protective effect of added potassium chloride. *Ann Intern Med* 47:263.

Meneely, G. R., Lemley-Stone, J., and Darby, W. J. (1961) Changes in blood pressure and body sodium in rats fed sodium and potassium chloride. *Am J Cardiol* 8:527.

Meneely, G. R., Tucker, R. G., Darby, W. J., and Auerbach, S. H. (1953) Chronic sodium chloride toxicity in the albino rat. II. Occurrence of hypertension and of a syndrome of edema and renal failure. *J Exp Med* 98:71.

Millar, J. A., Battistini, V., Cumming, R. L. C., Carswell, F., and Goldberg, A. (1970) Lead and aminolaevulinic acid dehydrase levels in mentally retarded children and in lead-poisoned suckling rats. *Lancet* 2:695.

Mullison, W. R. (1966) Some toxicological aspects of silvex. Paper presented at Southern Weed Conference, Jacksonville, Fla.

Murphy, R. J. F. (1950) The effect of "rice diet" on plasma volume and extracellular fluid space in hypertensive subjects. *J Clin Invest* 29:912.

NAS (1972) Airborne lead in perspective. The Committee on Biological Effects of Atmospheric Pollutants. Washington, D.C.

NAS (1977) Summary Report: Drinking Water and Health. National Research Council of the National Academy of Sciences. May 23, 1977.

National Institute of Occupational Safety and Health (NIOSH) (1973) Criteria document for inorganic lead. HEW, USPHS.

National Water Commission Report of 1973. Municipal and industrial water supply program, pp. 161–170.

Neubauer, O. (1947) Arsenical cancer: a review. *Br J Cancer* L:192–251.

Nielson, K., Kaempe, B., and Jensen-Holm, J. (1965) Fatal poisoning in man by 2,4-dichlorophenoxyacetic acid (2,4-D): determination of the agent in forensic materials. *Acta Pharmacol Toxicol* 22:224–234.

Oliver, W. J., Cohen, E. L., and Neel, J. V. (1975) Blood pressure, sodium and sodium related hormones in the Panoma Mo Indians, a "no-salt" culture. *Circulation* 52:146.

Page. L. B., Danion, A., and Moellering, R. C., Jr. (1974) Antecedents of cardiovascular disease in six Solomon Islands societies. *Circulation* 49:1132.

Parijs, J., Jossens, J. V., Vander Linden, L., Verstreken, G., and Amery, A. K. C. (1973) Moderate sodium restriction and diuretics in the treatment of hypertension. *Am Heart J* 85(1):22.

Patterson, C. C. (1965) Contaminated and natural lead environment of man. *Arch Environ Health* 11:344.

Personal Communication with Massachusetts Department of Public Health, Boston, Mass. April 2, 1977.

Petukhov, N. I. and Ivanov, A. V. (1970) Investigation of certain psychophysiological reactions in children suffering from methemoglobinemia due to nitrates in water. *Hyg Sanit* 35:29–31.

Prior, A. M., Evans, J. G., Harvey, H. P. B., Davidson, F., and Lindsey, M. (1968) Sodium intake and blood pressure in two Polynesian populations. *N Engl J Med* 279:515.

Public Health Service (April 10, 1925) The physical and chemical characteristics of acceptable water supplies. *Public Health Rep* 40:693.

Public Health Service (PHS) (1962) Fluoride. Public Health Service Drinking Water Standards. U.S. HEW PHS, pp. 41–42.

Public Health Service Drinking Water Standards (1943) *J Am Water Works Assoc* 35:96.

Public Health Service Drinking Water Standards adopted on September 25, 1942. (1943) *J Am Water Works Assoc* 35:96–104.

Public Health Service Drinking Water Standards (1946) *Public Health Rep* 61(11):-371–373.

Public Health Service Drinking Water Standards (1946) *J Am Water Works Assoc* 38:361–370.

Public Health Service Drinking Water Standards (1962) U.S. HEW PHS. Rockville, Maryland.

Ross, J. D. and Des Forges, J. F. (1959) Reduction of methemoglobin by erythrocytes from cord blood. Further evidence of deficient enzyme activity in newborne period. *Pediatrics* 23:218.

Rowe, U. K. and Hymas, T. A. (1954) Summary of toxicological information on 2,4-D and 2,4,5-T type herbicides and an evaluation of the hazards of livestock associated with their use. *Am J Vet Res* 15:622–629.

Russell, E. L. (1969) Sodium imbalance in drinking water. *J Am Water Works Assoc* 62(2):102.

Safe Drinking Water Act (December 16, 1974) Public Law, 93-523.

Sapirstein, L. A., Brandt, W. L., and Drury, D. R. (1950) Production of hypertension in the rat by substituting hypertonic sodium chloride solution for drinking water. *Proc Soc Exp Biol Med* 73:82.

Sasaki, N. (1964) The relationship of salt intake to hypertension in the Japanese. *Geriatrics* 19:735.

Sattlemacher, P. G. (1962) Methemoglobinemia from nitrates in drinking water. Schriftenveiche des vereins fur. Wasser, Boden and hufthygiene No. 20.

Schroeder, H. A. (1965) Cadmium as a factor in hypertension. *J Chron Dis* 18:647.

Schroeder, H. A. (1974) The role of trace elements in cardiovascular diseases. *Med Clin North Am* 58(2):381.

Scotch, N. (1960) A preliminary report on the relation of sociocultural factors to hypertension among the Zulu. *Ann NY Acad Sci* 84:1000.

Seaber, W. M. (1933) Barium as a normal constituent of Brazil nuts. *Analyst* 58:575–580.

Seabury, J. H. (1963) Toxicity of 2,4-dichlorophenoxyacetic acid for man and dog. *Arch Environ Health* 7:202–209.

Selye, H. (1943) Production of nephrosclerosis in the fowl by sodium chloride. *J Am Vet Med Assoc* 103:140.

Selye, H., Tuchwever, B., and Bertok, L. (1966) Effect of lead acetate on the susceptibility of rats to bacterial endotoxins. *J Bacteriol* 91:884.

Shaper, A. G. (1972) Cardiovascular disease in the tropics. III. Blood pressure and hypertension. *Br Med J* 3:805.

Shearer, L. A., Goldsmith, J. R., Young, C., Kearns, O. A., and Tamplin, B. (1971) Methemoglobin level in infants in an area with high nitrate water supply. Presented at the 99th annual meeting of the American Public Health Association, Minneapolis, Minn.

Simon, C., Mazke, M., Kay, H., and Mrowitz, G. (1964) Uber vorkommen Pathogenes and moglichkeiten zur Prophylaxe der durch nitrite verursachten Methamoglobinamie. *Z Kinderheilk* 91:124–138.

Sinnet, P. F. and Whyte, H. M. (1973) Epidemiological studies in a total highland population, Tukisenta, New Guinea. Cardiovascular disease and relevant clinical, electrocardiographic, radiological and biochemical findings. *J Chron Dis* 26:265.

Spencer, H., Lewin, I., Fowler, J., and Samachson, J. (1969) Effect of sodium fluoride on calcium absorption and balances in man. *Am J Clin Nutr* 22(4):381–390.

Springfield Daily News. State Plans Crackdown on Too Much Salt. Nov. 5, 1976.

State Office of Environmental Affairs, Massachusetts, Boston, MA (1977), Personal Communication.

Stein, A. A., Serrone, D. M., and Coulston, F. (1965) Safety evaluation of methoxychlor in human volunteers. *Toxicol Appl Pharmacol* 7:499.

Steinbach, M., Constantineau, M., Harnagea, P. et al. (1974) On the ecology of hypertension. *Rev Roum Med* 12/1(3–6) (Abstract).

Stofen, D. (1973) The maximum permissible concentrations in the U.S.S.R. for harmful substances in drinking water. *Toxicology* 1:187–195.

Stokinger, H. E. and Woodward, R. L. (April 1958) Toxicologic methods for establishing drinking water standards. *J Am Water Works Assoc* 50:515–529.

Subbotin, F. N. (1961) Nitrates in drinking water and their effect on the formation of methemoglobin. *Gig Sanit* 25:13–17.

Subrahmanyam, D. V. (Nov. 1975) Development and application of drinking water standards. Colloquium on Drinking Water Quality and Public Health, Water Research Center, United Kingdom.

Survey of Lead in the Atmosphere of Three Urban Communities, Bulletin 999 AP-12. Public Health Service, 1965.

Tabor, E. C. (1957) National air sampling network data. *U Missouri Bull* 60:9–15.

Taylor, E. W. (1958) Examination of water and water supplies, 7th ed., p. 841. Philadelphia: Blakiston.

Tobian, L., Jr. (1972) A viewpoint concerning the enigma of hypertension. *Am J Med* 52:595.

Toussaint, C., Wolter, R., et Sibelle, P. (1953) Effets de l'ingestion de grandes quantites de chlorure de sodium chez le rat. *CR Soc Biol* 147:1637 (cited in Freis, 1976).

Tuthill, R. W. (1976) Explaining variations in cardiovascular disease mortality within a soft water area. Water Resources Research Center, University of Massachusetts at Amherst. Publication No. 75.

U.S. National Research Council. (1953) Publication 294.

Vigil, J., Warburton, S., Haynes, W. S., and Haiser, L. R. (1965) Nitrates in municipal water supply cause methemoglobinemia in infants. *Public Health Rep* 80:1119–1121.

Walton, G. (1951) Survey of literature relating to infant methemoglobinemia due to nitrate-contaminated water. *Am J Public Health* 41:986–996.

Watkin, D. M., Fraeb, H. F., Hatch, F. T., and Gutman, A. B. (1950) Effects of diet in essential hypertension. II. Results with unmodified Kempner rice diet in fifty hospitalized patients. *Am J Med* 9:441.

Watson, R. L. and Langford, H. G. (1967) Sodium and socio-cultural effects on blood pressure of high school students. *Circulation* 36(Suppl I):246.

Weinsier, R. L. (1976) Salt and the development of essential hypertension. *Prevent Med* 5:7–14.

Wolf, H. W. and Esmond, S. E. (1974) Water quality for potable reuse of waste water. *Water Sewer Works* 121:2:48.

WHO (1956) Report of the European study group on standards of drinking water quality and methods of examination applicable to European countries (CMX/EUR/46.56) World Health Organization, Geneva.

WHO (1958) First International Standards for Drinking Water. World Health Organization, Geneva.

WHO (1963) Second Edition of the International Standards for Drinking Water, World Health Organization, Geneva.

WHO (1970) Second Edition of the European Standards for Drinking Water, World Health Organization, Geneva.

WHO (1972) Evaluation of certain food additives and of the contaminants mercury, lead and cadmium. 16th Report of the Joint FAO/WHO Expert Committee on Food Additives. Geneva, April 4–12.

7 Occupational Health Standards

HISTORICAL PERSPECTIVE

The rapid evolution of American economic life from a self-sufficient agricultural economy to an industrial economy developed through an amazing increase in technology during the eighteenth and nineteenth centuries. This process led to a centralization of resource development and the migration of laborers to urban factories and mills. As one might expect, the influx of workers was too rapid for the accompanying problems of industrial health and safety to be adequately dealt with through social legislation.

The legislative and court actions with respect to industrial health problems during this early industrial growth period posed no real threat to corporate interest. Since the government favored an expanding economy, judicial actions frequently reflected the attitude of many business leaders that progress brought with it certain inevitable costs. Therefore, many court decisions were made that not only shielded industry from the economic burdens of industrial accidents and health problems, but were also a precedent for legal defenses subsequently utilized by these industries (Ashford, 1976).

A number of factors contributed to this initial attitude and delayed the development of a safe and healthy work environment. Damage suits, which were trial oriented, were delayed for years or taken before an unsympathetic judge. The large number of immigrant laborers, with their language and social barriers, and the prevailing idea of a personal responsibility for success or failure likewise created conditions that contributed to labor exploitation (Page and O'Brien, 1973).

As the mechanization of industrial processes continued to accelerate, working conditions deteriorated, leading to even more widespread health and safety problems. The increasing number of compensation claims regarding industrial accidents resulted in an abundance of court suits. These cases helped publicize the existence of dangerous and inadequate working conditions. As a result of a gradual increase in the percentage of compensation claims won by disabled workers and the realization of the work, time, and money lost from industrial accidents, industry adopted a posture whereby it actually supported a modified or limited workmen's compensation law by which they assumed the liability, but not the fault. The end result was that the injured worker settled for comparatively small financial compensation, but it was theoretically received soon after the injury without the aggravation of ligitation. Thus, during the early 1900s safety legislation and workmen's compensation laws became prominent (Ashford, 1976).

In 1877 Massachusetts passed the first safety law. In 1908 the first state legislation was instituted for workmen's compensation. All but eight states enacted workmen's compensation laws by 1920. Jurisdiction, however, remained on the state level and was subject to influence from industrial lawyers and lobbyists (Ashford, 1976).

The early compensation laws were not backed by the tough enforcement standards necessary to counter the strengths of business. Also, these laws emphasized only accident-related injuries. Consideration of occupational diseases has developed more slowly, and even today the concept of compensable occupational diseases is still highly debatable (note the 1977 Senate hearings on Brown lung disease).

During this time*, the awakening of concern about occupational diseases prompted a number of researchers to initiate investigations of various substances to which workers may be exposed. According to Schrenk (1947), in 1912 Rudolf Kobert published one of the first tables concerning industrial toxicology. This table, entitled "The Smallest

* According to Blake et al. (1943) a group of engineers met in New York to form the National Council for Industrial Safety in 1913. The council's function focused primarily on the prevention of industrial accidents. In 1915 this group became the widely recognized National Safety Council.

Amounts of Noxious Industrial Gases Which are Toxic and the Amounts Which May Perhaps be Endured," addressed the acute toxic effects of 20 compounds. Some of these, hydrogen chloride, chlorine, bromine, and ammonia, agreed with later concentrations for repeated exposure or maximum acceptable concentrations (MACs). However, the values for the organic solvents were much greater than presently recognized as safe.

According to Cook (1956) and Schrenk (1947), a number of tables of standards were published during the 1920s and 1930s. These tables tended to focus on acute toxic effects. Information was also given on the least detectable odor and the least amount required to cause irritation. Included among these tabulations was that of Fieldner et al (1921), which was published in a Bureau of Mines Technical Paper. This table, which included 33 compounds, was one of the first published in the United States.

Other tables published during this period included Sayers (1927) which had 27 compounds, Henderson and Haggard's 1927 table listing 25 compounds, Sayers et al's 1929 table of 17 compounds, and an article entitled "Schaedliche Gase" published in 1931 that listed 20 compounds. The doses listed were based on symptoms after a given amount of time ranging from minutes to hours with effects ranging from minimum discomfort to death (Flury and Zernik, 1931).

The Henderson and Haggard table not only listed acute values, but for some substances gave a concentration limit for prolonged exposure. The values listed in this manner agree in dosage with later maximum acceptable concentration values. A table published a few years later by Sayers and Dalla Valle (1935) included 37 compounds. These compounds were listed with five concentrations and their subsequent effects. Four of those concentrations produced acute effects, and the fifth concentration gave a value for MACs reflecting prolonged exposure (Schrenk, 1947).

In contrast to previous reports, Schrenk noted that the tables published after 1935 did not provide a series of values for a range of acute effects. There was a shift in emphasis such that the reporting of the results yielded a single limit concerning repeated exposure. The tables referred to by Schrenk included those of Lehman and Flury (1938), Bowditch et al. (1940), Gafafer (1943), Cook (1945), and those of the American Conference of Governmental Industrial Hygienists (ACGIH) (1946). The extensive work by the ACGIH resulted in not only an expansion of the number of substances included in the lists, with approximately 140 listed by 1946, but also a reevaluation of the basis for many of the values. Prolonged exposure, reflecting true working conditions, was used as a basis for experimentation and reporting, as opposed to the earlier emphasis on single-dose acute effects.

Among other significant contributors to the advancement of industrial hygiene was the United States Public Health Service (USPHS). In 1943 this agency published a manual listing values based on past collective experience. In 1945 Cook made a landmark contribution by listing 141 values based on an analysis of suggested threshold limit values and new data. Finally, Smyth (1956) published the documentation for a considerably expanded list of toxic substances (approximately 240).

During these years, there was also a change in the types of organization created to deal with industrial health. As implied earlier, emphasis changed from exclusively accident prevention functions and safety criteria to include a more complete analysis of industrial hygiene and occupational diseases. Some organizations, which were based primarily on the former functions, included the United States Bureau of Standards and the Division of Labor Standards. However, two very influential groups that developed a strong orientation toward occupational disease problems were the ACGIH and the American Standards Association (ASA).

The increasing awareness of the importance of industrial hygiene was also reflected in several federal rulings that had an impact on working conditions. The Walsh-Healey Public Contracts Act of 1936 provided federal jurisdiction in the areas of regulation of working hours and also standards concerning the safety and health aspects of working conditions (Cralley and Konn, 1973). The control authorized in this act was limited to only those workers under contract to the government. Initially control affected only those companies with a contract with the federal government greater than $10,000.

ACGIH AND OCCUPATIONAL HEALTH

The role of the ACGIH in the development of occupational health standards in the United States and foreign countries has been so astonishingly influential that it is necessary to consider its historical foundations and present status with respect to occupational health standards. In fact, from about 1940 to the passage of the OSHAct in 1970, this organization was *the* leading force in the development of industrial health standards in the United States. Before 1936 there was very little official interest in the general area of industrial hygiene.

According to Bloomfield (1958), the extent of industrial hygiene activities before the mid 1930s consisted of the research activities of the USPHS, the United States Bureau of Mines, and several universities. Furthermore, only five states and one city were specifically involved with industrial hygiene work.

In 1936 Congress passed the far-reaching Social Security Act. Among the provisions of this act was the allocation of funds for the development and expansion of all aspects of public health activities. Consequently, the USPHS, along with the Conference of State and Provincial Health Authorities of North America, started a project whose purpose was to increase industrial hygiene activities in state and local health agencies (Bloomfield, 1958).

To carry out the objectives of the project, the USPHS developed a "short course" of industrial hygiene principles for personnel chosen by the participating state health departments. The first such course was conducted during the summer of 1936. The following year, a second short course was conducted, with the number of participants increasing from about 40 to over 100.

To provide more extensive educational experiences than could be offered via short courses, the USPHS developed cooperative field investigations in several states. For instance, field activities focusing on the problem of asbestosis were held in North Carolina and South Carolina. Several other states in which early cooperative activities were conducted included Connecticut (the hazards in the hatter's fur carroting and felt hat industries) and Utah (hazards of the metal and coal-mining industries) (Bloomfield, 1958).

Bloomfield stated that, during the final stages of the second short course (in 1937), it was felt that the courses should continue, but that it would be better to conduct them under the aegis of a nonofficial organization. The rationale behind this position was that more could be accomplished if they did not have to deal with the restrictions and pressures of formal governmental agencies. Several nonofficial organizations composed of government employees, such as the State of Provincial Health Authorities of North America, served to encourage this position, because a nonofficial group could make statements and take positions the governmental agency could not. As a result of such reasoning the ACGIH came into existence in 1938 and held its first conference June 27–29 in Washington, D.C.

The goals of the ACGIH as set forth in the first Constitution of 1938 were to "promote industrial hygiene in all its aspects and phases; to coordinate industrial hygiene activities . . . by official federal, state, local and territorial industrial hygiene agencies; to encourage the interchange of experience among industrial hygiene personnel in such official organizations; to collect and make accessible to all governmental industrial hygienists such information and data as may be of assistance to them in the proper fulfillment of their duties . . ."

During the early stages of its existence, the ACGIH developed standing committees intended to address some of the industrial health problems the membership thought to be most pressing. One of these committees was concerned with the development of threshold limit values (TLV) for toxic substances. The TLV committee, established in 1941, was made up of six nationally recognized industrial hygienists and toxicologists not employed by private industry. The initial TLV list, composed of 144 substances along with their MAC values, was published in 1946. The list was intended as a voluntary guide for industries. According to Stokinger (1970), a previous list of 45 substances was published by the Division of Industrial Hygiene of the USPHS in 1943. Stokinger noted that, even though suggested limits were classified MACs, they were TWAs and not a maximum ceiling value.

A second study group, the Committee of Uniform Codes, was also established and has served to complement the efforts of the threshold committee with respect to matters pertaining to standards development. It has been concerned with the development of a general industrial hygiene code, special exposures or industrial processes, standard methods of analysis, instrumentation, the uniform reporting of occupational disease, and so on.

The ACGIH has continued to review and extend the list of TLVs on a yearly basis. The TLVs were derived from the consensus of the judgment of expert committees. In 1955 the ACGIH produced formal documentation on which many TLVs were derived. Until 1962, this documentation was employed to assist the Committee in the revision of the TLVs. Since 1962 a widely circulated documentation has been published. At present, supplements to the documentation are published periodically by the ACGIH.

Additions have been made to improve on the annual publication. In 1962 an appendix was included to list carcinogens and substances that could not be given by a single value. In 1963 two more appendices were added. The first included the procedures used for determining the TLV of mixtures, and the second states the method employed in deciding if a substance rates a "C" or ceiling listing.

Changes have been brought about in the methods used by this group to propose and/or modify a standard. Committee actions to add a substance or to revise a standard must now be preceded by documentation. The evidence for the action is then referenced and submitted to a plenary committee. A Notice of Intent, instituted in 1964, requires that before any final action is taken, industry must be notified of TLV changes. This list is made available after the January meeting of the committee, at which time solicitation for comments and new substances for listing are made. Subsequent to the acceptance of the new annual list by the ACGIH, the Tentative Values column is used to note new additions and revisions of former standards. The values remain in this section for 2 years or until substantial evidence is collected on the substances that indicate no further unexpected results.

According to Stokinger (1964), concern had been expressed about the composition of the TLV committee, especially in view of the potential enforcement of ACGIH TLVs under the Walsh-Healey Public Contracts Act. Consequently, questions arose concerning this committee and its standards. The three principal areas of concern included (1) the bases on which the standards were established, (2) the degree of awareness of the committee of all data and experience available, and (3) the problem of establishing a TLV when only limited data are available.

Stokinger (1964, 1970) indicated that the ACGIH responded to such concerns by trying to improve the organizational framework whereby the standard derivation process could be more properly addressed. For instance, a change from eight to 15 persons was made in the committee to increase the coverage of industrial hygiene problems. The members of the committee represented complementary areas of expertise such as toxicology, engineering, industrial hygiene, analytical chemistry, and medicine. The TLV committee has five subcommittees that deal with insoluble dusts, economic poisons, hydrocarbons and halogenated compounds, oxygenated organic substances, and miscellaneous organic and inorganic compounds.

In addition to organizational improvement, the ACGIH decided to employ the use of both ceiling and TWA standards, since two standards based on different criteria provide greater safety. According to Smyth (1962), the ASA also considered this question, and set a precedent by naming two standards, each to guard against different injuries. A publication by Stokinger (1962) entitled "Threshold Limits and Maximum Acceptable Concentrations," supported by the ACGIH, led to this group's 1963 decision to adopt a ceiling limit for certain substances while retaining the time-weighted-average concept.

With regard to the rationale employed in establishing permissible limits, three basic health criteria were typically used by the ACGIH on which to base TLVs (Schrenk, 1947, 1955; Smyth, 1956). These include (1) pathological effects, (2) slight physiologic changes that result in no serious health effects, but could result in injury through loss of coordination and reaction time, and (3) discomfort or sensory effects. The possibility exists that the margin between a concentration that produces a mild response and a concentration that produces injury could be large, in which case exposure to a concentration slightly in excess of the TLV may not be injurious.

In 1962 the ACGIH published an article that attempted to provide a theoretical framework for interpreting the meaning of TLVs with regard to toxicity. The ACGIH considered the question of exceeding the standards and listed five points that should be analyzed: (1) the nature of the contaminant, (2) the acute effects that develop, (3) the cumulative effects that develop, (4) the frequency of the higher concentration, and (5) the duration of the exposure periods. Smyth (1962) was in agreement with these five criteria. After an analysis of 238 ACGIH TLVs and tentative TLVs, Smyth concluded that nine different adverse effects were guarded against and that comparison and useful interpretation of the values was difficult without a basis for understanding how the values were developed.

In a prepared discussion at a Symposium on Threshold Limits held in 1956, Smyth concluded that the TLVs of the ACGIH are the best available source for judgments on comparative toxicity of many industrial substances, although all the standards were not developed on the same basis. However, it should be noted that the TLVs were not to be used on a comparative basis, but have value in and of themselves.

THE THRESHOLD LIMIT VALUE

Initially the ACGIH designated its recommended limits as maximum allowable concentrations (MAC). As Dinman (1973) pointed out, the term MAC was a misnomer. For instance, the MACs were actually based on time-weighted average values and thus were not actually maximum ceiling values, as implied by the name. Furthermore, the use of the word "allowable" in the term MAC seemed to carry with it the impression of approval of concentrations below and up to the MAC level. To define more precisely the recommended standard, the term threshold limit value (TLV) replaced the term MAC in the 1960s.

Threshold limit values refer to time-weighted average concentrations of airborne substances and represent conditions under which it is thought that nearly all workers may be continuously exposed on a daily basis without harmful effects. The duration of exposure is assumed to be 8 hours/day, 40 hours/week for a working lifetime. Furthermore, it is recognized that certain individuals may be at high risk with respect to certain substances and may experience an adverse health effect of a variable nature at levels below the TLV (Dinman, 1973).

According to the 1976 Annual TLV listing booklet, there are three types of TLVs. The first is designated as the threshold limit value–time-weighted average (TLV-TWA). As previously noted, this represents the TWA concentration for a typical 8-hour workday in which the vast majority of workers can be continuously exposed without any harmful effect. The second type of TLV, or short-term exposure limit (TLV-STEL), is the maximum level that workers may be permitted exposure for a duration of up to 15 continuous minutes without experiencing (1) intolerable irritation, (2) chronic or irreversible tissue change, or (3) narcosis of sufficient magnitude to increase the likelihood of accidents, or to impair self rescue, or to significantly diminish work capacity, provided no more than four excursions* per day are allowed with at least 60 minutes between exposure period, and that the daily TLV-TWA is also not exceeded. According to the ACGIH document (1976), the STEL should be viewed as an absolute ceiling, not to be exceeding during the 15-minute excursion period. The final type of TLV or ceiling value (TLV-C) is the level that should not be exceeded even instantaneously. Only substances considered as fast acting and whose TLV is based on rapid response are given TLV-C limits. Examples of substances with a TLV-C limit include certain respiratory irritants (e.g., chlorine and formaldehyde), narcotic agents (e.g., methylene chloride), sensitizers (e.g., toluene, 2, 4-diisocyanate), and those substances which tend to accumulate rapidly in tissues (e.g., benzene). In contrast, those substances without a C-limit are usually characterized as effecting their toxic responses by cumulative, repeated exposure. Consequently, the use of a C limit for these substances would not be expected to prevent toxic effects.

The ACGIH has made contributions on a national and international level far beyond its inauspicious beginnings. It filled a niche that both government and industry found difficult to enter during the 1930s. Although the concept of the TLV may be challenged (or rejected as in

* Excursion values for substances without TLV ceiling values establish the degree of the maximum allowable level above the TLV-TWA. The approach adopted by the ACGIH involves the use of a general excursion factor for substances with similar TLVs. Thus substances with the following TLV ranges >0–1 ppm, >1–10 ppm, >10–100 ppm, and >100–1000 ppm have an excursion factor of 3, 2, 1.5, and 1.25, respectively. For example, a chemical substance with a TLV-TWA of 50 ppm may vary above the TLV, ultimately reaching a limit of 75 ppm for up to 15 minutes. The need to reduce the excursion factor as the TLV increases may seem odd at first. However, Dinman (1973) explained that if the excursion factors were not decreased as the TLV increased, extremely large absolute quantities would result.

the case of certain toxicologists, especially those in the Soviet Union), the orientation of the ACGIH toward trying to protect the health of the worker can only be praised. For nearly 35 years the ACGIH provided the leadership and council necessary to improve working conditions in the United States and many foreign countries. In fact, many western countries have adopted, almost in toto, the recommendations of the ACGIH. In 1971 OSHA promulgated the recommended TLVs of the ACGIH into law in the United States.

Another influential organization, the American National Standards Institute (ANSI), was organized in 1928. This institute was designed to set safety standards for a number of job-related items as well as the chemical compounds to which a worker may be exposed. This committee was composed of mostly industry-related professionals through whom major financial support was obtained. However, the organization did include a few representatives from government, trade associations, and private groups.*

Until 1969 the Member Body Council delegated committees that both set and approved standards. After that year, a separate Board of Standards Review was developed to rule on the standards on the basis of consensus.

The ASA used the idea of a maximum acceptable concentration (MAC) on which to base its standards. The definition of a MAC has been redefined through time. Initially the concept was of an average concentration to which an individual worker could be exposed for 8 hours daily for an indefinite period without injury or occupational disease. However, in 1957 the MAC of the ASA was redefined by their Z-37 committee. The MAC was then used "... to represent a limiting concentration, or ceiling, below which all values should fluctuate" (Stokinger, 1962). The redefinition of the MAC values was necessary because formerly the TLV was understood to have the same meaning as the MAC.

The MAC, according to Schrenk (1955), was derived from the belief that the body can detoxify a definite but limited amount of any potentially toxic substance. Most important, according to Elkins (1948), the term can be best understood graphically. By charting the concentration against the number of persons in a population affected at each particular dose, a bell-shaped curve is derived. This curve provides an indication of the probability of individuals affected by an increasing concentration of the substance. Elkins believed that this type of toxicity

* Since the ANSI Standards Committee is heavily weighted with respect to industrial representatives, Page and O'Brien (1973) strongly criticized the potential role of corporate influence in the development of the ANSI standards.

representation could be used in determining a level of socially acceptable exposure and thus serve as a basis for standard setting.

Irish (1965) presented an explanation of the concept of the MAC as envisioned by the Z-37 Committee of the ASA. The Z-37 Committee stated that each substance should have multiple limits depending on the extent and duration of exposure. The purpose of multiple limits on boundaries are to complement each other. Included are acceptable concentrations for repeated daily 8-hour time-weighted exposures, acceptable ceiling concentrations, short-term exposure limits, and minimum levels of sensory detection and avoidance of discomfort. Figure 7-1 shows the multiple limits (or boundaries) recommended by the ASA. This multiple limits approach of the ASA is in fundamental agreement with the more recent development by the ACGIH of the short-term exposure limits, (i.e., the excursion concept, and the ceiling limits).

Much controversy developed around the importance of these standards. Neither the ACGIH or ASA was formed through legislative action. Therefore, the standards were simply suggestions and carried no man-

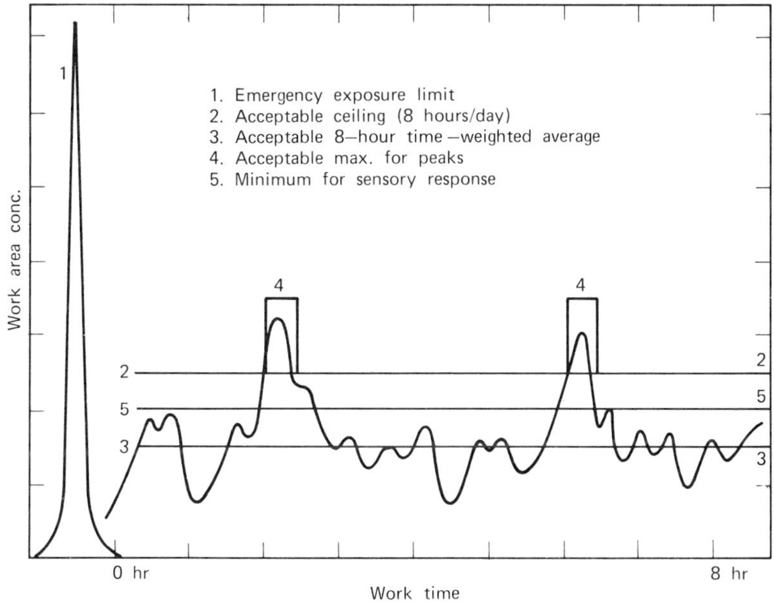

Figure 7-1. The five American National Standards Institute's acceptable concentrations expressed for an 8-hour day on a hypothetical substance.
Source: Irish, D. D. (1965) Concepts of standards. Arch Environ Health 10:546–549.

datory compliance terms. This lack of authority or enforcement ability was demonstrated, according to Page and O'Brien (1973), in a number of ways. One major revision was in the change of name to the American National Standards Institute (ANSI) from the United States of America Standards Institute (USASI). This change was made under pressure from the Federal Trade Commission, because the American National Standards Institute was formed by citizens and not by the government. The title originally used by this organization implied more authority than the group had.

ROLE OF THE FEDERAL GOVERNMENT

Federal support in the areas of industrial safety and health problems has developed slowly. The earliest program began with the passage in the 1880s of legislation concerning coal mines and railroad safety. The subsequent development of federal interest in this area, up through the early 1930s, was discussed previously. Following these earlier pieces of legislation, a lag period resulted during the latter part of the 1930s and continued through the 1940s. Interest shifted to the problems of employment and production, respectively, during the depression and World War II.

A certain amount of neglect continued in these areas, even after the war and its effects were over. However, three major pieces of legislation were enacted during the 1960s and early 1970s. These laws reflected an increased awareness of and interest in a program of industrial health and safety. The Metal and Nonmetallic Mine Safety Act of 1966, the Federal Coal Mine Health and Safety Act of 1969, and the most influential of all, the Occupational Safety and Health Act of 1970, were instituted.

The Occupational Safety and Health Act (OSHAct) was the culmination of a long series of attempts to pass a major piece of legislation that would reorganize and reshape the role of the federal government in the area of occupational safety and health. The act, proposed by Steiger of Wisconsin, was signed into law by President Nixon in December 1970.

The OSHAct established the National Institute of Occupational Safety and Health (NIOSH) whose functions include researching the effects of toxic substances in the workplace on health, developing criteria for assessing toxic substances, and recommending health standards to the Occupational Health and Safety Administration (OSHA). The role of the OSHA is to promulgate and enforce the standards.

According to Powell and Christensen (1975), the OSHAct provided the first opportunity to develop national occupational health standards that could be enforceable by law. Before the OSHAct of 1970, occupational health standards were, for the most part, a voluntary procedure. Standards prepared by the ACGIH, ANSI, or the National Fire Protection Associaton (NFPA) were typically employed by industry as guides to good work practice. Industries with government contracts greater than $10,000 were required to follow federal standards for safety and health under various legislative acts such as the Walsh-Healey Act of 1936. However, it was not until 1969 that the federal government adopted the TLV limits of the ACGIH as standards under the Walsh-Healey Act. By this time, the regulations applied to industries with contracts greater than $2,500. Despite this attempt of the government to be more involved with the regulation of occupational health and safety issues, none of the programs prior to the OSHAct was truly adequate in coping with widespread problems of worker safety and health in both a conceptual and practical manner (Ashford, 1976).

A second major thrust of the OSHAct included its development of multidimensional standards in contrast to previous narrow guidelines that include only environmental limits as a basis of judgment. Among some of the components of the proposed multidimentional standard are not only (1) the permissible limit of the contaminant in the workplace air, but also (2) the use of comprehensive medical examinations, (3) the designation of hazardous areas and containers, (4) the importance of work practices with the use of safety equipment, and (5) an ongoing monitoring and recording of environmental sampling and medical tests (Powell and Christensen, 1975).

One of the most important contributions of the OSHAct is the concept of thorough documentation of all standards. Inherent in this concept is the requirement that all data used to support the development of a standard must be publicly available. Previously, the recommended standards of the ANSI and the ACGIH depended to a certain extent on professional judgments and confidential data, although it should be reiterated that in 1962 the ACGIH began publishing the documentation from which their standards are derived. Nevertheless, Dinman (1973), commenting on Stokinger (1970), noted that as late as 1968, 24% of all TLVs published by the ACGIH were based on analogy.

The committee procedures by which standards are derived and reviewed differ from the consensus process of the ACGIH and especially the ANSI. A series of adversary-advocate reviews by three groups of people of differing backgrounds and experience serve to produce a final NIOSH proposed standard. The first group is drawn from senior NIOSH members, the second from the professional community, and the third from professional societies. It is important to note that individuals who help prepare the original criteria document are not members of any of these three review committees. Finally, in addition to the review process, NIOSH solicits comments from the interested general public via a request in the Federal Register.

Powell and Christensen (1975) listed a number of drawbacks this new method of standard setting must face. First, since documentation is essential, a disparity in the amount of information available on different substances reduces the criteria on which to base the values. Second, the time lag in the production of new information is a handicap.

The authors concluded by stating that a two-phased approach to standard setting would be initiated. The first phase encompasses the addition of extra information sections to the ACGIH and ANSI tables, including sections on field sampling, analytical methods, and so on as stated earlier. The second phase covers an extensive analysis of the OSHA standards in light of the information found in the supportive data and documentation reports.

Table 7-1 presents a list of the standards NIOSH has recommended to OSHA. It should be noted that NIOSH now bases its TWA on a 10-hour day, 40-hour week as compared to an 8-hour day, 40-hour week. As indicated previously, the additional research by NIOSH often addresses the same pollutants as did the ACGIH, and NIOSH may in fact recommend the same TWA, but the supporting documentation is now quite extensive. Although there is always room for improvement, the general thrust is a highly impressive and extremely thorough critical review of all available literature on the subject at hand.

FIRST TRULY NATIONAL FEDERAL STANDARDS

The OSHAct of 1970 provides the legal framework for the establishment of occupational standards. It does not list specific standards but only the methodology by which such standards may be promulgated. According to Ashford (1976), Section 4(b)(2) of the OSHAct (1970) permitted existing Federal standards covered by other acts (e.g., Walsh-Healey Service Contracts and National Foundation on the Arts and Humanities Acts) to be adopted automatically as occupational safety and health standards on the day the OSHAct was to be officially signed into law (April 28, 1971). However, it was not until approximately 1

Table 7-1. Summary of NIOSH Recommendations for Occupational Health Standards October 1976

Substance	Transmitted to OSHA	Current OSHA Environmental Standard	NIOSH Recommendation for Environmental Exposure Limit	Health Effect Considered	Comments and Basis for the Standard
Acetylene	July 1, 1976	2500 ppm (10% of lower explosive limit)	No exposure in excess of 2500 ppm	Indirect asphyxia	Employers to check for and inform employees of contaminants such as phosphine
Allyl chloride	Sept. 21, 1976	1 ppm 8-hr TWA	1 ppm TWA; 3 ppm ceiling (15 min)	Liver, kidney, lung effects	Urine, blood, and pulmonary function testing required
Ammonia	July 15, 1974	50 ppm 8-hr TWA	50 ppm ceiling (5 min)	Airway irritation	Hazardous liquid, eye damage
Arsenic, inorganic	Jan. 21, 1974 Revised June 23, 1975	0.5 mg As/cu m TWA	2 μg As/cu m ceiling (15 min)	Dermatitis, lung cancer	Chest x-rays required
Asbestos	Jan. 21, 1972	5 fibers/cc 8-hr TWA; 3 fibers/cc 8-hr TWA effective July 1, 1976; 10 fibers/cc ceiling	2 fibers/cc, 5 μ, 10 fibers/cc ceiling (15 min)	Asbestosis, lung cancer	Federal standard promulgated July 7, 1972, NIOSH reconsidering environmental limit

Table 7-1. *(Continued)*

Substance	Transmitted to OSHA	Current OSHA Environmental Standard	NIOSH Recommendation for Environmental Exposure Limit	Health Effect Considered	Comments and Basis for the Standard
Benzene	July 24, 1974 Revised Aug. 20, 1976	10 ppm 8-hr TWA, 25 ppm acceptable ceiling; 50 ppm maximum ceiling (10 min)	1 ppm ceiling (120 min)	Blood changes including leukemia	Urine monitoring required
Beryllium	June 30, 1972 Revised Dec. 10, 1975	2 μg/cu m 8-hr TWA , 5 μg/cu m acceptable ceiling, 25 μg/cu m maximum ceiling (30 min)	2 μg/cu m TWA, 25 μg/cu m ceiling (30 min)	Chronic lung disease (berylliosis), lung cancer	Chest x-ray required
Boron trifluoride	Oct. 4, 1976	1 ppm ceiling	0.25 mg/cu m TWA	Respiratory system effects	Adequate procedures for sampling and analysis not available; pulmonary function testing required

Cadmium	Aug. 23, 1976	0.1 mg/cu m 8-hr TWA ; 0.3 mg/cu m ceiling	40 μg Cd/cu m TWA; 200 μg Cd/cu m ceiling (15 min)	Lung and kidney effects	Urine and pulmonary function testing required
Carbaryl	Sept. 30, 1976	5 mg/cu m 8-hr TWA	5 mg/cu m TWA	Nervous system effects	Medical warnings of possible effects on reproductive system and minimum exposure during pregnancy required; skin and eye contact to be prevented
Carbon dioxide	Aug. 11, 1976	5000 ppm, 8-hr TWA	10,000 ppm TWA, 30,000 ppm ceiling (10 min)	Respiratory effects	
Carbon monoxide	Aug. 3, 1972	50 ppm 8-hr TWA	35 ppm TWA, 200 ppm ceiling	Heart effects	
Carbon tetrachloride	Dec. 22, 1975 Revised June 9, 1976	10 ppm 8-hr TWA, 25 ppm acceptable ceiling; 200 ppm maximum ceiling (5 min in 4 hr) hr)	2 ppm ceiling (60 min)	Liver cancer	
Chlorine	May 25, 1976	1 ppm 8-hr TWA	0.5 ppm ceiling (15 min)	Eye/airway irritation	Chest x-rays required

Table 7-1. *(Continued)*

Substance	Transmitted to OSHA	Current OSHA Environmental Standard	NIOSH Recommendation for Environmental Exposure Limit	Health Effect Considered	Comments and Basis for the Standard
Chloroform	Sept. 11, 1974 Revised June 9, 1976	50 ppm ceiling	2 ppm ceiling, 60 min	Liver or kidney tumors and central nervous system effects	Current federal standard should be TWA: published as "C" in error
Chromic acid	July 17, 1973	1 mg/10 cu m ceiling	0.05 mg CrO/cu m TWA, 0.1 mg CrO/cu m ceiling (15 min)	Nasal ulceration	
Chromium (VI)	Dec. 1, 1975	100 μg/10 cu m ceiling	1 μg/cu m for carcinogenic Cr (VI); 25 μg/cu m TWA for other Cr(VI); 50 μg/cu m ceiling (15 min)	Lung cancer, skin ulcers, lung irritation	Employer must demonstrate absence of carcinogenic Cr(VI); x-ray required
Coke oven emissions	Feb. 28, 1973	0.2 mg/cu m (coal tar pitch volatiles)	Work practices to minimize exposure to emissions	Tumors, primarily lung	Sputum cytology and chest x-ray required

Cotton dust	Sept. 26, 1974	1 mg/cu m (raw cotton dust)	0.2 mg/cu m lint-free cotton dust	Pulmonary disease (byssinosis)	Biologic (pulmonary function) testing required
Cyanide, hydrogen, and cyanide salts	Oct. 4, 1976	10 ppm 8-hr TWA , 8 hr TWA (alkali cyanides)	5 mg CN/cu m ceiling (10 min)	Thyroid, blood, and respiratory system effects	Concurrent measurement required for HCN when measuring for cyanide salt; trained first-aid personnel and first-aid kits to be available during use. Hazardous liquid, skin and eye
Epichlorohydrin	Sept. 17, 1976	5 ppm 8-hr TWA (20 mg/cu m)	2 mg/cu m TWA; 19 mg/cu m ceiling (15 min)	Skin, kidney, liver, and respiratory system effects	Medical warning of possible infertility effects required; hazardous liquid, skin

Table 7-1. *(Continued)*

Substance	Transmitted to OSHA	Current OSHA Environmental Standard	NIOSH Recommendation for Environmental Exposure Limit	Health Effect Considered	Comments and Basis for the Standard
Ethylene dichloride	March 9, 1976	50 ppm 8-hr TWA , 100 ppm acceptable ceiling; 200 ppm max. ceiling (5 min in 3 hr)	5 ppm TWA, 15 ppm ceiling, 15 min	Nervous system, respiratory, heart, and liver effects	Nursing infants at risk
Fluorides, inorganic	June 30, 1975	2.5 mg/cu m 8-hr TWA	2.5 mg F/cu m TWA	Kidney and bone effects	Urine monitoring required
Hot environments	June 30, 1972	None	Variable (sliding scale)	Heat stress	Acclimatization required
Hydrogen fluoride	March 9, 1976	3 ppm 8-hr TWA	2.5 mg F/cu m TWA; 5.0 mg/cu m ceiling (15 min, fluoride ion)	Skin/eye/airway irritation; bone effects	Pelvic x-ray (male) and urine testing required

Isopropyl alcohol	March 9, 1976	400 ppm 8-hr TWA	400 ppm TWA, 800 ppm ceiling (15 min)	Mucous membrane irritation; possible cancer threat in manufacturing process	More stringent work practices and medical surveillance for manufacturing workers
Kepone	Jan. 27, 1976	None	1 μg/cu m ceiling (15 min)	Nervous system effects; liver cancer	Liver function testing required
Lead, inorganic	January 5, 1973 Revised Aug. 4, 1975	0.2 mg/cu m 8-hr TWA	0.05–0.15 mg/cu m	Nervous system effects	Urine or blood monitoring required
Malathion	July 1, 1976	15 mg/cu m 8-hr TWA	15 mg/cu m TWA	Nervous system effects	Skin contact to be prevented; blood monitoring required
Mercury, inorganic	Jan. 5, 1973	0.1 mg/cu m ceiling	0.05 mg/cu m TWA	Central nervous system and mental effects	
Methyl alcohol	March 22, 1976	200 ppm TWA, 500 ppm ceiling	200 ppm TWA, 500 ppm ceiling (15 min)	Blindness; metabolic acidosis	

Table 7-1. *(Continued)*

Substance	Transmitted to OSHA	Current OSHA Environmental Standard	NIOSH Recommendation for Environmental Exposure Limit	Health Effect Considered	Comments and Basis for the Standard
Methyl parathion	Sept. 30, 1976	None	0.2 mg/cu m TWA	Nervous system effects	Skin contact to be prevented; blood monitoring required
Methylene chloride	March 9, 1976	500 ppm 8-hr TWA ; 1000 ppm acceptable ceiling; 2000 ppm max. (5 min in 2 hr)	75 ppm TWA; 500 ppm ceiling, (15 min) TWA to be lowered in presence of carbon monoxide	Central nervous system effects; carbon monoxide toxicity	Blood testing required
Nitric acid	March 9, 1976	2 ppm 8-hr TWA	2 ppm TWA	Dental erosion; nasal/lung irritation	Hazardous liquid, eyes and skin; chest x-rays required
Nitrogen, oxides	March 22, 1976	NO2: 5 ppm 8-hr TWA	NO2: 1 ppm ceiling	Airway effects	Pulmonary function testing required
		NO: 25 ppm 8-hr TWA	NO: 25 ppm TWA	Blood effects	

Noise	Aug. 10, 1972	90 dBA 8-hr TWA	85 dBA TWA; 115 dBA ceiling	Hearing damage	
Parathion	June 30, 1976	0.1 mg/cu m TWA	0.05 mg/cu m TWA	Nervous system effects	Skin contact to be prevented; blood monitoring required
Phenol	June 30, 1976	5 ppm 8-hr TWA (skin)	20 mg/cu m TWA; 60 mg/cu m ceiling (15 min)	Skin, eye, CNS, liver and kidney effects	Hazardous substance, skin and eyes
Phosgene	Feb. 23, 1976	0.1 ppm 8-hr TWA	0.1 ppm TWA; 0.2 ppm ceiling (15 min)	Airway effects	Pulmonary function testing and x-ray required
Silica, crystalline	Nov. 11, 1974	250/%SiO+5 in mppcf, or 10 mg/cu m/%SiO+2 respirable quartz	50 μg/cu m TWA, respirable free silica	Chronic lung disease (silicosis)	X-ray, pulmonary function testing required
Sodium hydroxide	Sept. 16, 1975	2 mg/cu m 8-hr TWA	2 mg/cu m ceiling (15 min)	Airway irritation	Hazardous liquid, eyes and skin
Sulfur dioxide	Feb. 11, 1974	5 ppm 8-hr TWA	2 ppm TWA	Pulmonary irritation	
Sulfuric acid	June 6, 1974	1 mg/cu m 8-hr TWA	1 mg/cu m TWA	Pulmonary irritation	Hazardous liquid, eyes and skin

Table 7-1. *(Continued)*

Substance	Transmitted to OSHA	Current OSHA Environmental Standard	NIOSH Recommendation for Environmental Exposure Limit	Health Effect Considered	Comments and Basis for the Standard
Tetrachloroethylene	July 2, 1976	100 ppm 8-hr TWA , 200 ppm acceptable max. ceiling; 300 ppm max. ceiling (5 min in 3 hr)	50 ppm TWA; 100 ppm ceiling (15 min)	Nervous system, heart, respiratory, and liver effects	Medical warning of possible congenital abnormalities required
Toluene	July 23, 1973	200 ppm 8-hr TWA ; 300 ppm acceptable ceiling; 500 ppm max. ceiling (10 min)	100 ppm TWA; 200 ppm ceiling (10 min)	Central nervous system depressant	
Toluene diisocyanate	July 13, 1973	0.02 ppm ceiling	0.005 ppm TWA; 0.02 ceiling (20 min)	Airway effects	Chest x-rays, blood tests, pulmonary function testing required

1,1,1,-trichloroethane	July 2, 1976	350 ppm 8-hr TWA	350 ppm ceiling (15 min)	Nervous system, liver, and heart effects	Action level set at 200 ppm TWA; medical warning of possible congenital abnormalities required
Trichloroethylene	July 23, 1973	100 ppm 8-hr TWA ; 200 ppm acceptable ceiling; 300 ppm max. ceiling; (5 min in any 2 hr)	100 ppm TWA, 150 ppm ceiling (10 min)	Central nervous system depressant	NIOSH reevaluating environmental limit
Ultraviolet radiation	Dec. 20, 1972	10 mW/cm averaged over any 1-hr period	1.0 mW/cm for over 1000 sec; 100 mW sec/cm for periods under 1000 sec	Skin and eye effects	
Vinyl chloride	March 11, 1974	1 ppm 8-hr TWA, 5 ppm ceiling, (15 min sample)	Minimum detectable level, 1 ppm ceiling (15 min)	Liver cancer	Standard promulgated Oct. 4, 1974; liver function testing required

Table 7-1. *(Continued)*

Substance	Transmitted to OSHA	Current OSHA Environmental Standard	NIOSH Recommendation for Environmental Exposure Limit	Health Effect Considered	Comments and Basis for the Standard
Xylene	May 23, 1975	100 ppm 8-hr TWA	100 ppm TWA, 200 ppm ceiling (10 min)	Central nervous system depressant; airway irritation	
Zinc oxide	Oct. 10, 1975	5 mg/cu m 8-hr TWA	5 mg/cu m TWA, 15 mg/cu m ceiling (15 min)	Metal fume fever	

NIOSH TWA recommendations based on up to a 10-hr exposure unless otherwise noted.

Source: NIOSH (1977). Cincinnati, Ohio.

month later (May 20, 1971) that OSHA proposed more than 400 pages of safety and health standards in the Federal Register. These standards included the consensus standards of the ANSI, the National Fire Protection Association, and the ACGIH. OSHA utilized section 6(a) of the OSHAct as the legal basis for the promulgation of the standards. Page and O'Brien (1973) strongly criticized OSHA's action, since Section 6(a) allows the Secretary of Labor up to 2 years after the date of enactment to promulgate the standards. Thus, instead of having effective federal standards as of April 29, OSHA opted for a different strategy. However, under considerable pressure to adopt these sets of standards, the Department of Labor acted to officially promulgate them in August 1971.

In the case of carcinogens, a different course of action occurred. First of all, according to Ashford (1976), the Labor Department established procedures such that the recommended values of the 1968 ACGIH TLV list were incorporated as standards under the Walsh-Healey Act. In their 1968 TLV report, the ACGIH listed suspected carcinogens and their recommended exposure limits (zero) in a special appendix. However, OSHA took the position that the appendix with the list of carcinogens was not incorporated within the Walsh-Healey Act, even though the Department of Labor had not originally recognized such an exception (Page and O'Brien, 1973). Consequently, the initial OSHA package of standards did not include the carcinogens.

As one could expect, OSHA's handling of the carcinogens became embroiled in controversy. In the midst of legal suits over which procedures to adopt while dealing with carcinogens, OSHA issued temporary emergency standards for 14 carcinogens on May 3, 1973. Two months later (July 16, 1973), OSHA proposed permanent standards for these 14 substances. Chapter 5 describes in detail OSHA's approach to the regulation of carcinogens in the workplace.

REFERENCES

ACGIH (1946) Threshold Limit Values for 1946. American Conference of Governmental Industrial Hygienists, Cincinnati, Ohio.

ACGIH (1962) Threshold Limit Values for 1962. Adopted at the 24th Annual Meeting of the American Conference of Governmental Industrial Hygienists, Washington D.C. May 13–15, 1962. *Indust Hyg J* 23:45–47.

ACGIH (1976) Threshold Limit Values for 1976. American Conference of Governmental Industrial Hygienists, Cincinnati, Ohio.

Ashford, N. A. (1976) *Crisis in the Workplace: Occupational Disease and Injury.* Cambridge: MIT Press.

Blake, R. P. et al. (1943) *Industrial Safety.* New York: Prentice-Hall.

Bloomfield, J. J. (1958) What the ACGIH has done for industrial hygiene. *Am Indust Hyg Assoc J* 19:338–348.

Bowditch, M., Drinker, C. K., Drinker, P., Haggard, H., and Hamilton, A. (1940) Code for safe concentrations of certain common toxic substances used in industry. *J Indust Hyg Toxicol* 22:251.

Cook. W. A. (1945) Maximum allowable concentrations of industrial atmospheric contaminants. *Indust Med* 14(11):936–946.

Cook. W. A. (1956) Symposium on threshold limits, present trends in MAC's. *Indust Hyg Q* 17:273–274.

Cralley, L. J. and Konn, W. H. (1973) The significance and use of guides, codes, regulations and standards for chemical and physical agents. In NIOSH, *The Industrial Environment: Its Evaluation and Control,* pp. 85–94.

Davies, D. M. (1975) The application of threshold limit values for CO under conditions of continuous exposure. *Ann Occup Hyg* 18:21–28.

Dinman, B. D. (1973) Principles and use of standards of quality for the work environment. In NIOSH, *The Industrial Environment: Its Evaluation and Control,* pp. 75–84.

Elkins, H. B. (1948) The case for MAC's. *Indust Hyg Q* 9(1):22–25.

Fieldner, A. C., Katz, S. H., and Kinney, S. P. (1921) Gas masks for gases met in fighting fires. *Bureau of Mines and Tech. Paper* 248:56 (cited in Schrenk, 1947).

Flury, P. and Zernik, P. (1931) *Schadliche Gase,* p. 453. Berlin (cited in Schrenk, 1947).

Gafafer, W. M. (1943) *Manual of Industrial Hygiene and Medicine Service in War Industries,* p. 508. Philadelphia.

Henderson, Y. and Haggard, H. (1927) Noxious gases and the principles influencing their action. New York (cited in Schrenk, 1947).

Irish, D. D. (1965) Concepts of standards. *Arch Environ Health* 10:546–549.

Kobert, R. (1912) Kompendium der praktischen toxikologie zum Gebrauche fur Artze, Studierende und Medizinalbeamte p. 45 Shuttgart (cited in Schrenk, 1947).

Lehman, K. B. and Flury, F. (1938) *Toxikologie and Hygiene der technischen Losungsmittel.* Berlin.

Page, J. and O'Brien, M. V. (1973) *Bitter Wages,* New York: Grossman.

Powell, C. H. and Christensen, H. E. (1975) Development of occupational standards. *Arch Environ Health* 30:171–173.

Sayers, R. R. (1927) Toxicology of gases and vapors: International criteria tables of numerous data. *Physics, Chemistry and Technology.* Volume 2, New York. p. 318–321.

Sayers, R. R. and Dalla Valle, J. M. (1935) Prevention of occupational diseases other than those that are caused by toxic dust. *Mech Eng* 57:230–234.

Sayers, R. R. Yant, W. P., Thomas, B. G. H., and Berger, L. B. (1929) Physiological response attending to exposure to vapors of methyl bromide, methyl chloride, ethyl bromide and ethyl chloride. *U.S. Public Health Service,* No. 185. 56 p

Schrenk, H. H. (1947) Interpretation of permissible limits. *Am Indust Hyg Assoc Q* 8(3):55–60.

Schrenk, H. H. (1955) Pitfalls in using maximum allowable concentrations in air pollution. *Indust. Hyg. Quart.* 16:230–234.

Silk, S. J. (1955) The threshold limit value for CO. *Ann Occup Hyg* 18:29–35.

Smyth, H. F. (1956) Improved communication-hygiene standards for daily inhalation. *Indust Hyg Q* 17:129–185.

Smyth, H. F. (1962) A toxicologist's view of threshold limits. *Indust Hyg Q* 23:37–43.

Sterner, J. H. (1956) Methods of establishing threshold limits. *Indust Hyg Q* 17:280–284.

Stokinger, H. E. (1962) Threshold limits and maximum acceptable concentrations: their definition and interpretation 1961. *Indust Hyg J* 23:45–47.

Stokinger, H. E. (1964) Modus operandi of the threshold limits committee of the ACGIH. Threshold limits committee of the ACGIH. *Indust Hyg J* 25:589–594.

Stokinger, H. E. (1970) Criteria and procedures for assessing the toxic responses to industrial chemical in permissible levels of toxic substances in the working environment. ILO Occupational Safety and Health Series, No. 20, Geneva, Switzerland.

8 Biological Indicators of Pollutant Exposure: Their Role in Occupational Standard Setting

BIOLOGIC MONITORING IS PRImarily considered as a vehicle by which medical and/or health personnel can carefully watch (i.e., in a quantitative manner) the level of exposure to a specific toxic substance. In most cases, it is not possible to measure the amount of toxicant at the site of action. For example, although lead is primarily feared for its neurotoxic effects, it is not yet possible to measure the amount of exposure in the specific neurons thought to be affected without possibly injuring these cells. Consequently, to determine what level of exposure may be tolerated without producing an adverse health effect (i.e., neurologic disorders), it is often necessary to develop an indirect indicator of exposure, that is, a measurement of pollutant level in one segment of the body (e.g., blood) that accurately predicts the level of pollutant in the area in which the most important toxic response may occur (i.e., neurons in the case with lead). Since the numerous systems of the body are usually in dynamic equilibrium with each other for most chemical substances (Joselow, 1976), the theoretical basis for indirect indicators of pollutant exposure is firmly established.

The process by which investigators have gained information concerning biomonitoring often resulted from unfortunate situations. For example, if a worker experienced neurologic disorders as a result of lead intoxication, the worker was usually given a blood or urine analysis. The level of lead in the blood or urine was then associated with a degree of toxicity. As a result of screening many workers a dose-response relationship between levels of lead in the blood and urine and neurologic effects has developed. Historically, this has been the pattern by which much of our knowledge concerning lead toxicity in workers evolved. The next step was to try to establish a level of atmospheric lead that could accurately predict the lead level in the blood or urine. Once this is developed, the derivation of a possible air standard can be more easily pursued. The following scheme represents the interrelationships of air pollutant levels, bioindicators, and toxic effects.

Air Levels → Must predict pollutant level in body (bioindicator) → This bioindicator must predict the response in the critical organ system.

This chapter discusses the concept of biomonitoring of pollutant exposure and its application and uses in industrial and community settings and in the development of occupational and environmental health standards.

SPECIFICITY AND SENSITIVITY

There are several concepts central to an understanding of the role or value of diagnostic testing in the process of biomonitoring of pollutant exposures. The two principal concepts in diagnostic testing are specificity and sensitivity. Specificity refers to the ability of a test to detect those people with exposure or the site of exposure while simultaneously excluding those persons or organs without exposure. Thus a diagnostic test is termed highly specific when it can distinguish between true positives and true negatives. As implied earlier, specificity is applied in both individuals and organ systems.

The response indicator should also co-vary consistently in a quantitative manner with the degree of exposure. Consequently, the ideal indicator changes relatively soon after exposure and exhibits a dose-response relationship (sensitivity). Thus an ideal biologic indicator indicates "what, where, and how much."

Most bioindicators are not ideal in the sense that they are both highly specific and highly sensitive. Under these circumstances, the indicator should be used concurrently with other complementary bioindicators and/or environmental monitoring.

Evolving naturally from the concepts of specificity and sensitivity are the so-called "selective and nonselective" bioindicator tests. Selective tests are usually based on the concept of pollutant specificity; that is, if the presence of a specific pollutant can be detected by a certain test, the diagnostic test is considered selective. Thus selective tests are based on the measurement of the concentration of a particular foreign compound or its metabolic derivative(s) in body or excretory fluids (e.g., determination of heavy metals in the blood or urine). Tests based on symptoms that are characteristic of certain toxic substances are also selective [e.g., determination of the activity of acetylcholinesterase in plasma after exposure to organophosphate insecticides (Ariens et al., 1976)].

Nonselective tests are actually bioindicator tests of low specificity with respect to the pollutant. Nonselective testing usually indicates exposure to a toxic substance whose characteristics are shared by a group of substances such that the specific pollutant cannot be definitely determined. Consequently, a nonselective test is actually a very general test for specificity. However, having only weak properties of specificity does not preclude a loss of sensitivity. For example, although the specific pollutant may not be precisely known, the extent of its toxic effects may be directly or indirectly monitored to a fairly precise degree (sensitivity). Nonselective but reasonably sensitive bioindicator tests may include the quantitative determination of glucuronides, mercapturic acid derivatives, and glycine conjugates in the urine. Also, greyscale ultrasonography, which is also considered a nonselective diagnostic test for a number of chemicals that cause liver damage, can be used to determine the extent or degree of dysfunction (Williams et al., 1976; Taylor, 1977).

BLOOD AND URINE TESTING

The use of norms offers the industrial hygienist the opportunity to make comparisons between the values of an individual worker in relation to the general population. If an individual's urine, blood, or enzymatic parameters deviate from what would be expected to be "normal," the hygienist is alerted to the possibility that the worker may be exposed to excessive levels of a certain toxic substance. This is a very crude approach to trying to determine if a worker has had unusual exposure to a pollutant. However, it is a useful initial screening process, which identifies some adversely affected individuals. Of course, it also brings with it the possibility of many false positives, since certain people may be very healthy yet their individual norms fall completely out of the range of the general population. Normal in this sense must be viewed as only a statistical expression of a large number of individual responses. This process of comparing the individual values to general norms may also result in many false negatives. For example, a person's values may usually fit on the low edge of the norm; however, following exposure to a toxic substance, his/her values may increase to a certain extent but may still be considered in the normal range. Here is an individual who exhibits a value in the normal range, yet he/she is adversely affected but not detected. The solution to this problem appears to be that each person's values at different times should be compared to determine whether an adverse effect is occurring. Thus the use of statistical norms in the process of biomonitoring must be recognized for its definite, albeit limited, value and also as one of a series of important potential screening tools, which, when interpreted by trained personnel, may offer an important component in disease prevention.

The evidence discussed in the preceding paragraphs strongly suggests that the utilization of biologic monitoring must take into account the concept of biochemical individuality. Overwhelming biochemical evidence has shown that each person has his/her own characteristic metabolic pattern, identifiable and recognizable by the activities of the enzymes involved.

To illustrate this phenomenon, Williams (1956) tested the activities of the serum enzymes phosphohexose isomerase (PHI), glutamic-oxalacetic transaminase (GOT), glumatic-pyruvic transaminase (GPT), lactic dehydrogenase (LD), malic dehydrogenase (MD), alkaline phosphatase (AKL), leucyl aminopeptidase (LAP), aldolase (ALD), and acetycholinesterase (ACE). Sera from seven pairs of identical twins, siblings and parents of the twins, and 50 other healthy individuals were assayed at various intervals from 2–10 days for up to 6 months. The results indicated that serum enzyme patterns are highly specific and are biochemical characteristics of the individual. Each person not only has his/her unique biochemical profile but also responds in an individualistic fashion to various toxic agents.

Frajola et al. (1960) tested this hypothesis in persons poisoned by a number of toxic chemicals, including trichloroethylene, benzene, phosphorus, carbon monoxide, decaborane, and carbon tetrachloride, among others. Figure 8-1 shows the typical biochemical profiles of the normal male, the normal female, and an individual poisoned by decaborane—the individuality of each pattern is apparent. Biochemical profiles obtained by radial plotting of the activity of a number of enzyme systems is a natural extension of the use of such indicators in biomonitoring. Metabolic disturbances have a greater probability of being detected if a number of parameters are simultaneously determined. This procedure offers the possibility of identifying the causative agent more readily than any single determination. With preemployment profiles of normal levels for a worker, biochemical profiles in situations of exposure would be highly useful tools in biomonitoring (Stokinger, 1962).

The most commonly used biologic and excretory indicators of pollutant exposure are blood and urine analyses, which can offer either selective or nonselective indicator tests, depending on the pollutant considered. The role of blood and urine in biomonitoring of pollutant exposure is considered in greater detail in the subsequent pages. Table 8-1 shows the "normal" urinary parameters and Tables 8-2 and 8-3 show the "normal" hematological parameters for human adults. Analyses of urine and/or blood parameters are good starting points in biomonitoring, since they are useful in revealing evidence of early poisoning.

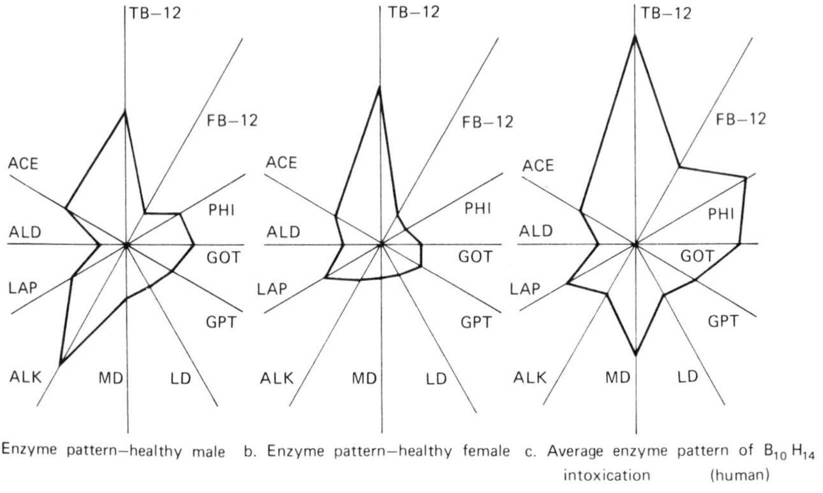

Figure 8-1. Biochemical profiles of a normal male, normal female, and an individual poisoned by decaborane. Source: Stokinger, H. E. (1962) New concepts and future trends in toxicology. Am Indust Hyg Assoc 23:879.

Table 8-1. Normal Values of Urine

Test	Range of Normal Values	Significance
Color	Pale straw to deep amber	Low specific gravity usually associated with pale color, high specific gravity with a deeper color.
Turbidity	Usually clear, if specimen freshly voided	Turbidity not necessarily indication of abnormality.
Acidity	pH 4.8–7.5	Fresh urine usually slightly acid on standing, may become alkaline due to decomposition with ammonia formation.
Specific gravity	1.001–1.030	Depends on fluid intake. Fixed at 1.010 in certain kidney diseases.
Sugar (glucose or reducing bodies)	None	A meal high in carbohydrate may give transient rise. Presence does not necessarily mean diabetes.
Albumin (protein)	None by ordinary methods; 2–8 mg/100 ml by quantitative method	Presence denotes kidney disease. Occasionally appears in urine of normal persons following long standing (postural albuminuria).
Casts: RBCs, WBCs, and epithelial cells	0–9,000 in 12 hr 0–1,500,000 in 12 hr 32,000–4,000,000 in 24 hr	In routine tests of urine a few casts and blood cells may be found in a normal specimen.

Table 8-1. *(Continued)*

Test	Range of Normal Values	Significance
Porphyrins	Trace qualitative; 0.001–0.010 mg/100 ml quantitative	May increase with lead exposure.
Concentration	Specific gravity above 1.020	Rough, but useful measure of kidney function.
Dilution	Specific gravity 1.002	Same.
Dye excretion	15 min 30–50% 30 min 15–25% 60 min 10–15% 120 min 3–10% Total of 70–80% in 2 hr	Values given are based on IV injection of phenolsulfonphthalein (PSP). Low dye excretion means poor kidney function.

Source: Goldwater, L. J. (1968) In N. Irving Sax., ed., *Dangerous Properties of Industrial Materials,* pp. 1–29. New York: Van Nostrand Reinhold.

Table 8-2. Normal Blood Values

Test	Normal Range	Significance
Red blood cells		
Males	4.5–6.0 million/cmm	
Females	4.0–5.0 million/cmm	Values for U.S.
Hemoglobin		
Males	14–18 g/100 cc	Values given in percent are meaningless unless the gravimetric equivalent of 100% is stated.
Females	12–15 g/100 cc	
White blood cells		
Total	5,000–10,000/cu mm	Total white cell counts fluctuate widely from hour to hour. Differential counts remain fairly static.
Differential		
Band forms	0–5%	
Segmented	35–70%	
Lymphocytes	20–60%	
Monocytes	2–10%	
Eosinophiles	0–5%	
Basophiles	0–2%	

Table 8-2. *(Continued)*

Test	Normal Range	Significance
Platelets	200,000-500,000/cu mm	
Reticulocytes	0.1-0.5%	High values indicate active blood regeneration.
Erythrocyte sedimentation rate (ESR)	Wintrobe method Males 0-9 mm in 1 hr Females 0-20 mm in 1 hr Westergren method Less than 20 mm in 1 hr	Values differ in males and females.
Hematocrit	40-48% cell volume	Affected by dehydration.
Mean corpuscular volume (MCV)	80-94% cu μ	Average size of RBCs.
Mean corpuscular Hb (MCH)	27-32 uug	Average Hb content per cell.
Mean corpuscular concentration Hb (MCC)	32-38%	Average concentration of Hb in RBCs.
Red cell fragility	Hemolysis starts in solution of 0.42% NaCl, complete in 0.32% NaCl	
Bleeding time	Less than 3 min	Capillary bleeding.
Coagulation time	6-10 min	Venous blood.
Prothrombin time	12-15 sec	
Appearance of RBCs	Uniform size and staining, appear circular	Parasites such as those of malaria may be found in stain-blood smears.

Source: Goldwater, L. J. (1968) In N. Irving Sax., ed., *Dangerous Properties of Industrial Materials*, pp. 1-29. New York: Van Nostrand Reinhold.

Before moving on to other important, although less used bioindicators of pollutant exposure, several comments are necessary concerning methodologic considerations during urine analysis. The value of urine as a biologic sample is increased by adjusting for creatinine or specific gravity (Elkins et al., 1974; Elkins and Pagnotto, 1965). Adjustment is necessary because of the variation in individual fluid intake and the rate of urine excretion, which cause the concentration of the excreted substance of concern to vary (Elkins and Pagnotto, 1965). Elkins et al. (1974) found that neither the specific gravity nor the creatinine adjustment consistently superior to the other, but the adjusted values were better than unadjusted ones. Creatinine adjustment was better for very dilute or very concentrated samples. However, because most early research reported in the literature was adjusted for specific gravity, adjustment of specific gravity to 1.024 would result in a good reference by allowing comparability. The NIOSH criteria document for occupational exposure to benzene (1974) recommends a specific gravity adjustment to 1.024.

As indicated above, with knowledge of the normal urinary values for a worker, it is possible to gain presumptive evidence of poisoning; that is, results of such testing are sensitive to abnormality, but are nonspecific as to cause. For example, the presence of elevated levels of albumin in the urine is indicative of kidney damage that may be caused by a number of diverse occupational poisons, including arsenic, carbon disulfide, chlorobenzene, and DDT, among others (Goldwater, 1968).

Table 8-3. Normal Blood Chemistry Values

Test	Normal Range	Significance
Blood urea	20–35 mg/100 cc	Increased in kidney disease.
Urea nitrogen	9–17 mg/100 cc	
Nonprotein nitrogen (NPN)	25–40 mg/100 cc	Increased in kidney disease.
Total protein (LF)	6.0–8.0 g/100 cc	
Serum albumin (LF)	3.5–5.5 mg/100 cc	
Serum globulin	2.0–3.0 mg/100 cc	
Albumin-globulin ratio (A/G) (LF)	1.5:1 to 2.5:1	
Uric acid	2–7 mg/100 cc	Females slightly lower.
Total cholesterol	150–250 mg/100 cc	Increased in biliary obstruction, decreased in liver disease.
Cholesterol esters (LF)	60–75% total	Decreased in liver disease.
Alkaline phosphatase (LF)	1.0–4.0 Bodansky units/100 cc of serum	High values—biliary obstruction, little change—intrinsic liver disease.

Table 8-3. *(Continued)*

Test	Normal Range	Significance
Acid phosphatase	0.5–2.0 Gutman units/ 100 cc serum	Not related to liver disease.
Blood sugar	80–120 mg/100 cc blood	
Icterus index (LF)	2–9 U	Index of jaundice, unreliable because of interfering substances.
Calcium	8.5–11.5 mg/100 cc serum	4.5–5.7 meq/liter
Chlorides as NaCl	350–400 mg/100 cc plasma 100–110 meq/liter	
Creatinine	1–2 mg/100 blood	
Oxygen capacity	18–24 cc/100 cc blood	Arterial blood.
CO_2 combining power	45–70 vol/100 cc 21–30 meq/liter	
Cholinesterase	RBC 0.67–0.86 pH units/hr	Michel Method.
	Plasma 0.70–0.97 pH units/hr	Decreased in organic phosphate poisoning
Potassium	16–22 mg/100 cc serum 4.0–6.0 meq/liter	
Sodium	315–340 mg/100 cc serum 136–145 meq/liter	
Phosphorus, inorganic	3–4 mg/100 cc serum	1.5–2.8 meq/liter
Magnesium	1.5–2.4 meq/liter	

Tests marked (LF) are used in liver function studies.

Source: Goldwater, L. J. (1968) Toxicology. In N. Irving Sax, ed., *Dangerous Properties of Industrial Materials.* pp. 1–29. New York: Van Nostrand Reinhard.

OTHER INDICATORS OF EXPOSURE

The use of saliva, fingernails, hair, placental tissue, sweat, milk, and feces may also be particularly useful in determining the extent of pollutant exposure. These types of bioindicators are not commonly used, but in specific cases their role as a predictor of exposure is highly significant. The two subsequent examples (saliva and fingernails) should provide the reader with the notion of how certain "uncommon" bioindicators are uniquely suited for certain pollutants.

Joselow et al. (1969) studied the use of parotid salivary fluid in biochemical monitoring. For mercury, parotid fluid showed highly significant correlations with blood concentrations. Denson et al. (1967) and Barylko-Pikielna and Pangborn (1968) found parotid saliva very useful in determining the extent of cyanide exposure via cigarette smoke as indicated by thiocyanate (SCN) levels. The use of parotid gland saliva was found to be a more reliable indicator of cyanide exposure than urine, blood, or submaxillary saliva thiocyanate levels. Such a test may be particularly useful in biomonitoring the effects of cigarette smoking on nonsmokers with respect to cyanide exposure.

The use of fingernail composition as an indirect measure of vanadium exposure has been suggested as a specific and sensitive bioindicator (Mountain et al., 1955). In fact, the Environmental Health Resource Center for the State of Illinois has recommended an ambient air standard for vanadium based in part on the predictive potential of the fingernail as a bioindicator. The rationale for this vanadium bioindicator is:

> "Subclinical depression of sulfur metabolism is thought to represent the earliest indication of vanadium intoxication. Cystine, an essential sulfur-containing amino acid, is a readily available indicator for analysis of sulfur metabolism, since it is contained in the hair and fingernails. An abnormally low amount of cystine indicates inhibited sulfur metabolism, which has serious implications for such physiological functions as cell membrane integrity, the breakdown of fat in the liver, the degradation of specific hormones, and the prevention of red blood cell agglutination. The cystine content of fingernails has been found (Mountain et al., 1955) to be reduced in workers exposed to vanadium pentoxide dust. An inverse relationship exists between vanadium urinary excretion levels and percentage of cystine in the fingernails; as the concentration of vanadium in the urine increases, the cystine content decreases. This inverse relationship was used by the Environmental Health Resource Center of the state of Illinois as a basis for calculating an ambient air quality standard for vanadium. In short, a reduced fingernail cystine content is of importance not only because it is a direct indication of a metabolic abnormality involving an essential amino acid, but also because it is easily measured" (EHRC, 1974).

LIVER AND KIDNEY FUNCTION TESTS

Liver and kidney function tests are of limited value in biomonitoring for two reasons: (1) Disturbances in function may not be present until extensive damage has occurred. (2) Normal results do not necessarily mean the liver or kidneys are without damage (Goldwater, 1968). Table 8-4 lists clinical liver function tests that may be useful in conjunction with other tests. Table 8-5 lists occupational poisons that may produce abnormalities in liver function. The pollutant specificity of liver function tests is low as implied by the diversity of toxic substances that cause abnormalities.

Table 8-4. Liver Function Tests (LF)

Test	Normal Range	Comments
Bromosulfonephthalein test (BSP)	Less than 5% retention after 45 min	Disease of liver tissue, biliary obstruction, or circulation may cause greater retention.
Cephalin flocculation	0–1+-flocculation in 48 hr	Increased flocculation found in disease of liver tissue, associated with abnormalities of serum proteins.
Serum bilirubin	Total 0.2–1.0 mg/100 ml Direct 0.1–0.7 mg/100 ml Indirect 0.1–0.3 mg/100 ml	
Thymol turbidity	0–4 Maclagan units	Significance same as cephalin flocculation.
Urine urobilinogen	0.5–2.0 mg/24 hr or dilution of 1:4 to 1:30	May be abnormal in biliary obstruction or disease of liver tissue.

Source: Goldwater, L. J. (1968) Toxicology. In N. Irving Sax, ed., *Dangerous Properties of Industrial Materials*, pp. 1–29. New York: Van Nostrand Reinhold.

Table 8-5. Occupational Poisons That May Produce Abnormalities in Liver Function

Acrylonitrile
Antimony
Arsenic[a]
Beryllium[a]
Cadmium[a]
Carbon disulphide[a]
Carbon tetrachloride[a]
Chlordane
Chlorinated diphenyls[a]
Chlorinated naphthalenes[a]
Chloroform
Cobalt
Cycloparaffins
DDT
Diethylene dioxide[a]
Dimethylformate
Dinitrophenol[a]
Diphenyl
Ethylene chlorohydrin
Ethylene dichloride
Methyl bromide
Methyl chloride[a]
Methyl formate
Methylene chloride
Nitrobenzol[a]
Phenol[a]
Phenylhydrazine
Phosphorus[a]
Tetrachlorethane
Trichlorethylene
Trifluorochloroethylene
Trinitrotoluene[a]
Uranium[a]

[a] Substance known to produce liver injury.

Source: Goldwater, L. J. (1968) Toxicology. In N. Irving Sax, ed., *Dangerous Properties of Industrial Materials,* pp. 1–29. New York: Van Nostrand Reinhold.

A relatively new and potentially useful liver response test is the determination of serum ornithine carbamyl transferase (OCT) levels. The OCT test is an organ-specific and -sensitive indicator of liver dysfunction (Drotman, 1975; Divincenzo and Krasavage, 1974). According to Divincenzo and Krasavage (1974), OCT is a mitochondrial enzyme that catalyzes the condensation of L-ornithine and carbamyl phosphate to L-citrulline and inorganic phosphate. This metabolic transformation occurs in the urea cycle of the liver. In mammals, OCT is present nearly exclusively in the liver. Consequently, an increase in OCT activity in blood serum is viewed as possible liver damage and serves as the basis for a bioindicator of liver dysfunction (Table 8-6).

As previously noted, another potentially encouraging bioindicator of liver or spleen function is greyscale ultrasonography. Greyscale ultrasonography has been successfully employed in the detection of liver disorders among vinyl chloride workers. Williams et al. (1976) reported that five out of 12 vinyl chloride workers diagnosed as normal by traditional methods exhibited modified liver texture, splenic vein enlargement, or increased spleen size. It is hoped that greyscale ultrasonography will be successfully used in the detection of early and definite pathological changes in liver and spleen structure.

Serum enzyme isozyme profiles offer organ-specific and -sensitive

Table 8-6. Relative Order of Hepatotoxicity[a]

Low	Moderate	High
Amyl acetate	Chloroform	Bromotrichloromethane
Butyl acetate	Hexane	Carbon tetrachloride
Butyl alcohol	Tetrachloroethylene	
1,2-dichloroethane	1,1,2-Trichloroethane	
Diethyl ether	Vinyl acetate	
Methyl alcohol		
Methyl ethyl ketone		
Methyl isobutyl ketone		
Xylene		

[a] Low indicates an elevated OCT at a dose greater than 500 mg/kg. Moderate indicates an elevated OCT at a dose of 50–500 mg/kg. High indicates an elevated OCT at a dose less than 50 mg/kg.

Source: Divincenzo, G. D. and Krasavage, W. J. (1974) Serum ornithine carbamyl transferase as a liver response test for exposure to organic solvents. *Am Indust Hyg Assoc J* 35:21–29.

methods of hepatic and nephrotic biomonitoring. Serum lactic dehydrogenase (LDH) isozyme patterns are markedly different in response to liver and kidney damage. For example, kidney damage results in increased LDH-1 and LDH-2, whereas liver damage causes increased LDH-5 (Grice et al., 1971; Cornish et al., 1970).

BIOINDICATORS FOR INORGANIC AND ORGANIC SUBSTANCES

This section contains a discussion of the relative usefulness of biomonitoring for representative inorganic (lead) and organic (benzene) toxic substances. A representative of both inorganic and organic substances was selected because organic substances offer a somewhat different set of biochemical problems (e.g., metabolic transformation into chemical derivatives of the parent or original toxic substance) than inorganic substances such as lead* and are thus complementary examples. These substances have also been extensively studied and tested for many years in industrial settings and therefore provide the most reliable (although not without controversy and disagreement) source of data from which new research and technical developments may evolve. Finally, all the major principles inherent in biologic monitoring [i.e., specificity, sensitivity, and direct and indirect indicators (metabolic transformations, multiple systems approach, etc.)] can be amply demonstrated by one or the other toxic substance.

Lead

The absorption of lead into the body results from respiratory exposure and gastrointestinal ingestion of food and drinking water. Absorption efficiency rates are approximately 30% for respiratory exposure and only 10% via the gastrointestinal tract (EPA, 1975). Once absorbed, lead is generally distributed throughout the various body tissues. According to Figure 8-2, most lead is bound to hard (i.e., dense bone, hair, teeth, etc.) and soft tissue (brain, kidney, bone marrow, etc.) (Joselow, 1976). Quantities of lead bound to the erythrocyte con-

* This is not to imply that inorganic substances do not undergo metabolic transformations. For instance, tetraethyl lead is metabolized to triethyl lead in the liver (Williams, 1963). However, the clear emphasis in biologic monitoring for organics concerns detecting the presence of biochemical derivatives in contrast to determining the total amount of an inorganic substance regardless of its various forms.

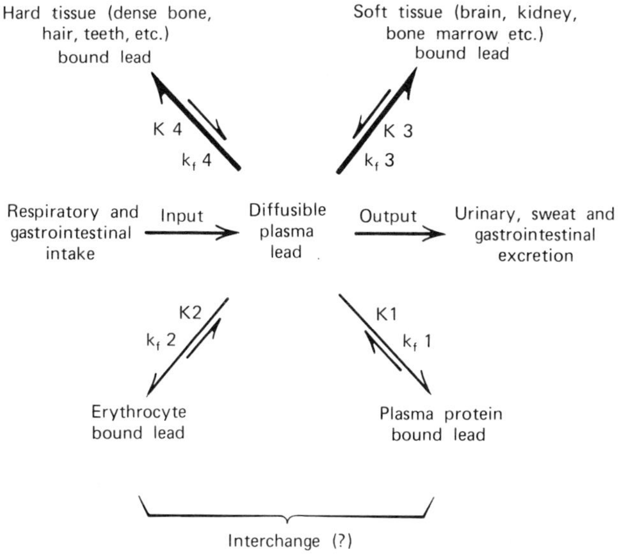

Figure 8-2. The dynamic interchange of the body lead pool. Source: Baloh, R. W. (1974) Laboratory diagnosis of increased lead absorption. Arch Environ Health 28:198–208.

stitute only a very small percentage (about 2%) of the total body burden. Despite the fact that lead is bound to different tissues of the body, this bound state is not permanent but is in a dynamic equilibrium between previously mentioned organ systems; consequently, there exists an exchangeable body lead pool. As indicated by the figure, each segment of the body with bound lead has an interchange of lead with the plasma, although the equilibrium constants are usually markedly shifted toward the tissues that bind the lead as compared to the diffusable plasma. However, it is precisely the concept of a dynamic interchangeable pool of body lead that establishes the basis on which a biologic indicator of lead body burden may be derived.

Thus, when one determines the level of lead via any type of biologic or excretory sample, predictions of the total lead body burden and levels in specific body organs can be made. Traditionally, levels of lead have been measured in urine and blood to obtain an estimation of overall lead exposure. Certain indirect diagnostic tests for lead exposure have been developed that determine the extent of lead present by the degree of activity of certain enzymes that lead is known to inhibit. The subsequent section evaluates the state of the art with respect to biologic indicators of lead exposure and their role in standard setting.

The present occupational health standard for lead is based primarily on preventing neurological and kidney damage in humans (NIOSH, 1972). At this juncture in our biomedical knowledge of the health effects of lead, it is not possible to say whether kidney tissue is more or less sensitive than nerve cells. The adverse health effects on which the standards are based are historically derived, resulting from numerous cases of lead-induced neurological and kidney damage. Because the standards are based on these two health effects does not mean that lead does not cause toxicity in other organ systems and at lower concentrations than in kidney and nerve cells. What it does mean is that, for the present time, these two adverse health effects are what the standard is designed to prevent. The questions then are, what biologic and/or excretory index or indices can be effectively used to determine if lead is being absorbed? How much is being absorbed? What is the danger of developing kidney and neurological damage?

A large number of biologic/excretory tests have been developed to monitor lead exposure (Table 8-7). Three general approaches for biomonitoring lead exposure are (1) the different biologic indices of exposure including the various urinary and hematopoietic indicators that denote a lead interference with heme synthesis, (2) the indices that indicate lead absorption, deposition, and excretion, and (3) other supplementary indices of lead exposure, including bone density, renal tubular function, neurobehavioral function, and others.

Figure 8-3 shows the steps in the biosynthesis of heme, as well as the known and possible sites of lead interference. Lead interference is known to result in both the qualitative and quantitative alteration of several of the intermediates involved in the synthesis of heme (Joselow, 1976). Interference at a particular site in heme biosynthesis results in an accumulation of precursors that may be excreted in the urine.

More specifically, it is known that minute amounts of absorbed lead are able to inhibit the activity of σ-aminolevulinic acid dehydrase (ALAD), which is needed to synthesize porphobilinogen from two molecules of σ-aminolevulinic acid (ALA) (Blejer, 1976). With inhibition by lead, ALAD levels in blood decrease, resulting in an increase in blood ALA levels (ALA-B) and later an increase in urinary ALA (ALA-U). Further along the heme biosynthesis scheme lead is thought to be inhibitory at stages 4 and 5 (Figure 8-3), thereby leading to (1) the accumulation of coproporphyrin III (CP) in red blood cells (CP-B) and its later excretion in urine (CP-U) and (2) the accumulation of free erythrocyte protoporphyrin (FEP), nonheme iron, and zinc-bound protoporphyrin (ZP) in red blood cells.

How well do these biochemical indices actually predict lead ex-

Table 8-7. Inorganic Lead: Biologic Indices of Exposure, by Type

I. INTERFERENCE WITH HEME SYNTHESIS	II. LEAD ABSORPTION/ DEPOSITION/ EXCRETION
A. Urinary Index	Index
Delta-Aminolevulinic acid (ALA)	Blood lead
Coproporphyrin	Urinary lead
Uroporphyrin	Hair lead
Porphobilinogen	Nail lead
	Bone lead
B. Hematopoietic Index	III. OTHER TYPES OF INDICES
Blood ALA	Bone density (radiographic)
Erythrocyte ALA-dehydratase activity	Electromyography
Erythrocyte nonheme iron	Serum proteins
Erythrocyte protoporphyrin	Renal tubular function
Erythrocyte zinc protoporphyrin	Endocrine function
Bone marrow sideroblasts	Neurobehavioral function
Reticulocytes (immature erythrocytes)	
Punctate Basophilia ("stipple" cells)	
Erythrocyte life span (increased fragility)	
Hemoglobin-hematocrit	

Source: Blejer, H. P. (1976) Inorganic lead: biological indices of absorption—biological threshold limit values. In B. W. Carnow, ed., *Health Effects of Occupational Lead and Arsenic Exposure—A Symposium*, pp. 165-178. Rockville, Md.: USHEW NIOSH.

posure? Are they both highly specific and highly sensitive? ALAD activity is acknowledged to be highly specific for lead, with only ethyl alcohol capable of effecting a decrease in activity that lasts for only a short time (Tola et al., 1973). However, despite its high specificity for lead, ALAD has marked limitations in providing an adequate dose-response relationship. ALAD is highly valuable in indicating low-level lead exposure, but it is a very poor indicator of prolonged, relatively elevated lead levels (Goldstein et al., 1975). These qualities make ALAD potentially attractive for monitoring lead exposure in environments in

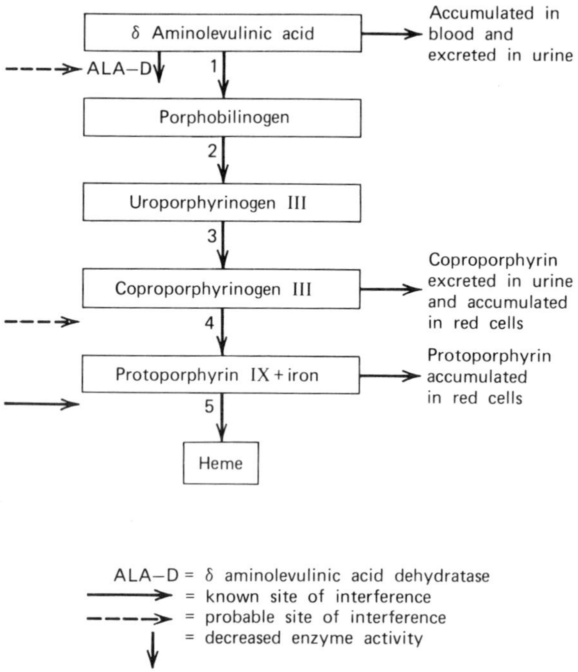

Figure 8-3. Scheme for heme biosynthesis. Source: Chisholm, J. J., Jr. (1971) Lead poisoning. Scientific American 224:15–33.

which lead exposure is thought to be at a potentially low level, but its usefulness in occupational settings is quite limited.

As for ALA-B, it becomes noticeably increased in concentration when Pb-B levels begin to equal or exceed 30 μg/100 g. As concentrations of Pb-B start to exceed 40 μg/100 g, a good correlation exists between ALA-U and Pb-B (Blejer, 1976).

According to Blejer, CP-U is nonspecific and subject to extreme variability, thereby restricting its efficiency in biomonitoring. Exposures to arsenic, ether, aniline, and other compounds are known to cause increases in CP-U (Ornosky, 1968).

As previously indicated, protoporphyrins increase in the red blood cell when the insertion of iron into the porphyrin ring to form heme is inhibited by lead. Joselow (1976) indicated that the increases in the protoporphyrin level in blood is "roughly proportional" to the Pb-B level; since the protoporphyrin in FEP is usually bound to zinc, ZP has been found to be a more accurate and convenient predictor of lead toxicity than FEP. This test has been successfully applied to children and occupationally exposed workers.

From the preceding discussion of biochemical indices of lead interference with heme synthesis, it can be seen that there are certain limitations for all the possible bioindicators. However, it appears that indicators such as ALAD and ALA and ZP do complement each other. For example, although ALAD is very sensitive to low-level exposure and inadequate to predict higher lead exposure levels, ALA and ZP offer reliable dose-response relationships as the amount of absorbed lead increases.

The use of multiple diagnostic tests that complement each other should be encouraged. Reliance on any one particular bioindicator should be interpreted in light of that test's limitations. It should be noted that the ALA-U and ZP tests were evaluated in light of their close correlation with Pb-B levels. Since their usefulness is thus related to the validity of the Pb-B values, the use of Pb-B as a bioindicator of lead exposure is considered. Pb-B levels are a measure of the amount of lead absorbed and do not directly determine its toxic effects (Benson et al., 1976; Elkins and Pagnotto, 1965). Pb-B levels have been used by industrial hygienists for years as the primary basis in monitoring workers' lead exposure. The NIOSH criteria document for inorganic lead adopted the use of Pb-B as the most reliable indicator of lead exposure. NIOSH (1972) considers Pb-B and Pb-U as good indicators of current lead absorption, but not particularly accurate predictors of a lead body burden or of the state of the individual's health. NIOSH suggested that bone lead is probably a better predictor of total body burden than Pb-B; however, a bone sample as a routine procedure is not possible. NIOSH considered other possible indicators, such as CP-U, ALA-U, ALAD-B, cell stippling, FEP, and ZP, as less reliable than blood and urinary lead determinators for estimating the absorption of lead. The validity of Pb-B as an accurate predictor of lead toxicity has been challenged by Vitalli et al. (1975), who reported the coexistence of increased Pb body burdens as evidenced by EDTA mobilization tests with Pb-B levels below 80 μg/100 ml and the occurrence of lead nephropathy in some of these individuals.

Table 8-8 shows the results of multiple lead screening tests for 30 workers (aged 24–62 years) in the lead refining, cutting, and welding occupations in New Jersey. Two of the 30 workers exhibited Pb-B levels in excess of the 80 μg/100 ml limit, whereas only one of 25 workers had Pb-U values greater than 150 μg/100 ml. Since these values represent the BLVs* adopted and recommended by NIOSH, it is expected that individuals with lower values did not have "officially" unacceptable

* Biological limit values

Table 8-8. Lead Screening Tests on 30 Lead Workers

Subject	Occupa-tion	Bpb (μg/100 ml)	ALA-D (μg/100 ml RBC)	FEP (μg/100 ml)	Uala (mg/liter)	Upb (μg/liter)	24-hr Urine Execretion: Control Copro (μg/day)	Control Pb (μg/day)	EDTA Pb (μg/day)
H.Z.	LT	29	66	71	7	138	127	135	976
M.B.	LC	94	43	129	26	—	420	—	2922
J.H.	LC	70	95	147	7	—	350	—	2176
J.P.	LC	64	57	107	3	—	610	—	1794[c]
J.B.	LC	34	97	4	5	—	48	—	227[c]
R.V.	LC	47	120	11	4	—	673	—	673
J.Bo.	LB	38	88	73	4	73	—	91	1051
R.A.	LB	48	43	95	5	68	—	81	—
G.B.	LB	68	47	242	—	66	—	145	3375
R.R.	LB	45	44	124	7	82	—	84	1477
S.N.	LB	53	50	125	2	116	—	112	—
M.S.	LB	50	27	51	—	72	—	72	1881[c]
C.U.	LB	44	45	101	4	44	—	103	1153[c]
T.G.	LB	32	88	54	4	70	—	—	—
M.A.	LB	38	77	50	5	84	8	65	819
J.Ba.	LB	46	64	45	5	106	—	127	—
S.D.	LB	46	80	71	9	142	—	138	2053
G.H.	LB	54	65	134	4	96	—	102	1988

F.C.	LB	59	52	163	15	180	—	136	2810
C.K.	LB	40	65	59	5	123	—	149	2294
R.F.	LB	52	69	138	5	92	20	43	530
J.Zi.	LB	41	51	63	5	128	—	116	2401
A.B.	LB	40	—	—	5	57	—	58	1776
O.D.	LB	39	74	151	3	84	16	99	2068
J.T.	LB	48	116	63	6	128	—	129	1793
C.E.[a]	LB	51	82	64	5	80	7	53	1134
R.S.[a]	LB	66	78	26	6	86	24	99	1590
S.B.[a]	FR	98	67	77	18	334	737	474	4018[c]
F.C.[b]	PB	35	88	3	2	70	13	112	990
J.Z.[a]	SC	48	42	29	70	65	757	305	5200

[a] Chronic lead nephropathy.
[b] Renal disease and hypertension of uncertain etiology.
[c] Underestimation of 24-hr Pb excretion. Creatinine excretion < 1.2 g/24 hr.
LT = lead tin solder worker.
LC = lead cutter.
LB = lead burner.
FR = firing range sweeper.
PB = painted steel burner.
SC = solder cream worker.

Source: Vitalli, L. F., Joselow, M. M., Wedeen, R. D., and Pawlow, M. (1975) Blood lead—an adequate measure of occupational exposure. *J Occup Med* 17:156.

levels. However, it was pointed out that three of the workers with values below 80 μg/100 ml had evidence of lead nephropathy, whereas four of eight workers from this group of 30 who volunteered for renal function studies had reduced kidney function. Additionally, three of the four workers with impaired kidney function had renal biopsies that revealed histological abnormalities. The use of the other biologic indicators of lead exposure as predictors of lead intoxication revealed that eight of 28 workers had abnormal ALA-U values (>6 mg/liter), 27 of 29 workers had abnormal ALAD (<120 μg/100 ml RBC), 28 workers had abnormal FEP values (>25 μg/100 ml RBC), and 23 of 26 workers had excess lead excretion (>0.50 μg/100 ml) following EDTA mobilization tests.

At this point it should be recognized that no single indicator of lead exposure is a truly excellent predictor of toxicity. However, when used in collaboration, biologic indicators such as Pb-B, ALAD, FEP, ALA-U, Pb-U, and CP are able to compensate for the inherent weaknesses that their complements lack (Table 8-9). Different investigators have suggested various indices of occupational exposure that may indicate excessive lead absorption (Table 8-10).

As previously indicated, Vitalli et al. (1975) have shown the value of multiple indices of exposure in dealing with heavy occupational exposures. Reliance on any one indicator is actually based on a false optimism not supported by current data. Although more research is needed to determine the predictive potential of various combinations of indicator tests, it is clear that industrial hygienists should be biomonitoring with at least two tests that tend to complement each other.

One of the most significant problems with respect to the adequacy of the lead bioindicators is a general lack of data that correlate air concentrations and exposure to the lead. According to Blejer (1976), the "only fairly good correlations" have been presented by Williams et al. (1969). They reported that at the present standard of 150 μg Pb per m^3 of air, the blood lead levels were generally around 60 μg/100 ml. Since 80 μg/100 ml is a blood lead level at which the incidence of toxic effects begins to increase markedly, Blejer concluded that there was only a small safety factor built into the standard. Lerner (1976, discussion) has been highly critical of the Williams et al. study on the basis of (1) its short duration (2 weeks), (2) its small sample size of 39 individuals, and (3) the wide variability (up to 400%) of air measurements taken by personal monitors. Since NIOSH has established environmental monitoring in contrast to biologic monitoring as the primary basis for occupational health standards, more information is needed to associate an air level with the biologic indicator value.

Table 8-9. Inorganic Lead: Biologic Indices of Occupational Exposure by Type and Suggested Usefulness

	I. Interference with Heme Synthesis		
	Increased Concentration		Decreased
	Erythrocyte protoporphyrin[a]		Erythrocyte ALA-dehydratase activity[a]
	Erythrocyte zinc protoporphyrin[a]		Hemoglobin[d]
	Urinary σ-aminolevulinic acid (ALA)[b]		Increased Count
	Urinary coproporphyrin[c]		Stipple cells (punctate basophilia)[e]
II.	Lead Absorption	III.	Other Types of Indices
	Increased concentration		Electromyography[g]
	Blood lead[f]		Motor nerve conduction slowness—extensor muscle weakness
	Urinary lead[f]		Renal tubular function[g]
			Proximal tubular dysfunction

[a] Specific for biochemical effect of absorption.
[b] Semispecific for biochemical effect of absorption.
[c] Semispecific, as above; valuable for screening.
[d] Nonspecific; corroborates overabsorption.
[e] Nonspecific; too variable and not useful.
[f] Specific for degree of lead absorption.
[g] Nonspecific; useful in early detection and diagnosis.

Source: Blejer, H. P. (1976) Inorganic lead: biological indices of absorption—biological threshold limit values. In *Health Effects of Occupational Lead and Arsenic Exposure—A Symposium*, pp. 165–178. USHEW NIOSH.

Table 8-10. Multiple Indicators of Lead Exposure Suggesting Excessive Lead Absorption According to Different Researchers

Biologic/ Excretory Indicators	Tsuchiya and Harashima 1965	Vitalli et al 1975	Zielhuis 1971	Expert Group of 18 1968	NIOSH 1972
Pb-B			>80 μg/100 g	>80 μg/100 ml liter	>80 μg/100 g
Pb-U	>150 μg/liter		>150 μg/liter	>150 μg/liter	>200 μg/liter
Copro-U	> 50 μg/liter		>80 μg/liter	>50 μg/liter	
ALA-U		>6 mg/liter	>20 mg/liter	>20 mg/liter	
ALAD		<120 μg/100 ml RBC			
FEP		>25 μg/100 ml RBC			
ZnProtop.		>50 μg/100 ml			
EDTA-Pb		>650 μg/day			
Baso. St.	0.3/1000				
Pb-Plasma		>10 μg/100 ml			

The role of biomonitoring in the context of industrial hygiene has a well-established history. Advances in the development of new diagnostic tests have helped to overcome deficiencies in the specificity and/or sensitivity of early tests. Biomonitoring has become a highly controversial area in occupational health because OSHA has made the legally binding standard one of environmental monitoring; that is, if an air concentration exceeds a specific level, a violation may be called even if biologic monitoring does not support the contention of excessive pollutant exposure (Nelson, 1973). The question is, on what should the actual standard be based—a biologic level of exposure, an environmental level, or some combination of biologic and environmental monitoring? Nelson (1973), in his article entitled "The Place of Biological Measurements in Standard Setting Concepts," directly addressed this issue. He stated that there are five minimum requirements that must be met for a biologic monitor to be preferred over environmental sampling. These include

1. The air contaminant must not provoke irreversible biochemical changes or health effects in work exposures usually encountered;
2. levels of the contaminant measured in exhaled air, blood or urine must be quantitatively related to an adverse health effect;
3. measurements of metabolites or biochemical changes resulting from contaminant inhalation must also be quantitatively related to an adverse effect;
4. biological monitoring must yield information of potential health risks equal or superior to the information obtained by air sampling; and,
5. facilities for accurate biological measurements must be readily available at costs below or not greatly exceeding costs for air sampling.

Nelson (1973) suggested that, in addition to lead, arsenic, cadmium, selenium, tellurium, and thallium could satisfy these minimum requirements.

As can be seen from the previous discussion, there is considerable disagreement over the issue of whether biologic or environmental monitoring should serve as the basis for OSHA air standards. Since this issue is central to the role of bioindicators in standard setting, a comparison of the strengths and weaknesses of these respective positions is presented. Table 8-11 compares the views of representative leading

Table 8-11. Conflicting Statements Of Leading Theoreticians Concerning The Use Of Biologic Monitoring vs. Environmental Monitoring For An Occupational Health Standard For Inorganic Lead

Comments Supporting Biologic Monitoring	Comments Supporting Environmental Monitoring	Comments Supporting Integration of Biological and Environmental Monitoring
Lead Industries Association—1972 ". . .a biological standard, based on blood lead determinations, provides the best means of protecting the worker and determining compliance under the OSHAct of 1970. Biochemical indices provides a much more accurate assessment of possible hazard to lead exposure than do air concentrations. It is recommended that air sampling be used only to indicate the necessity to institute biological monitoring and to evaluate engineering controls."	Dr. Samuel Epstein—1976 ". . .the only really effective and accurate way of determining the extent of worker exposure to lead is by environmental monitoring, as opposed to biological monitoring. Biological monitoring is useful for back up purposes, but to suggest that this is the most accurate way of assessing exposure is. . .inconsistent with the fact." Dr. Epstein indicated that as a result of the prophylactic use of versenate, and the lack of reliability of blood lead levels to reflect long-term exposure, frequent monitoring of blood levels would be needed to provide valid estimations of lead exposure.	Dr. Robert Kehoe—1947 ". . .making the lead trades safe, not for a few weeks or years, but for a lifetime. . . The hazard of lead poisoning. . .is not controlled by medical practice [i.e., prophylaxis]. . .(but by) controlling the environment. . .What the industrial physician does is to act as the policeman in determining the extent to which environmental conditions are satisfactorily controlled. If he acts in any other capacity, he is missing the point of his efforts."
Dr. Jerome F. Cole—1976 Listing of some inadequacies of air monitoring to evaluate human exposure to lead. "1. Air sampling for lead does not represent the amount of lead		Dr. Robert Kehoe—1963 ". . .the comparative mildness of lead intoxication, as it is seen in American industry. . .contributes to the . . .irresponsible . . .view that there is no need to

absorbed into the body because of differences in particle sizes and solubilities;

2. Air samples represent only a small fraction or aliquoit of the total volume of air inhaled by an individual;
3. Air sampling does not consider the potential exposure from ingestion; Biological indicators measure lead from all sources."

He concluded that biochemical indices offer a more accurate assessment of possible hazard to lead exposure than does air monitoring. He recommended that the BLV and OSHA complicance standards should be a blood lead concentration of 80 μg/100 g of blood. Also, he recommended the adoption of certain action levels, means of reduction, and frequency of sampling.

Dr. Robert Eckardt—1976

"I would like to support. . .(the) concept of biomonitoring by giving a brief example. While performing monthly urinary lead levels on

Timothy F. Cleary—1976

The U.S. Circuit Court of Appeals for the Eighth Circuit held that "while biological monitoring, including urine and blood sampling, is also accurate,. . .air sampling is the most efficient and practical way for the Secretary of Labor and employers to determine air borne lead levels for the purposes of enforcement actions under the OSHAct."

achieve complete control of the hazard. This view, it seems, is especially prevalent in some of the long-established lead trades in which conditions are much better than they used to be while failing to meet adequate hygienic standards. . .What is required is. . .not. . .prophylactic medical therapy, but. . .the application of orthodox engineering principles and equipment coupled with satisfactory medical supervision."

Dr. Hector Blejer—1976

"Although primary prevention can be achieved only by engineering controls, the medical use of biological monitoring indices is still needed to protect workers occupationally exposed to inorganic lead. Biologic monitoring may be employed in an occupational medical program for lead-exposed workers as follows: (1) assists medical placement, (2) serves as a check on the efficiency of engineer con-

Table 8-11. *(Continued)*

Comments Supporting Biologic Monitoring	Comments Supporting Environmental Monitoring	Comments Supporting Integration of Biological and Environmental Monitoring
lead refinery workers, a general upward shift in urinary lead excretion was noted. Further investigations revealed that these men were working a 12-hour or 7 day/week because a large building program was occurring within the refinery. Without the biological monitoring this upshift would not have been detected."		trols and air monitoring, (3) prevents both acute and chronic exposure to lead, and (4) determines the extent of recovery of lead-affected workers when they can be permitted a safe return to a healthy work environment."
Mr. Kenneth Nelson—1973 "Biological measurements have several important advantages (over environmental measurement). They eliminate time-consuming jobs of air sampling, particle-sizing and time studies. The interpretation of blood lead levels is also much easier than trying to guess what airborne lead concen-		Dr. Sidney Lerner—1976 "I would have a false-negative sense of confidence if I simply went on the lead in air [analysis]. So I think we really have to use both tools, neither to the exclusion of the other, for the ultimate determinant of whether there is an effect. After all, we are trying to protect man, so let us study man."

trations really mean. The cost to industry would be considerably reduced if biologic monitoring were adopted."

Mr. David Cameron—1976

"Engineering controls, environmental sampling, medical monitoring and good personal hygiene all must be employed in order to control occupational exposure in a lead environment."

Dr. Bertram W. Carnow—1976

"Biological monitoring alone is not an appropriate way to assure safety and health for workers. Both environment and hosts must be monitored to assure low levels of contaminants for the continued health of workers."

theoreticians on the subject of biologic indicators as industrial standards versus environmental monitoring standards. Every effort has been made not to take statements out of context.

Benzene

Benzene is among the most widely used industrial chemicals as a result of its properties as an organic solvent and its relatively low cost (Browning, 1965). Benzene is easily absorbed via the skin, intestinal tract, and lungs (Gerarde, 1966; Browning, 1965). Most of the toxic effects of benzene (thrombocytopenia, erythopoiesis, and leukopoiesis) are related to its phenolic metabolites (Parke, 1968; Gerarde, 1966; NIOSH, 1974), since benzene itself is thought to have a low toxicity because it is rapidly excreted via the lungs as a result of its insolubility in blood.

Considerable research has been directed toward developing a dose-response relationship with respect to the adverse health effects benzene exhibits on both animal models and humans. Animal studies have revealed that when rats are exposed to 40–50 ppm, leukopenia is induced (Deichmann et al., 1944). At higher concentrations (80–88 ppm) Wolf et al. (1956) noted the occurrence of leukopenia and significantly increased numbers of nucleated cells in the bone marrow of animal models (rats, rabbits, and guinea pigs). Industrial exposures of workers to 60 ppm have resulted in hematologic changes for total RBCs and WBCs, Hb, polymorphonuclears, lymphocytes, and eosinophils (Hardy and Elkins, 1948). Thus it was felt that exposures should be kept below 25 ppm.

To develop a specific and sensitive indicator test for benzene, it is necessary to consider how benzene is metabolized by the body. In contrast to bioindicators for lead, which are based on the actual measurement of lead in blood and urine and on enzyme or precursor levels associated with the biosynthesis of heme, the potential bioindicators for benzene are either the level of benzene in expired air or the level of a benzene metabolite in the urine. As previously mentioned, benzene is insoluble in blood and is readily excreted in expired breath. In fact, Stewart et al. (1965) were able to detect benzene in expired air by vapor phase chromatography with a minimum sensitivity of 0.1 ppm. However, the urinary diagnostic tests have been adopted by both governmental and industrial personnel as the most reliable and convenient bioindicator. Yet, to understand how a specific and sensitive bioindicator for benzene can be developed, it is necessary to understand the metabolic transformations of benzene.

Because of its lipid solubility, benzene tends to accumulate in adipose tissue, with this tissue acting as a reservoir, releasing the benzene very slowly (Ariens et al., 1976; Browning, 1965). The metabolism of benzene within the body involves its biotransformation to phenolic derivatives, excreted as ethereal sulfates or glucuronic acid conjugates or phenylmercapturic acid derivatives (Parke, 1968; Gerarde, 1966; Browning, 1965; Fishman, 1961; Yant et al., 1936; Laham, 1970).

In rabbits orally exposed to benzene, approximately 45% of benzene was eliminated via expired air in 2 days (43% benzene and 1.5% CO_2), and 35% was eliminated via the urine as metabolites (23% conjugated phenol, 4.8 quinol, 2.2% catechol, 0.3% hydroxyquinol, 0.5% phenylmercapturic acid, and 1.3% *trans*-muconic acid) (Figure 8-4). Thus respiratory and urinary excretion accounted for nearly 80%, with the remainder in the adipose tissue being slowly excreted (Laham, 1970; Browning, 1965; Fishman, 1961; Walkley et al., 1961). Most of these metabolic pathways are functional in humans, with about 40% of the absorbed benzene also being excreted in the urine as phenol.

There has been considerable discussion concerning the choice of a benzene bioindicator; should it be urinary phenol or the urine sulfate ratio? The normal urinary inorganic to organic sulfate ratio is approximately 85:15. Exposure to benzene decreases this ratio by increasing the organic sulfate levels relative to inorganic levels. Ratios of less than 70:30 are considered abnormal and may be indicative of benzene poisoning. Ratios of less than 60:40 indicate dangerous exposure (Djerassi and Lumbroso, 1958; Gerarde, 1966, Walkley et al., 1961; Yant et al., 1936).

Teisinger and Fiserova-Bergerova and medical personnel from the Bethlehem Steel Corp reported that the determination of urinary phenol is a better indicator of benzene exposure than urine sulfate ratios (Utidjian, 1976). The urinary sulfate ratio is considered a poor index of benzene exposure because concentrations as high as 40 ppm are needed to effect a change in the sulfate ratio. In contrast, benzene air levels of 25 ppm have been reported to cause a urinary phenol level of 170–225 mg/liter in exposed individuals (Utidjian, 1976; Walkley et al., 1961; Doctor and Zeilhuis, 1967) compared to the normal phenol levels of 5–42 mg/liter in nonexposed people (Doctor and Zielhuis, 1967; Deichmann and Schafer, 1942; Walkley et al., 1961). An exposure of 10 ppm benzene has been found to produce an intermediate urine level of 70–80 mg/liter.

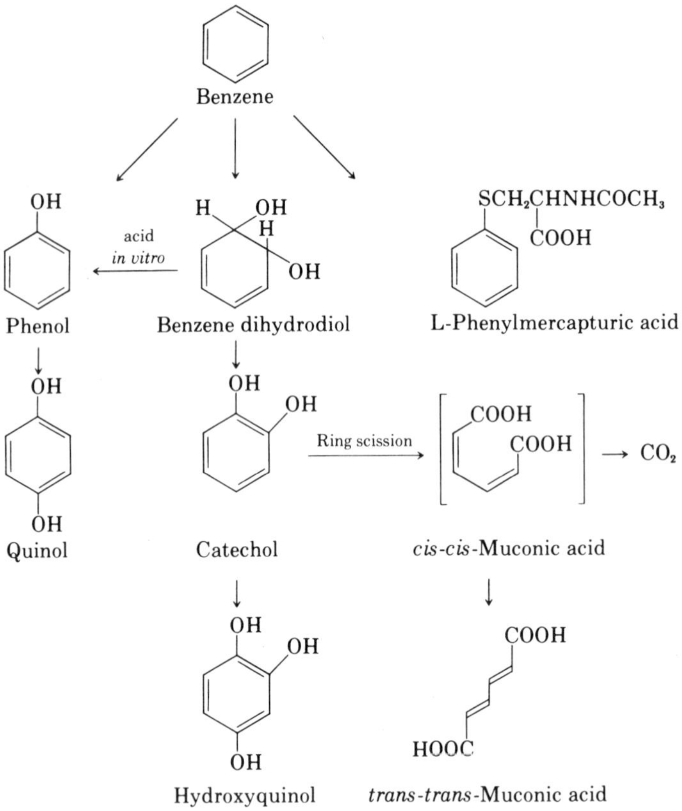

Figure 8-4. Metabolic transformations of benzene. Source: Parke, D. V. (1968) The Biochemistry of Foreign Compounds, p. 216. New York: Pergamon Press.

NIOSH has recommended to OSHA that occupational exposure to benzene should not be greater than 10 ppm based on a TWA for a 10-hour day and 40-hour work week; a ceiling value of 25 ppm has also been established. However, if the workers are exposed to benzene levels above one-half the recommended standard (i.e. the action level), a program of biomonitoring must be adopted that would involve the determination of total urinary phenol content in an effort to help prevent the absorption of "unacceptable" levels of benzene. The value designated as unacceptable with respect to benzene is $\geq$75 mg phenol per liter of urine (with the urine specific gravity corrected to 1.024). The biologic monitoring requirement of the standard involves a quarterly urine sample for those "exposed to benzene" (i.e. exceeding the action level). If the levels of benzene, as detected by environmental monitoring, are equal to or greater than the standard, the frequency of biologic monitoring is speeded up so that a urinary phenol analysis is conducted every 2 weeks. This rate of biomonitoring is to be maintained for at least 2 months after the high environmental exposure was determined (NIOSH, 1974).

Furthermore, if the urinary phenol level of a worker equals or exceeds the 75-mg concentration, two follow-up urine samples (one at the start and one at the finish of the work day) must be determined within 1 week after first recognizing the elevated phenol level. If these elevated levels are maintained, various efforts must be promptly made to reduce absorption. These efforts may include improvement in environmental controls, personnel protection, personal hygiene or administrative control. The heavy reliance of urinary phenol levels in the biomonitoring of benzene exposure as required by the NIOSH recommendation was criticized by Fishbeck et al. (1975), who indicated that nonprescription medication such as peptobismol and chloraseptic lozenges cause a marked elevation in urinary phenol levels. Consequently, the test for phenol in urine can be reasonably criticized for a lack of specificity with respect to the pollutant. Such confounding factors must be considered by industrial hygiene personnel, especially in the retesting of individuals who show high phenol levels.

The role of environmental and biologic monitoring with respect to NIOSH recommendations can be seen as a complementary process. The primary concern is given to environmental monitoring. In fact, if action level concentrations are not reached, biomonitoring is not implemented.

REFERENCES

Ariens, E. J., Simonis, A. M., and Offermeier, J. O. (1976) *Introduction to General Toxicology*, New York: Academic.

Baloh, R. W. (1974) Laboratory diagnosis of increased lead absorption. *Arch Environ Health* 28:198–208.

Barylko-Pikielna, N. and Pangborn, R. M. (1968) Effect of cigarette smoking on urinary and salivary thiocyanates. *Arch Environ Health* 17:739–45.

Benson, G. I., George, W. H. S., Litchfield, M. H., and Seaborn, D. J. (1976) Biochemical changes during the initial stages of industrial lead exposure. *Br J Indust Medl* 33:29–35.

Blejer, H. P. (1976) Inorganic lead: biological indices of absorption-biological threshold limit values. In Bertram Carnow, ed., *Health Effects of Occupational Lead and Arsenic Exposure—A Symposium*, pp. 165–178. Rockville, Md: USHEW.

Browning, E. (1965) *Toxicity and Metabolism of Industrial Solvents.* New York: Elsevier.

Cameron, D. (1976) Discussion comment. *Health Effects on Occupational Lead and Arsenic Exposure: A Symposium.* USHEW NIOSH Publication Number (NIOSH) 76-134, p. 200.

Carnow, B. W. (1976) Discussion of TLV's for lead. *Health Effects of Occupational Lead and Arsenic Exposure: A Symposium.* USHEW. NIOSH Pub. No. 76-134, pp. 186–190.

Chisholm, J. J., Jr. (1971) Lead poisoning *Scientific American* 224:15–33.

Cleary, T. F. (1976) Panel discussion of TLV's for lead. *Health Effects of Occupational Lead and Arsenic Exposure: A Symposium.* USHEW. NIOSH Pub. No. 76-134, pp. 162–164.

Cole, J. F. (1976) Occupational health standard for lead. In Bertram W. Carnow, ed., *Health Effects of Occupational Lead and Arsenic Exposure: A Symposium,* pp. 179–185. Rockville, Md.: USHEW.

Cornish, H. H., Barth, M. L., and Dodson, V. N. (1970) Isozyme profiles and protein patterns in specific organ damage. *Toxicol Appl Pharmacol* 16:411–23.

Deichmann, W. B., Kitzmiller, K. V., and Witherup, B. S. (1944) Phenol studies. VII. Chronic phenol poisoning, with special reference to the effects upon experimental animals of the inhalation of phenol vapor. *Am J Clin Pathol* 14:273–277.

Deichmann, W. and Schafer, L. J. (1942) Phenol studies. I. Review of the literature. II. Quantitative spectrophotometric estimation of free and conjugated phenol in tissues and fluids. III. Phenol content of normal human tissues and fluids. *Am J Clin Pathol* 12:129–143.

Denson, P. M., Davidow, B., Bass, H. E., and Jones, E. W. (1967) A chemical test for smoking exposure. *Arch Environ Health* 14:865–74.

Divincenzo, G. D. and Krasavage, W. J. (1974) Serum ornithine carbamyl transferase as a liver response test for exposure to organic solvents. *Am Indust Hyg Assoc J* 35: 21–29.

Djerassi, L. S. and Lumbroso, R. (1968) Carbon disulphide poisoning with increased ethereal sulphate excretion. *Br J Indust Med* 25:220–22.

Doctor, H. J. and Zielhuis, R. L. (1967) Phenol excretion as a measure of benzene exposure. *Ann Occup Hyg* 10:317–326.

Drotman, R. B. (1975) A study of kinetic parameters for the use of serum ornithine carbamyl transferase as an index of liver damage. *Food Cosmet Toxicol* 13:649–51.

Eckardt, R. (1976) Discussion comment. *Health Effects of Occupational Exposure to Inorganic Lead and Arsenic.* USHEW, NIOSH Pub. No. 74-134, p. 197.

Elkins, H. B. and Pagnotto, L. D. (1965) Is the 24-hour urine sample a fallacy? *Am Indust Hyg Assoc J* 26:456–60.

Elkins, H. B., Pagnotto, L. D., and Smith, H. I. (1974) Concentration adjustments in urinalysis. *Am Indust Hyg Assoc J* 35:559–65.

Environmental Health Resource Center (EHRC) (1974) Health effects and recommended standard for vanadium. Illinois Institute for Environmental Quality, Chicago, Ill.

Epstein, S. (1976) Discussion comments. *Health Effects of Occupational Exposure to Inorganic Lead and Arsenic.* USHEW, NIOSH Pub. No. 74–134, pp. 105–106.

Fishbeck, W. A., Langner, R. R., and Kociba, R. J. (November 1975) Elevated urinary phenol levels not related to benzene exposure. *Am Ind Hyg Assoc* 36:820–824.

Fishman, W. H. (1961) *Chemistry of Drug Metabolism,* Springfield, Ill.: Charles C. Thomas.

Frajola, W., Meyer-Arendt, T. E., and Waltz, J. (1960) Serum enzymes and biochemical individuality. *Fed Proc* 19:46.

Gerarde, H. W. (1966) The aromatic hydrocarbons. In David W. Fassett and Don D. Irish, eds., *Industrial Hygiene and Toxicology, Volume II, Toxicology,* pp. 1219-1240. New York: Interscience Publishers.

Goldstein, D. H., Kneip, T. J., Rulon, V. P., and Cohen, N. (1975) Erthrocytic amino-levulinic acid dehydratase (ALAD) activity as a biologic parameter for determining exposures to lead. *J Occup Med* 17:157-62.

Goldwater, L. J. (1968) Toxicology. In N. Irving Sax, ed., *Dangerous Properties of Industrial Materials,* pp. 1-29. New York: Van Nostrand Reinhold.

Grice, H. C., Barth, M. L., Cornish, H. H., Foster, G. V., and Gray R. H. (1971) Correlation between serum enzymes, isozyme patterns and histologically detectable organ damage. *Food Cosmet Toxicol* 9:847-55.

Group of 18 (Land, R. E., Hunter, D., Malcolm, D., Williams, M. K., Hudson, T. G. F., Browne, R. C., McCallum, R. I., Thompson, A. R., deKretser, A. J., Zeilhuis, R. L., Cramer, K., Barry, P. S. I., Goldberg, A., Beritic, T., Vigliani, E. C., Truhaut, R., Kehoe, R. A., and King, E.) (1968) Diagnosis of inorganic lead poisoning: a statement. *Br Med J* 4:501.

Hardy, H. L. and Elkins, H. B. (1948) Medical aspects of maximum allowable concentrations: benzene. *J. Indust Hyg Toxicol* 30(3):196-200.

Joselow, M. M. (1976) Biological monitoring—problems of blood lead levels. In *National Conference on the Health Effects of Occupational lead and arsenic exposure: A Symposium.* USHEW USPHS NIOSH pp. 27-33.

Joselow, M. M., Ruiz, R., and Goldwater, L. J. (1969) The use of salivary parotid fluid in biochemical monitoring. *Am Indust Hyg Assoc J* 30:77-82.

Kehoe, R. A. (November 14, 1947) Safe occupational lead exposure. Presented before the 23rd Annual Meeting of the Association of American Battery Mfrs., Inc.

Kehoe, R. A. (1963) Industrial lead poisoning. In Frank A. Patty, ed., *Industrial Hygiene and Toxicology,* 2nd revised ed., vol. II. Interscience Publishers.

Laham, S. (1970) Metabolism of industrial solvents. 1. The biological biotransformation of benzene and benzene substitutes. *Occup Health Rev* 21:24-28.

Lead Industries Association, Inc. (November, 1972) Industrial inorganic lead poisoning, A Program for prevention. Position paper.

Lerner, S. (1976) Discussion comments. *Health Effects of Occupational Exposure to Inorganic Lead and Arsenic.* USHEW, NIOSH Pub. No. 76-134, p. 201.

Mountain, J. T., Stockell, F. R., and Stokinger, H. E. (1955) Studies in vanadium toxicology. III. Fingernail cystine as an early indicator of metabolic change in vanadium workers. *Arch Indust Health* 12:494.

Nelson, K. W. (1973) The place of biological measurements in standard setting concepts. *J Occup Med* 15:439-40.

NIOSH (1972) Criteria for a recommended standard for occupational exposure to inorganic lead. USHEW.

NIOSH (1974) Criteria for a recommended standard for occupational exposure to benzene. USHEW.

Ornosky, M. (1968) Coproporphyrinuria and urine-lead findings: fifteen years of experience. *Am Indust Hyg Assoc J* 29:228-32.

Parke, D. V. (1968) *The Biochemistry of Foreign Compounds.* New York: Pergamon Press.

Stewart, R. D., Dodd, H. C., Erley, D., and Holder, B. B. (1965) Diagnosis of solvent poisoning. *JAMA* 193:1097–1100.

Stokinger, H. E. (1962) New concepts and future trends in toxicology. *Am Indust Hyg Assoc J* 23:8–19.

Stokinger, H. E. (1972) Concepts of thresholds in standards. *Arch Environ Health* 25:-153–57.

Taylor, K. J. W. (1977) Defects of liver and spleen pathology after vinyl chloride exposure—potential of grey-scale ultrasound and doppler techniques. In *Proceedings—Toxic Substances in the Air Environment Specialty Conference* held Nov., 1976, sponsored by New England Section of the Air Pollution Control Association.

Teisinger, J. and Srbora, J. (1955) Elimination of benzoic acid with the urine and its relation to the maximum tolerable concentration in the air. *Arch Med Prof* 16:216–220 (French).

Tola, S., Hernberg, S., Asp, S., and Nikkanen, J. (1973) Parameters indicative of absorption and biological effect in new lead exposure: a prospective study. *Br J Indust Med* 30:134–41.

Tsuchiya, K. and Harashima, S. (1965) Lead exposure and the derivation of maximum allowable concentrations and threshold limits values. *Brit. J. Indus. Med.* 22:181–186.

Utidjian, J. M. (1976) Criteria documents. I. Recommendations for a benzene standard. *J Occup Med* 18(7):499–504.

Vitalli, L. F., Joselow, M. M., Wedeen, R. D., and Pawlow, M. (1975) Blood lead—an inadequate measure of occupational exposure. *J Occup Med* 17:155–56.

Walkley, J. E., Pagnotto, L. D., and Elkins, H. B. (1961) The measurement of phenol in urine as an index of benzene exposure. *Am Indust Hyg Assoc J* 22:362–67.

Williams, D. M. J., Smith, P. M., Taylor, K. J. W., Crossley, I. R., and Duck, B. W. (1976) Monitoring liver disorders in vinyl chloride monomer workers using grey-scale ultrasonography. *Br J Indust Med* 33:152–57.

Williams, R. (1956) *Biochemical Individuality,* New York: Wiley.

Williams, R. T. (1963) Metabolic fate of foreign compounds of toxicity. *Arch Environ Health* 7:612–620.

Williams, M. K., King, E., and Walford, J. (1969) An investigation of lead absorption in an electric accumulator factory with the use of personnel samplers. *Br. J Indust Med* 26:202–216.

Wolf, M. A., Rowe, V. K., McCollister, D. D., Hollingsworth, R. L., and Oyen, F. (1956) Toxicological studies of certain alkylated benzenes and benzene. *Arch Ind Health* 14:387–398.

Yant, W. P., Schrenk, H. H., Sayers, R. R., Horvath, A. A., and Reinhart, W. H. (1936) Urine sulphate determination as a measure of benzene exposure. *J Indust Hyg Toxicol.* 18(1):69–90.

Zielhuis, R. L. (1971) Interrelationship of biochemical responses to the absorption of inorganic lead. *Arch Environ Health* 23:299–311.

9 Novel Work Schedule TLVs

IN RECENT YEARS, NUMEROUS industries have instituted novel work schedules for a variety of reasons, among them greater energy conservation and more efficient use of manpower. These novel schedules may include such variations as four 10-hour workdays per week, a 6-week cycle of three 12-hour workdays for 3 weeks followed by four 12-hour workdays for 3 weeks, and numerous other variations. In light of these possible work schedules, several recent papers have proposed that new occupational exposure limits should be developed for such novel work schedules (Brief and Scala, 1975; Mason and Dershin, 1976). The new exposure limit, or modified TLV, must consider not only the increased hours of exposure per day, but also the decreased hours of recovery (Brief and Scala, 1975). Accordingly, Brief and Scala (1975) devised a mathematical formula* for the derivation of a TLV reduction factor that can be applied to such novel situations.

To develop a meaningful dialogue concerning this limited yet important area of industrial hygiene, the original concept by Brief and Scala (1975) will be extended by providing some necessary supplements to the general formula approach (GFA).

* Reduction of TLV = effect of increased exposure time multiplied by effect of decreased exposure-free time: TLV reduction factor (RF) = $8/h \times [(24 - h)/16]$; h = hours worked per day, (RF) (TLV) = novel work schedule TLV.

Note: This chapter was previously published within the American Industrial Hygiene Association Journal 38:443–446, 1978 (E. J. Calabrese) and is reprinted with permission of the publisher.

The GFA is highly appealing because it places nearly all pollutants on a similar basis; thus reduced TLVs can be quickly and easily determined. In light of the approximately 600 substances with TLVs, this is highly desirable. However, it is important to realize that our primary concern is not with efficiency and simplicity, as desirable as they may be, but with the maintenance and enhancement of human health. Before the GFA becomes a widely accepted protocol for the determination of novel work schedule TLVs, several relevant factors must be considered.

LIMITATIONS OF THE GENERAL APPROACH

The GFA does not sufficiently discriminate among individuals, pollutants, and existing TLVs, yet it is known that the differential susceptibility of adults to stressor agents, the toxicologic activities of these stressor agents, and the degree of safety built into TLVs are highly variable. The remainder of this chapter discusses these three limitations of the GFA and how they may be dealt with in the development of more appropriate TLVs for novel work schedules.

Differential Susceptibility

High-Risk Groups

The GFA does not take into account individuals who may be at higher risk with respect to the toxicologic activity of the pollutant in question. The concept of differential susceptibility provides the fundamental basis for explaining the phenomenon of human high-risk groups. People may be at high risk to the toxic or carcinogenic effects of a pollutant for a variety of reasons, including developmental/aging processes, genetic factors, nutritional inadequacies, disease conditions, or behavioral activities (Calabrese, 1978).

An example of such enhanced hypersusceptibility has been reported with respect to the exposure of individuals with a deficiency of the enzyme glucose-6-phosphate dehydrogenase (G-6-PD) to cyanogenic compounds such as aniline and nitrobenzene (Linch, 1974). The data of Linch indicate that the incidence of cyanosis among those with the G-6-PD deficiency was greater than three times that of the general worker population. The enhanced susceptibility of G-6-PD-deficient individuals to the oxidant stress of aniline and nitrobenzene-like compounds is in remarkable agreement with previous research (Beutler,

1957; Zinkham et al., 1958) which indicated that human G-6-PD-deficient red blood cells are approximately four times more susceptible to oxidative stress (as measured by the GSH stability test). These results, in part, served as the basis for a recent theoretical study by Calabrese et al. (1977) that indicated enhanced susceptibility of G-6-PD-deficient individuals to ambient ozone at concentrations significantly lower than would be expected to adversely affect the general population.

Based on the considerable volume of information concerning high-risk groups (Calabrese, 1978; Linch, 1974; Calabrese et al., 1977), it seems highly questionable to assume that an individual at increased risk to a given pollutant will respond similarly to a normal individual when both switch to novel work schedules involving unusually long workdays (e.g., up to 12 hours/day) in light of the inherently diminished adaptive capacity of the hypersusceptible worker. Thus, before developing modified TLVs for novel work schedules, a characterization of the high-risk groups with respect to the pollutant(s) considered should be undertaken. This is necessary because these are the individuals who will be affected first and most adversely by either the increased exposure or lack of recovery period or both.

Circadian Rhythms

The enhanced susceptibility of individuals to environmental stressors as a result of disruptions of normal circadian rhythms and the adoption of rotation schedules should be considered in the derivation of novel work schedule TLVs. Circadian rhythms have been reported in cell growth, hormone levels, and body temperature (Luce, 1970), as well as in susceptibility to various stressor agents, including bacteria (Halberg et al., 1960), noise (Halberg et al., 1955, 1958), alcohol (Hans and Halberg, 1959; Hans et al., 1959), and the carcinogen benzo(a)pyrene (Mottram, 1945). The disruption of normal social routine by shifting work hours is known to cause restlessness, nervousness, headache, gastrointestinal tract irregularities, fatigue, slower reaction time, and error proneness. According to Felton and Patterson (1971), when individuals change their regular 8-hour schedule and go abruptly into novel work hours, it takes time for their body rhythms to make the adjustment, and in some cases they may not adjust at all. Although research has not yet established definitive relationships linking phase shifting and disruption of circadian rhythms to increased susceptibility of humans to environmental stressors, this definitely appears to be an area of potential concern for the development of novel work schedule standards.

Pollutant Toxicities

Differential Toxic Responses

Another factor to consider is the type of toxic response (i.e., chronic toxicity, acute toxicity, narcosis, irritation, asphyxiation, hemolysis, cancer, allergic reactions, and others) the specific TLVs are designed to prevent (Smyth, 1956). Four specific examples (carcinogens, heavy metals, hemolytic chemicals, and irritants) are discussed briefly to illustrate the concept of differential toxic responses of pollutants and their relevance for novel work schedule TLVs.

How novel work schedules will affect the human response to carcinogens is, for the most part, unknown. It has been pointed out that the latent periods for the development of radiation and several chemical-induced tumors varied with the inverse cube root of the dose (Jones and Grendon, 1975). Thus, as the dose increases, the latent period decreases; conversely, as the dose decreases, the latent period increases. Since the data concerning carcinogens dealt with variable doses over extended time periods and not with daily variations in exposure, they are not readily applicable to evaluating the carcinogenic risk associated with novel work schedules. Yet it is obvious that the total dose over extended periods of time is an important criterion of the carcinogenic response. However, if novel work schedules only increased the daily but not weekly, monthly, and yearly exposure to carcinogens, would they be inherently more dangerous than the traditional 8-hour day? The only available information that relates to this problem, albeit indirectly, has indicated that short-term excursions of elevated exposures, despite a low TWA, markedly enhanced the carcinogenic response to asbestos (Enterline, 1976). Although this does not precisely focus on the concept of novel work schedules, it does indicate that carcinogenic responses may be affected by highly variable carcinogen exposure of a short-term nature. Such information infers that highly variable exposure of a daily nature, as may occur during novel work schedules, may also be of concern. Consequently, the adoption of novel work schedules without an appropriate reduction in the TLV would not take into account the possible, although presently unsubstantiated, enhanced carcinogenic potential of highly variable daily exposure to carcinogens.

The exposure of individuals to variable daily amounts of heavy metals such as lead or mercury as a result of novel work schedules should not cause any increase in the body burden that may result in enhanced susceptibility to chronic toxicity in such critical areas as the central nervous system and/or kidney as long as weekly exposures are not increased. This is due to the relatively long biologic half-life of these metals in the body.

For potentially hemolytic chemicals that are rapidly metabolized by the body, it is critical to control daily, as compared to weekly exposures so as not to exceed a potentially critical level. In these cases we are not concerned with a chronic response resulting from many months or years of accumulating an excessive body burden; however, the most important consideration is the prevention of a rather rapidly developing hemolytic response.

In light of the paucity of information regarding the responses of individuals, including potential high-risk groups, to irritant chemicals during novel work schedules, the suggestion by Brief and Scala (1975) that TLVs for irritants would probably not be reduced seems premature. In fact, of all types of chemicals, those with the smallest built-in safety factors are usually the chemical irritants (Smyth, 1956; Finklea et al., 1974). Furthermore, several of the well-known irritant oxidants (e.g., ozone and nitrogen dioxide) have been recently found to have systemic effects [e.g., red cell membrane peroxidation (Fletcher and Tappel, 1972; Chow et al., 1974) and induced chromosomal aberrations in human circulating lymphocytes (Merz et al., 1975)] that may be even more serious than the strictly irritant effects. These additional factors certainly preclude the previously noted suggestion.

As a result of novel work schedules, individuals should not be exposed to greater than normal pollutant levels based on either daily (e.g., carcinogens, hemolytic chemicals, and irritants) or weekly (heavy metals) levels, depending on the pollutant.

Differences in Metabolism

With substances that have a long biologic half-life (e.g., lead, mercury), the daily recovery phase should not be considered as strongly as compared to those with a short half-life. A recent study by Mason and Dershin (1976) recognized this inconsistency in the GFA. They proposed a new model for the derivation of novel work schedule TLVs that takes into account the kinetics of accumulation and metabolism of individual compounds and represents an improvement on the GFA. Thus it does not appear logical to consider the diminished recovery periods for all pollutants in a similar manner. Knowledge of the absorption, distribution, storage, site of action, metabolism, and excretion are necessary to ensure the derivation of appropriate novel schedule TLVs.

Inconsistent Safety Factors

It may be instructive to consider Table 2 of Smyth (1956), which summarizes the toxicologic basis on which the TLVs were based. Of particular interest are the columns of predicted effects at twice and ten times the TLV for an 8-hour day. In many cases there is no predicted toxic effect at even twice the TLV; in other cases highly toxic responses ensue. It appears that the margin of safety in many TLVs, therefore, is highly variable, being less than a factor of 2 in numerous cases and usually no greater than a factor of 10. It is important to realize that any "safety factor" built into an existing TLV may already be considerably reduced with respect to those persons at high risk, since TLVs are not designed to specifically protect the hypersusceptible worker. By reducing the TLV in a general and nondiscretionary manner (i.e., by a general formula methodology), it is evident that those at high risk are not specifically considered despite their enhanced susceptibility. It should be realized that the GFA is very similar to the toxicologist applying a "safety factor" of 10, 100, or 1000 for often vague and indeterminate reasons.

RECENT NIOSH POLICY

It should be noted that NIOSH addressed the novel work schedule TLVs in the development of criteria documents for different toxic substances. The initial criteria documents (e.g., inorganic lead and others) were based on the traditional exposure of 8 hours/day and 40 hours/week. However, the criteria document for inorganic lead (NIOSH, 1972) indicated that in cases in which the weekly lead exposure exceeded 40 hours, a proportionate reduction in average concentrations should occur. For example, to maintain a TWA of 0.15 mg/m^3, the average should be 0.15 mg/m^3 for a 40-hour week and 0.12 mg/m^3 for a 50-hour week. This, of course, is quite similar to the GFA with its inherent limitations. At present, the new recommended occupational standards of NIOSH are intended for a 10-hour/day exposure for a 40-hour week in marked contrast to the earlier 8 hour/day approach (NIOSH, 1976).

REFERENCES

Beutler, E. (1957) The glutathione instability of drug-sensitive red cells. A new method for the *in vitro* detection of drug sensitivity. *J Lab Clin Med* 49:84.

Brief, R. S. and Scala, R. A. (1975) Occupational exposure limits for novel work schedules. *Am Indust Hyg Assoc J* 36:467.

Calabrese, E. J. (1978) *Pollutants and High Risk Groups*. New York: Wiley.

Calabrese, E. J., Kojola, W. J., and Carnow, B. W. (1977) Ozone: A possible cause of hemolytic anemia in G-6-PD deficient individuals. *J Toxicol Environ Health* 2: 909–912.

Chow, C. K., Dillard, C. J., and Tappel, A. L. (1974) Glutathione peroxidase system and lysozyme in rats exposed to ozone or nitrogen dioxide. *Environ Res* 7:311.

Enterline, P. E. (1976) Pitfalls in epidemiological research. An examination of the asbestos literature. *J Occup Med* 18(3):150.

Felton, G. and Patterson, M. G. (1971) Shift rotation is against nature. *Am J Nursing* 71:4.

Finklea, J. F., Shy, C. M., Moran, S. B., Nelson, W. C., Larsen, R. I., and Akland, G. C. (1974) The role of environmental health assessment: The control of air pollution. Paper presented to the 67th Meeting of the American Institute of Chemical Engineers, Houston, Texas.

Fletcher, B. L. and Tappel, A. L. (1972) Protective effects of dietary α-tocopherol in rats exposed to toxic levels of ozone and nitrogen dioxide. *Environ Res* 6:165.

Halberg, F., Bittner, J. J., Gully, R. J., Albrecht, P. G., and Brackney, E. L. (1955) 24-hour periodicity and audiogenic convulsions in I-mice of various ages. *Proc Soc Exp Biol Med* 88:169.

Halberg, F., Jacobsen, E., Wadsworth, G., and Bittner, J. J. (1958) Audiogenic abnormality spectra, twenty-four hour periodicity and lighting. *Science* 128:657.

Halberg, F., Johnson, E. A., Brown, B. W., and Bittner, J. J. (1960) Susceptibility rhythm to *E. coli* endotoxin and bio-assay. *Proc Soc Exp Biol Med* 103:142.

Hans, E. and Halberg, F. (1959) 24-hour rhythm in susceptibility of C mice to a toxic dose of ethanol. *J Appl Physiol* 14:878.

Hans, E., Hanton, E. M., and Halberg, F. (1959) 24-hour susceptibility rhythm to ethanol in full fed, starved and thirsted mice and the lighting regimen. *Physiologist* 2:54.

Jones, H. B. and Grendon, A. (1975) Environmental factors in the origin of cancer and estimation of the possible hazard to man. *Food Cosmet Toxicol* 13:251.

Linch, A. L. (1974) Biological monitoring for industrial exposure to cyanogenic aromatic nitro and amino compounds. *Am Indust Hyg Assoc J* 35:426.

Luce, G. G. (1970) Biological rhythms in psychiatry and medicine. *USHEW USPHS Pub. No. 2088*, pp. 1–15.

Mason, J. W. and Dershin, H. (1976) Limits to occupational exposure in chemical environments under novel work schedules. *J Occup Med* 18(9):603.

Merz, T., Bender, M. A., Kerv, H. D., and Kuller, T. J. (1975) Observations of aberrations in chromosomes of lymphocytes from human subjects exposed to ozone at a concentration of 0.5 ppm for 6 and 10 hours. *Mutat Res* 31:299.

Mottram, J. L. (1945) A diurnal variation in the production of tumors. *J Pathol Bacteriol* 57:265.

NIOSH (1972) Criteria for a recommended standard: Occupational exposure in inorganic lead. *USHEW USPHS NIOSH.*

NIOSH (1976) Criteria for a recommended standard: Occupational exposure to nitric acid. *USHEW USPHS NIOSH.*

Smyth, H. F. (1956) Hygienic standards for daily inhalation. *Am Indust Hyg Assoc J* 17(2):129.

Zinkham, W. H., Lenhard, R. D., and Childs, B. (1958) A deficiency of glucose-6-phosphate dehydrogenase activity in erythrocytes from patients with favism. *Bull Johns Hopkins Hosp* 102:169.

10 National Ambient Air Quality Standards

AIR POLLUTION IN INDUSTRIALized countries has long been known as a serious aesthetic and health problem. In fact, as far back as 1273, unsuccessful attempts to prohibit the burning of coal in London were made. In an interesting chapter on the history of air pollution, Stern et al. (1973) quoted Sir Hugh Beaver (1955) concerning the early approaches to controlling the air pollution problem in Great Britain.

> By 1819, there was sufficient pressure for Parliament to appoint the first of a whole dynasty of committees "to consider how persons using steam engines and furnaces could work them in a manner less prejudiced to public health and comfort." This committee confirmed the practicality of smoke prevention, as so many succeeding committees were to do, but as was often again to be experienced, nothing was done.
>
> In 1843, there was another Parliamentary Select Committee, and in 1845, a third. In that same year, during the height of the great railway boom, an act of Parliament disposed of trouble from locomotives once and for all (!) by laying down the dictum that they must consume their own smoke. The Town Improvement Clauses Act two years later applied the

Table 10-1. Acute Air Pollution Episodes the World Has Experienced

Place	Time	Effect and Number	Conditions and Probable Cause
Meuse Valley, Belgium; coke ovens, blast furnaces, steel, glass, zinc, and sulfuric acid plants	Dec. 1–6, 1930	60 deaths, "thousands ill," coughing, breathlessness, chest pain, eye and nose irritation experienced.	Inversion, stagnation in 15-mile river valley for 1 week; smoke and irritant gases; sulfur oxide, sulfuric acid mist, and fluorides suspected; estimated sulfur dioxide 25–100 mg/cu m (10–40 ppm).
Donora, Pa., U.S.; zinc smelter, wire coating mill, steel mill, sulfuric acid plant	Oct. 27–31, 1948	6000 of 14,000 population ill, 1400 sought medical care, 17 died; coughing, sore throat, chest constriction, burning and tearing eyes, vomiting, nausea, excessive nasal discharge.	Temperature inversion and fog along horseshoe-shaped valley of Monongahela River; sulfur oxides, smoke, and zinc compound particulates present; sulfuric acid mists likely; estimated sulfur dioxide of 1.5–5.5 mg/cu m (0.5–2 ppm).
Poza Rica, Mexico; petrochemical plant, hydrogen sulfide recovery system	4:45–5:10 a.m., Nov. 24, 1950	22 deaths; 320 hospitalized, acute hydrogen sulfide poisoning, unconsciousness, vertigo, severe irritation of respiratory tract, loss of sense of smell.	Low inversion layer, fog, weak winds; hydrogen sulfide released when burner on 4-day-old sulfur recovery plant failed under increased hydrogen sulfide flow rate; release lasted for only 25 min.

London, England	Dec. 5–9, 1952	3500–4000 deaths in week of Dec. 5–12 in excess of expected norm of like weeks; causes of death: chronic bronchitis, bronchopneumonia, and heart disease; increased hospital admissions for respiratory and heart disease.	"Pea soup" fog and temperature inversion covered most of the U.K.; smoke and sulfur dioxide accumulations in stagnated air; reported smoke highs of 4.5 mg/cu m and sulfur oxide highs of 3.75 mg/cu m (1.4 ppm).
London, England	Jan. 1956	1000 excess deaths charged to a pollution episode.	Extended fog conditions similar to 1952 episode; resulted in Parliament's passing Clean Air Act.
London, England	Dec. 5–7, 1962	700 excess deaths and increased illness charged to a pollution episode; emergency medical care plan functioned.	Severe fog and inversion; sulfur dioxide levels higher than in 1952, but particulates were lower; alert system operated.

From Chanlet, E. T. (1973) *Environmental Protection*, p. 236. New York: McGraw-Hill. Used with permission of McGraw-Hill Book Company.

> same panacea to factory furnaces. Then 1853 and 1856 witnessed two acts of Parliament dealing specifically with London and empowering the police to enforce provisions against smoke from furnaces, public baths, and washhouses and furnaces used in the working of steam vessels on the Thames.

However, the problem of air pollution was not to be solved in London by this series of Parliamentary acts. In fact, for Great Britain and the other industrial countries of the world, the situation would "get worse before it got better."

Table 10-1 lists a series of air pollution episodes throughout the world that focused the public's attention on the health implications of airborne pollution. As a result of the air pollution episode in London in 1952, the British Clean Air Act of 1956 was passed. The first air pollution legislation passed in the United States was a California law in 1947. It should be noted that the infamous Los Angeles smog preceeded the 1947 California law by several years (Stern et al., 1973). The first national ambient air standards were adopted in 1951 by the Soviet Union (Izmerov, 1973).

The first federal air pollution legislation in the United States was adopted in 1955. This legislation granted the federal government the authority to conduct research, training, and technical assistance programs. The concept of national ambient air standards was still a long way off (Stern et al., 1973).

Since the passage of the original legislation in 1955, Congress has passed and the President has signed into law a variety of bills: the Clean Air Act of 1963, amended in 1965, 1966, 1967 (with the 1967 amendments generally known as the Air Quality Act), and 1970 (The Clean Air Amendments of 1970). The 1970 Amendments required the EPA to develop and promulgate national ambient air quality standards for pollutants for which health criteria had been issued. This represented a significant change from the 1967 Act, which provided that the states adopt their own air quality standards. To help achieve compliance with the national ambient standards, the EPA was also granted authority to establish emission standards for hazardous substances and performance standards.

On January 30, 1971 the EPA published its proposed ambient air quality standards in the Federal Register. Three months later (April 30, 1971) the final standards were promulgated (Federal Register 1971). Table 10-2 compares the proposed and the promulgated ambient air quality standards.

Table 10-2. US EPA National Ambient Air Quality Standards

Substance	Air Quality Standard, μg/cu m (ppm) Primary (human health)	Secondary (all other effects)
Sulfur dioxide		
Annual arithmetic mean	80 (0.03 ppm)	60 (0.02 ppm)
24-hr max[a]	365 (0.14 ppm)	260 (0.01) ppm)
3-hr max[a]	—	1300 (0.5 ppm)[b]
Particulate matter		
Annual geometric mean	75	60
24-hr max[a]	260	150
Carbon monoxide		
8-hr max[a]	10,000 (9 ppm)	—
1-hr max[a]	40,000 (35 ppm)	—
Oxidant		
1-hr max[a]	160 (0.08 ppm)[c]	—
Nitrogen oxides (e.g., NO_2)		
Annual arithmetic mean	100 (0.05 ppm)	—
Hydrocarbons		
3-hr max (6–9 a.m.)	160 (0.24 ppm)[d]	—

[a] Not to be exceeded more than once per year.
[b] The 3-hour maximum secondary SO_2 standard was not in the originally proposed set of standards as issued by the EPA on January 30, 1971. However, it was later adopted for the finally promulgated standards to minimize the effect of emissions from large sources during short periods of adverse weather conditions.
[c] The standard proposed on January 30, 1971 by the EPA was 125 μg/cu m (0.06 ppm). However, the standard was increased to 160 μg/cu m (0.08 ppm) because of evidence indicating that 125 μg/cu m may actually be the natural oxidant level in certain areas of the country.
[d] The standard proposed on January 30, 1971 was 125 μg/cu m (0.19 ppm). However, it was subsequently revised to achieve consistency with the related standards for photochemical oxidants.

This chapter critically considers the biomedical basis on which several of the national air quality standards (oxidants, nitrogen dioxide, sulfur dioxide, and carbon monoxide) have been derived.* The reader should see the Criteria Documents for SO_2, CO, oxidants, hydrocarbons, NO_2, and particulates to review the summarized data on

* On December 14, 1977 the EPA proposed a national ambient air quality standard for lead of 1.5 g/m^3, monthly average (see page 316 of Chapter 10).

which the EPA based their standards. These documents represent comprehensive reviews of the literature consisting of approximately 150–200 pages in most cases. In addition to reviewing the data in the 1971 Criteria Documents, research published after 1971 is also considered.

AMBIENT AIR QUALITY STANDARDS

A study of the derivation of ambient air standards offers several unique situations compared to industrial health and drinking water standards. For instance, the ambient air standards must be established to protect the public from 24-hour exposure of each day for the lifetime of the entire population, including the very young, the elderly, and all other potential high-risk groups. In contrast, industrial health standards apply to 8 hours/day, 40 hours/week exposure for a working lifetime. Further, industrial health standards apply only to those aged 18–65 years who are physically and psychologically capable of sustaining employment. With regard to drinking water standards, there is the obvious difference in the route of exposure and continuous versus noncontinuous exposure. These differences between ambient air pollutants and their standards and both industrial air and drinking water standards create several variations within the general scheme for the derivation of standards.

The difference in health status between the general public and the working population has been found to be quite substantial in many studies and has led to the development of the concept commonly referred to as "the healthy worker effect." This refers to the fact that in mortality studies of various industries, the standardized mortality ratios are usually much lower (during certain age ranges) compared to the general population. These data clearly imply that the general population is not as healthy (or adaptive) as those in the work force (see Gaffey, 1976; Goldsmith, 1975; Enterline, 1975; Redmond and Breslin, 1975).

PRIMARY AND SECONDARY STANDARDS

There were two types of general ambient air quality standards promulgated by the EPA in 1971. The first type, primary air standards, is designed to protect human health. The second type, secondary air standards, is intended to protect human welfare, including those things which humans value and do not want degraded or injured such as buildings, materials, plants, and so on. Table 10-2 lists the secondary air quality standards for sulfur dioxide and particulates. The remaining four pollutants (CO, O_3, HC, and NO_2) for which there are primary air standards do not have secondary standards because their principal adverse effects are thought to be with human health. The secondary standards for SO_2 and particulates are more strict than the primary standards, indicating that human health is not the most sensitive "valued possession" with respect to SO_2 and particulate toxicity. Stokinger (1972) suggested that there is a need for a secondary standard for photochemical oxidants, since very low levels of ozone (0.04 mg/m^3) are known to adversely affect tobacco leaves. Also, he pointed out that NO_2 is known to be corrosive to metal surfaces at concentrations much lower than the 0.1 mg/m^3 primary standard.

EPISODE STANDARDS

Another type of standard developed by the EPA is the episode control standard series. These series consist of standards called alert, warning, emergency, and significant harm levels (Table 10-3). They are designed to prevent the occurrence of air pollution episodes such as the ones in Donora, London, and so on. The significance of the episode standards is that there is a progressively more stringent set of implementation controls scheduled for activation at each specific episode standard violation. Table 10-4 is a schematic representation of a general type of episode implementation plan (Stern et al., 1973; JAPCA, 1974).

The prevention of the occurrence of air pollution episodes is a highly controversial component of any state clean air program. The disagreements do not really focus on the actual standards selected as much as the required implementation plans. The issue is obviously one of economic importance. For example, in hearings during 1975 and 1976 in Illinois concerning the state's ozone episode implementation strategy, the Illinois EPA proposed an implementation plan that would require closing all parking lots with a capacity greater than 200 automobiles when the ozone level reached 0.30 ppm. Other suggestions included closing O'Hare International Airport when the ozone concentration reached the emergency level (0.5 ppm).

SELECTED AMBIENT AIR STANDARDS

There has been considerable controversy concerning the legitimacy of each of the adopted ambient air quality standards. When one considers

Table 10-3. United States Alert, Warning, and Emergency Level Criteria[a]

- Alert level criteria
  - SO_2
    - 800 μg/cu m (0.3 ppm), 24-hr average
  - Particulate
    - 3.0 COH's or 375 μg/cu m, 24-hr average
  - SO_2 and particulate combined
    - Product of SO_2 ppm, 24-hr average, and COH's equal to 0.2; or product of SO_2 μg/cu m, 24-hr average and particulate μg/cu m, 24-hr average, equal to 65×10^3
  - CO
    - 17 mg/cu m (15 ppm), 8-hr average
  - Oxidant (O_3)
    - 200 μg/cu m (0.1 ppm), 1-hr average
  - NO_2
    - 1130 μg/cu m (0.6 ppm), 1-hr average; 282 μg/cu m (0.15 ppm), 24-hr average
- Warning level criteria
  - SO_2
    - 1600 μg/cu m (0.6 ppm), 24-hr average
  - Particulate
    - 5.0 COH's or 625 μg/cu m, 24-hr average
  - SO_2 and particulate combined
    - Product of SO_2 ppm, 24-hr average, and COH's equal to 0.8; or product of SO_2 μg/cu m, 24-hr average, and particulate μg/cu m, 24-hr average, equal to 261×10^3
  - CO
    - 34 mg/cu m (30 ppm), 8-hr average
  - Oxidant (O_3)
    - 800 μg/cu m (0.4 ppm), 1-hr average
  - NO_2
    - 2260 μg/cu m (1.2 ppm), 1-hr average; 565 μg/cu m (0.3 ppm), 24-hr average
- Emergency level criteria
  - SO_2
    - 2100 μg/cu m (0.8 ppm), 24-hr average
  - Particulate
    - 7.0 COH's or 875 μg/cu m, 24-hr average
  - SO_2 and particulate combined
    - Product of SO_2 ppm, 24-hr average, and COH's equal to 1.2; or product of SO_2 μg/cu m, 24-hr average, and particulate μg/cu m, 24-hr average, equal to 393×10^3
  - CO
    - 46 mg/cu m (40 ppm), 8-hr average

Table 10-3. (*Continued*)

Oxidant (O_3)
1200 μg/cu m (0.6 ppm), 1-hr average
NO_2
3000 μg/cu m (1.6 ppm), 1-hr average; 750 μg/cu m (0.4 ppm), 24-hr average

[a] Federal Register, Vol. 36, No. 206, October 23, 1971, 15593.

Source: Stern et al. (1973) *Fundamentals of Air Pollution,* p. 48. New York: Academic.

the economic, social, and health implications the air standards are known to affect, controversy surrounding their adoption and implementation is certainly expected. Some of the more critical papers that do not support certain of the EPA-adopted positions with respect to the ambient air standards include those of Heuss et al. (1971), Ross (1971), Grennard and Ross (1974), and Megonnell (1975).

Ozone

Among the most controversial of the ambient air standards is that of the photochemical oxidants (usually measured as O_3—ozone). There is little disagreement that ozone is a potent irritant and may adversely affect the respiratory system. In fact, the TLV for occupational exposure is 0.1 ppm for an 8-hour daily exposure. This TLV was designed to reduce, in large measure, irritation of the respiratory tract.

The originally proposed ambient air standard by EPA was 125 mg/m^3 (0.06 ppm). However, the question of whether this differed substantially from natural (uncontrollable?) levels was raised. Levels of ozone vary naturally from 0.01 ppm up to approximately 0.05–0.06 ppm, depending on a variety of atmospheric conditions. The formation of ozone is dependent, in part, on several chemical precursors (e.g., NO_x and certain hydrocarbons). Since there are natural sources of these chemical precursors, the strictest environmental controls over stationary and nonstationary sources of NO_x and hydrocarbons may not be able to effect the achievement of a 0.06 ppm standard. Thus the levels of ozone in the breatheable atmosphere already have a "head start" from natural sources. Consequently, if the standard was placed at 0.06

Table 10-4. Air pollution episode control scenario. (a) See Table 10-3. Source: Fundamentals of Air Pollution, p. 47. New York: Academic.

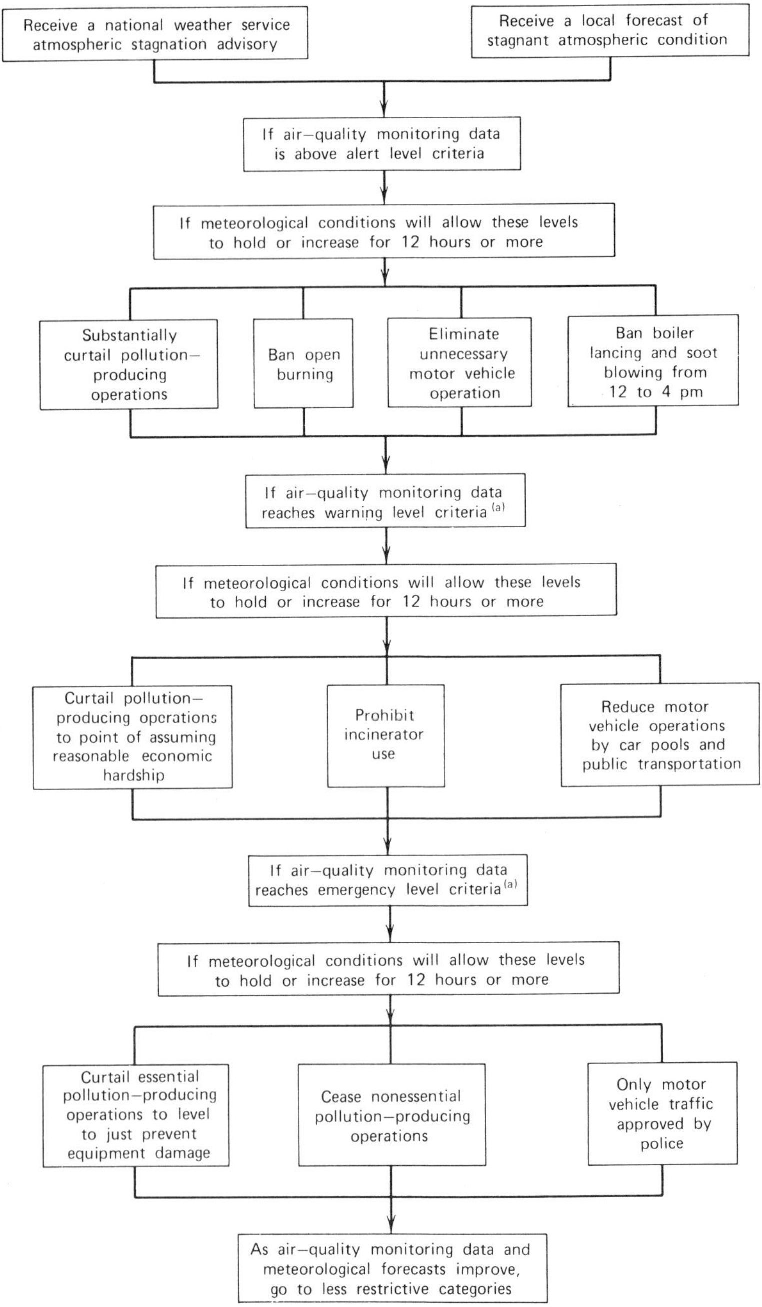
Receive a national weather service atmospheric stagnation advisory
Receive a local forecast of stagnant atmospheric condition
If air–quality monitoring data is above alert level criteria
If meteorological conditions will allow these levels to hold or increase for 12 hours or more
Substantially curtail pollution–producing operations
Ban open burning
Eliminate unnecessary motor vehicle operation
Ban boiler lancing and soot blowing from 12 to 4 pm
If air–quality monitoring data reaches warning level criteria (a)
If meteorological conditions will allow these levels to hold or increase for 12 hours or more
Curtail pollution–producing operations to point of assuming reasonable economic hardship
Prohibit incinerator use
Reduce motor vehicle operations by car pools and public transportation
If air–quality monitoring data reaches emergency level criteria (a)
If meteorological conditions will allow these levels to hold or increase for 12 hours or more
Curtail essential pollution–producing operations to level to just prevent equipment damage
Cease nonessential pollution–producing operations
Only motor vehicle traffic approved by police
As air–quality monitoring data and meteorological forecasts improve, go to less restrictive categories

ppm, many regions of the country would exceed the standard on a daily basis during the summer months. Furthermore, in recent years, the concept of rural oxidant transport has complicated the oxidant problem immeasurably. Ozone may be transported several hundred miles in air currents from highly polluted areas to minimally polluted locations. As a result of the problem of elevated natural levels of ozone, the EPA decided to increase the standard from 0.06 to 0.08 ppm. This was still thought to promote an adequate safety factor for most individuals. Table 10-5 presents a summary of the literature on ozone health effects according to a dose-response relationship.

It should be emphasized that the standard was based primarily on preventing respiratory irritation. However, it has been reported that

Table 10-5. Best Judgment Exposure Thresholds for Adverse Effects Due to Photochemical Oxidants (Short Term)

	Threshold	
Effect	μg/cu m	ppm
Aggravation of asthma	500	0.25
Aggravation of chronic lung disease	<500	<0.25
Aggravation of certain anemias	400–500	0.20–0.25
Irritation of eyes	200–300	0.10–0.15
Irritation of respiratory tract in otherwise healthy adults	500–600	0.25–0.30
Decreased cardiopulmonary reserve in healthy adults	240–740	0.12–0.51
Increased susceptibility to acute respiratory disease[a]	160[b]	0.08[b]
Risk of mutations[a]	400–600[b]	0.20–0.30[b]
Impaired fetal survival[a]	200–400[b]	0.10–0.20[b]
Decreased visual acuity[a]	400–1000	0.20–0.50
Present standard (1 hr)	160	0.08

[a] Involves exposures of 3–7 hr daily for up to 3 weeks.
[b] Based solely on animal studies.

From Finklea et al. (1977) The role of environmental health assessment in the control of air pollution. In J. N. Pitts and R. L. Metcalf, eds., *Advances in Environmental Sciences and Technology*, vol. 7, p. 340. New York: Wiley.

there is a natural adaptation to ozone toxicity (Stokinger, 1960, 1965). It may be argued that, unless people can adapt to ozone stress, how can one explain the nonepisodic responses of residents of Los Angeles to ozone levels that may occasionally exceed 0.6 ppm (greater than seven times the ambient air standard). Studies by Buckley et al. (1975) and Hackney et al. (1977) clearly support the concept of biochemical adaptive processes (e.g., increase in G-6-PD levels) during ozone exposure. In fact, in the Hackney et al. study, Canadians who are normally exposed to low levels of ozone exhibited greater respiratory stress at 0.37 ppm ozone (chamber study) than Southern Californians who normally live in a high ozone environment—suggesting an adaptive response. It may be argued that the standard should not be as low as it is, since individuals can adapt to the stress.

However, the argument against such reasoning is that there may be certain subsegments of the general public who lack the capability of adapting to the ozone stress for one or more of a variety of reasons. These people are referred to as high-risk groups and are discussed in Chapter 3. With respect to ozone, it has been reported that one of the adaptational changes is an increase in the activity of the pentose phosphate shunt, especially the activity of G-6-PD. If a person has a G-6-PD deficiency is it not possible that his adaptational capability may be diminished to a certain (as yet undefined) extent? This is the subject of a theoretical paper by Calabrese et al. (1977), who suggested that those with this enzyme deficiency may be at increased risk for the development of acute hemolytic anemia caused by elevated levels of ozone.

This leads directly into biomedical evidence of ozone toxicity which has been gathered subsequent to the Criteria Document for Photochemical Oxidants. Since 1969, a growing number of studies (Merz et al., 1976; Calabrese et al., 1977) suggests that ozone not only effects the respiratory tract, but also has systemic effects. The two most important of the potential systemic effects involve the influence of ozone on producing chromosomal aberrations in circulating lymphocytes in laboratory animals (0.2 ppm) and humans (0.5 ppm) and the possibility of inducing hemolytic responses. These relatively new biomedical effects could markedly influence the assessment of risk associated with ozone exposure and thus affect future revisions in the oxidant standard. Zelac et al. (1971) even suggested that the industrial health standard take the mutagenic effects of ozone into consideration since it is based solely on respiratory effects. At present, however, the effect of ozone on chromosome breaks in humans seems to occur at levels significantly higher than those which induce respiratory symptoms (approximately 0.5 ppm vs. 0.1 ppm). Consequently, a standard that prevents respiratory symptoms should also protect against mutagenic activities of ozone.

The derivation of the ozone standard presents a variety of interesting factors, especially with respect to episode standards. First, ozone produces multiple adverse health effects at different levels of exposure. Consequently, the adoption of air quality standards for various episodic stages (alert, warning, emergency, and significant harm) may be based on different adverse health effects (e.g., irritation vs. hemolysis vs. chromosomal breaks). Second, an unique group is at high risk for each adverse health effect.

The levels of ozone, as indicated earlier, are dependent in part on the occurrence of NO_x and hydrocarbon precursors. In fact, the ambient air standard for hydrocarbons is not based on their direct adverse health effects, but on their role in the formation of photochemical oxidants. A similar attempt toward standard setting for ambient manganese levels was adopted by the Environmental Health Resource Center of the Illinois Institute for Environmental Quality (1975). In the case of manganese, its recommended standard was based on minimizing the catalytic effect of manganese on the conversion of SO_2 to SO_3.

Nitrogen Dioxide

The ambient air quality standard for nitrogen dioxide is based on both animal and human toxicologic studies as well as epidemiologic research. The studies given the strongest consideration in the derivation process were the human epidemiologic studies. In a review of the basis of the nitrogen dioxide standard, particular emphasis is directed toward these epidemiologic studies (EPA, 1971).

The nitrogen dioxide standard was based, to a rather large extent, on the so-called "Chattanooga Study." This investigation considered the health implications of nitrogen dioxide from a large stationary source (i.e., a TNT plant). According to Barth et al. (1971), this represented a truly unique situation to consider the influence of nitrogen dioxide without many of the confounding covariables normally found in urban areas. In fact, these authors indicated that the opportunity to verify their results via replicate studies in other communities has not yet been possible.

The study consisted of testing the ventilatory function (over a 24-week period) of school children living in close proximity to various air monitors in four different areas near Chattanooga. Furthermore, the families of the children were questioned concerning their incidence of acute respiratory illnesses within the previous two weeks. Finally, a retrospective epidemiologic investigation concerning the incidence of acute lower respiratory tract infections of elementary school children over 3 years was carried out.

Heuss et al. (1971) were extremely critical of the accuracy of the NO_2 monitoring technique and the interpretation of the medical results of the Chattanooga study as presented by Shy et al. (1970, 1970a) and Pearlman et al. (1971). The first criticism involved the accuracy of the Jacobs-Hochheiser monitoring methods. Since this book is primarily concerned with the health effects of pollutants, we recommend that the reader refer to the works of Heuss et al. (1971) and Barth et al. (1971) and their references for an initial consideration of this controversial issue.

Heuss et al. (1971) also offered very strong criticism concerning some of the methodologic procedures employed during the analysis of the medical data from the Chattanooga study. For instance, in the high nitrogen dioxide area, three schools were used as sampling sites. However, one of the schools (school 3) had a mean nitrogen dioxide level of 0.062 ppm, which was nearly identical with control area 1 (0.063 ppm). If, during the data analysis, school 3 was included with the controls, the difference in ventilatory function between the treatment and control groups would have been diminished and, according to Heuss et al., may not have been statistically significant. Furthermore, although Shy et al. considered control 1 as a control group during the ventilatory function study, control 1 was considered as an intermediate group during the 3-year study of bronchitis among school children. According to Heuss et al., if Pearlman et al. (1971) used the same comparisons of Shy et al. (i.e., high NO_2 groups vs. controls), the variations between the study groups would not have been statistically significant. There are other challanges to the validity of the Chattanooga study presented in the paper by Heuss et al., which should be consulted by the interested reader. Based on their criticisms, Heuss et al. suggested that the conclusions of the Chattanooga study were more equivocal than presented by the authors. Finally, Heuss et al. suggested that the EPA standard for NO_2 (Table 10-2) be changed to 0.25 ppm for 1 hour until the Chattanooga study is validated.

Because of the controversial nature of the paper by Heuss et al., the Air Pollution Control Association Journal permitted a rebuttal of the criticism by representatives of the EPA (Barth et al., 1971). They suggested that the placement of school 3 in the high nitrogen dioxide group was justified on the basis of its high standard deviation even though its average value was nearly identical to control 1. Somewhat higher nitrate levels in the area of school 3 (3.8 $\mu g/m^3$) as compared to control 1 (2.8 $\mu g/m^3$) supported the view that school 3 belonged in the high nitrogen dioxide group. Furthermore, the research design of the study by Pearlman et al. was independent of the methodologic approaches of the studies by Shy et al. and should not be criticized *a priori*. Although the response of Barth et al. diminished the impact of some of the criticisms of Heuss et al., the need to validate this study emerges as increasingly important.

Barth et al. tried to diminish the importance of the Chattanooga study as the only basis for the nitrogen dioxide standard. They indicated that, in addition to the Chattanooga study, toxicologic studies have been conducted that support the present EPA standard. The reader is referred to two extensive literature reviews of the biologic effects of nitrogen dioxide with respect to air quality standards (Cooper and Tabershaw, 1966; Morrow, 1975).

Based on a comprehensive review of the literature, Morrow (1975) concluded that the lowest NO_2 level not causing adverse health effects in a variety of species is approximately 0.6 ppm (1200 $\mu g/m^3$). If one accepts this as a reliable figure reasonably applicable to humans, the question arises, what safety factor should be applied? If a safety factor of 10 is used, a value of 0.06 ppm would be derived, which is, of course, nearly identical to the standard adopted by the EPA.

Other criticisms of the Chattanooga study were presented by Warner and Stevens (1973) and Morrow (1975). Warner and Stevens noted that Shy et al. and Pearlman et al. did not adequately consider other air pollutants that would have been expected to be present in the atmosphere. Even though Shy et al. did consider the effects of total suspended particulates, Warner and Stevens (1973) indicated that sulfuric and nitric acid vapors (which were not considered) could have been serious contributors to whatever respiratory symptoms were reported. Consequently, these authors concluded that the EPA standard of 0.05 ppm was based on an inadequate study design and was more pessimistic than realistic.

In addition to the criticisms of Heuss et al. (1971) and Warner and Stevens (1973), the National Academy of Sciences (1973) indicated that the data on which the nitrogen dioxide standard is based are not as well demonstrated as the other ambient standards. Furthermore, the NAS stated that the Chattanooga study had required more adequate atmospheric analyses in agreement with the Warner and Stevens (1973) report.

Thus, in retrospect, an analysis of the data supporting the nitrogen dioxide standard is rather striking in the sense that the data concerning human epidemiologic studies are extremely minimal as compared to those for other pollutants such as ozone or sulfur dioxide. Perhaps one of the comforting situations that a decision maker faces in dealing with setting standards for pollutants such as sulfur dioxide, ozone, and nitrogen dioxide is that there is considerable past industrial experience. Thus the underlying feeling is that one cannot be too far off if an appropriate safety factor is included. Since the TLV for nitrogen dioxide is 5 ppm, the adoption of a 0.05 ppm ambient air standard represents a difference of a factor of 100 (realizing of course that the TLV is an 8-hour TWA, and the environmental standard is an annual arithmetic average).

Sulfur Dioxide—Particulates and Particulate Sulfates

Numerous toxicologic (animal and human) and epidemiologic studies have been directed toward trying to provide an understanding of the effects of sulfur dioxide on humans. The principal effect of sulfur dioxide on normal humans is upper respiratory tract irritation resulting in reflex bronchoconstriction and increased pulmonary flow resistance (Stokinger, 1972). Stokinger pointed out that most of the toxicologic studies with animals and humans were of limited value in the derivation of the primary air quality standard for sulfur dioxide because (1) most animal species employed in these studies were not considered as ideal models for extrapolation to human high-risk groups such as those with chronic obstructive pulmonary disease and cardiorespiratory difficulties; (2) human toxicologic studies are conducted with healthy volunteers—therefore, the relevance of these studies as accurate predictors of adverse health effects in susceptible segments of the population is limited.

As a result of the limitations inherent in the application of toxicologic data in the derivation of the sulfur dioxide standard, primary emphasis was placed on epidemiologic data of urban smog episodes in various cites in Europe (London, Rotterdam, and Ruhn) and the United States (Denora, Pennsylvania; New York; Chicago). The data that emerged from these studies included increased mortality, hospital admissions of older persons with respiratory diseases, and morbidity such as bronchitis. An important study published subsequent to the Criteria Document noted an increase of 10–20 deaths per day in New York City on days when sulfur dioxide concentrations increased from 0.2 ppm (0.57 mg/m^3) to 0.4 ppm (1.14 mg/m^3) or greater (Stokinger, 1972). Table 10-6 summarizes the nine most relevant epidemiologic studies with respect to developing the EPA-promulgated sulfur dioxide standard. These studies indicate that adverse health effects may be initiated at concentrations as low as 105 mg/m^3 (0.037 ppm) on an annual average and at 300–500 mg/m^3 (0.11–0.19 ppm) on a 24-hour average. According to Stokinger (1972), the primary standard of 80 mg/m^3 of an annual mean with a 24-hour maximum of 365 mg/m^3 represents a sensible standard that will afford protection to most high-risk segments of the population with chronic obstructive pulmonary disease and asthma.

Since the promulgation of the sulfur dioxide ambient air quality standard, there has been a growing consensus that sulfur dioxide may not be the culprit it had been thought. Particular emphasis has been directed toward the role of suspended sulfates as the causative agent in much of the respiratory irritation that was once thought to be principally related to sulfur dioxide exposure.

The investigations of the EPA, under the Community Health and Environmental Surveillance System (CHESS) (1974), attempted to distinguish and rank the relative contributions of sulfur dioxide, particulates, and suspended sulfates to a variety of adverse respiratory health effects in a number of sections of the United States. The results indicated consistent positive relationships only between the ambient levels of suspended sulfates and the daily aggravation of symptoms among individuals with general cardiopulmonary and asthmatic conditions. Delineation of the health effects of sulfur dioxide and particulates proved to be inconsistent and at times contradictory. The investigators concluded that "while adverse effects were occurring at daily concentrations below the National Daily Primary Air Quality Standard for sulfur dioxide and total suspended particulates . . . these effects . . . [should be attributed] to suspended sulfate concentrations on those days rather than to the [sulfur dioxide and particulate] pollutants." According to Battigelli and Gamble (1976), subjective adverse health effects noted in these CHESS studies could not be positively related to sulfur dioxide levels; consequently, the EPA "shifted the blame" to suspended sulfates.

Charles and Menzel (1975) and Menzel (1976) provided a theoretical explanation for the observations that sulfate salts rather than sulfur dioxide are the principal irritants. Charles and Menzel reported that ammonium sulfate exhibited extreme potency with respect to inducing the release of histamine stores of the lungs of guinea pigs. Subsequent *in vivo* studies in which rats were administered 1 mole of ammonium sulfate resulted in the release of most of the histamine stores of the lung along with a reduction in tidal volume. According to Menzel, these results implied that much of the bronchoconstriction caused by sulfate salts (see Amdur and Corn, 1963) results from the release of histamine.

Table 10-6. Epidemiologic Studies of SO_2 Toxicity Ranked in Importance As Reported in the Criteria Document of 1971

Sulfur Dioxide Concentration	Time Average	Particulate Concentration	Health Effects	Country	Reference
1500 μg/cu m (0.52 ppm)	24-hr	7 6 COH	Increased mortality	U.S.A.	McCarroll and Bradley, 1966
715 μg/cu m (0.25 ppm)	24-hr	Smoke 750 μg/cu m	Increased mortality	England	Lawther, 1963
500 μg/cu m (0.19 ppm)	24-hr	Low particulate	Increased mortality	Netherlands	Wilkins, 1954
300–500 μg/cu m (0.11–0.19 ppm)	24-hr	Low particulate	Increased hospital admissions of older persons with respiratory diseases	Netherlands	Brasser et al., 1967
715 μg/cu m (0.25 ppm)	24-hr	Particulates present—not precisely measured	Increase in illnesses in bronchitics older than 54 yr	U.S.A.	Carnow et al., 1968

600 μg/cu m (0.21 ppm)	24-hr	Smoke 300 μg/cu m	Increased symptoms in those with chronic lung disease	England	Lawther, 1958
105–265 μg/cu m (0.037–0.092 ppm)	Annual mean	Smoke 185 μg/cu m	Increased frequency of respiratory symptoms and lung disease	Italy	Petrilli et al., 1966
120 μg/cu m (0.046 ppm)	Annual mean	Smoke 160 μg/cu m	Increased frequency and severity of respiratory disease in school children	England	Lunn et al., 1967
115 μg/cu m (0.04 ppm)	Annual mean	Smoke 160 μg/cu m	Increased mortality	England	Wicken and Buck, 1964

Menzel concluded that these studies provide support for the epidemiologic findings that suspended sulfate rather than sulfur dioxide is the principal respiratory irritant.

It is interesting to note that the previously sited epidemiologic studies on which the EPA based the sulfur dioxide standard did not monitor suspended sulfates. Could it be possible that the adverse health effects attributed to sulfur dioxide are actually related to another pollutant (i.e., possibly suspended sulfates)? It is quite possible that the ambient levels of sulfur dioxide are reasonably good predictors of suspended sulfate levels. If this is the case, the sulfur dioxide standards may be doing their job—but for the wrong reason.

On a similar theoretical plane, Schimmel (1976) indicated that when sulfur dioxide is employed as an indicator of air pollution, one should not consider sulfur dioxide as the cause of the adverse effects because one is measuring the influence of all substances associated with sulfur dioxide levels. He noted that most investigations that have employed sulfur dioxide as the only indicator of air pollution erroneously concluded that sulfur dioxide is the cause of the described mortality and morbidity.

Since a number of epidemiologic studies reported the occurrence of increased mortality in association with elevated levels of sulfur dioxide, Schimmel and Murawski (1976) investigated whether the major reduction in sulfur dioxide levels, which had taken place since 1969, was associated with a concomitant decrease in adverse health effects (mortality).* This type of study offers a certain type of validation of the previous studies that were used in the derivation of the sulfur dioxide standard. It also provides for an evaluation of the assumed societal benefits inherent in implementation programs that have forced the adoption of low sulfur fuels.

The results of Schimmel and Murawski (1976) did not indicate any concomitant decrease in mortality as the sulfur dioxide level declined. Thus the hypothesis that a substantial decline in ambient sulfur dioxide levels should result in a decrease in mortality rates was not supported by the data. Schimmel and Murawski theorized that it was possible that the decline in sulfur dioxide levels was not sufficient to affect mortality patterns.

On the other hand, they felt that sulfur dioxide levels should be employed primarily as indicators of the presence of other pollutants that are known to occur along with sulfur dioxide. These other pollutants are most likely the principal causes of the increased

* Since the SO_2 standard was promulgated, the levels of sulfur dioxide in the ambient air of New York City have decreased by 85% (Schimmel and Murawski, 1976).

mortality. Thus the authors concluded that "the absolute levels of SO_2 do not appear, of themselves, to be significant." Such data seem consistent with mortality and morbidity studies in England that have reported a reduction of morbidity and mortality incidence with a marked decline in particulate levels but only slight changes in sulfur dioxide levels (Schimmel and Murawski, 1976).

The array of conflicting reports with respect to the toxicity of sulfur dioxide clearly calls into question the validity of the basis of the original standard. It also suggests that considerable effort must be directed toward a more precise delineation of the relationship of sulfur dioxide to particulates and suspended sulfates with respect to their occurrence and health effects. In light of this present controversy and our past course of action, we would do well to recall Stokinger's (1971) first commandment for sanity in research and evaluation of environmental health: "Standards must be based on scientific facts, realistically derived, and not on political feasibility, expedience, emotion of the moment, or unsupported information." Stokinger further stated, "In selecting criteria on which to base standards for the prevention of diseases caused by more than one factor, the temptation to ascribe an effect of questionable significance and doubtful connection to a pollutant merely because that pollutant was selected for investigation must not be yielded to. All too often in the past, this commandment has not been followed, particularly in those cases in which a disease entity was not sharply defined or when there was a paucity of data on it."

Carbon Monoxide

The development of the national ambient air quality standards for carbon monoxide was based exclusively on human exposure studies. Perhaps the unique aspect of developing a standard for carbon monoxide as compared to gases such as sulfur dioxide, nitrogen dioxide, and ozone is that carbon monoxide exposure can be measured in terms of percentage COHb formation, whereas the other gases are primarily irritants. Consequently, COHb is an excellent bioindicator of CO exposure, whereas gaseous irritants generally do not have good biologic indicators of exposure. The importance of COHb as an indicator of carbon monoxide exposure in the standard-setting process can be seen in the statement of the EPA in the Federal Register when the carbon monoxide standard was promulgated. The carbon monoxide standard is "intended to protect against the occurrence of COHb levels above 2 percent" (Federal Register, 1971).

The formation of COHb occurs because carbon monoxide has greater affinity for the hemoglobin molecule than does oxygen. More specifically, although oxygen combines somewhat more quickly with hemoglobin as compared to carbon monoxide, the strength of the bond between carbon monoxide and hemoglobin is approximately 200–300 times greater than that of oxygen (NIOSH, 1972). As the level of COHb increases, a decrease in the oxygen-carrying capacity of the blood develops. For this reason carbon monoxide has long been known as a dangerous asphyxiant, causing a deprivation of oxygen for physically active tissues, especially those of the heart and brain.

The primary ambient air standard for carbon monoxide is currently 35 ppm (40 mg/m^3) for a maximum 1-hour exposure or 9 ppm (10 mg/m^3) for a maximum 8-hour period. Each of these standards may be exceeded once per year. The Criteria Document for carbon monoxide (1971), from which these standards were derived, summarized the four most influential studies with respect to the carbon monoxide ambient standard.

1. Experimental exposure of nonsmokers to a concentration of 35 mg/m^3 (30 ppm) for 8–12 hours has shown that an equilibrium value of 5% COHb is approached in this time; about 80% of this equilibrium value (i.e., 4% COHb) is present after only 4 hours of exposure. These experimental data verify formulae used for estimating the equilibrium values of COHb after exposure to low concentrations of CO. These formulae indicate that continuous exposure of nonsmoking sedentary individuals to 23 mg/m^3 (20 ppm) results in a blood COHb level of about 3.7%, and an exposure to 12 mg/m^3 (10 ppm) results in a blood COHb level of about 2%. (From Criteria Document for carbon monoxide, 1971; data from McIlvaine et al., 1969).
2. Experimental exposure of nonsmokers to 58 mg/m^3 (50 ppm) for 90 minutes has been associated with impairment in time-interval discrimination. This exposure produces an increase of about 2% COHb in the blood. This same increase in blood COHb occurs with continuous exposure to 12–17 mg/m^3 (10–15 ppm) for 8 or more hours (From the Criteria Document for carbon monoxide, 1971; data from Beard and Wertheim, 1967).
3. Experimental exposure to carbon monoxide concentrations sufficient to produce blood COHb levels of about 5% (level producible by exposure to about 35 mg/m^3 for 8 or more hours) has provided evidence of impaired performance on certain other psychomotor tests and impairment in visual discrimination (From the Criteria Docu-

ment for carbon monoxide, 1971; data from Beard and Grandstaff, 1970).

4. Experimental exposure to carbon monoxide concentrations sufficient to bring blood COHb levels to about 5% (a level producible by exposure to 35 mg/m^3 or more for 8 or more hours) has produced evidence of physiologic stress in patients with heart disease. (From the Criteria Document for carbon monoxide, 1971; data from Ayres et al., 1969).

Based on these conclusions of the Criteria Document, the ambient standard for carbon monoxide was derived. The adverse health effects on which the standard was based centered on the impairment of psychomotor function in normal adults and the production of physiologic stress in patients with heart disease. The change in the psychomotor responses, however, was reported to be more sensitive to COHb levels than the stress phenomenon in the heart patients; that is, adverse behavioral changes were first recorded when COHb levels reached about 2% COHb, but it was not until COHb levels reached 5% that the heart patients experienced stress. Heuss et al. (1971) concluded that the carbon monoxide standard was strongly influenced by the research of Beard and Wertheim, which demonstrated impaired time-interval discrimination following 90 minutes of exposure to 50 ppm carbon monoxide and a COHb level of about 2%. These are the results on which the desire to prevent the COHb level from exceeding 2% are presumably based. Thus, if one can prevent the impairment of psychomotor activities, the heart patients should also be protected.

As a result of its singular importance, a closer look at the study by Beard and Wertheim (1967) is in order. This study attempted to test the hypothesis that low levels of carbon monoxide cause impairment in the ability to discriminate between short time intervals. The study involved the exposure of 18 young adults who were nonsmokers to carbon monoxide. While in an audiometer booth, these 18 subjects were presented a pair of pure tones, and each subject was to determine if the second tone was shorter, equal to, or longer than the first. There was minimal physical activity by the participants, and any questions were answered except whether they were being subjected to carbon monoxide. Blood samples were taken before and after each exposure, but accurate COHb levels were not obtained. However, based on the exposure levels, a COHb saturation of 2.5% was assumed. The tones were presented in blocks of 25 trials with a 13-minute rest period between blocks. Six hundred trials were run in each session, and each subject participated in at least 15 sessions. Carbon monoxide concentrations of 0, 50, 100, 175, and 250 ppm were used. The results indicated that 30 minutes after exposure, the ability to discriminate between tones decreased. Furthermore, as the concentration increased, the percentage of correct responses decreased in a linear fashion. One implication drawn from this study was that drivers exposed to relatively low levels of carbon monoxide could be predisposed to involvement in traffic accidents as a result of decreased ability to distinguish between time intervals.

The study of Beard and Wertheim (1967) was critized by Heuss et al. (1971) for its lack of quantitative determination of COHb levels and the single-blind* nature of the study. In their highly controversial article, Heuss et al. quoted Beard and Wertheim's conclusion that "we do not suggest the immediate application of these observations to the establishment of new air quality standards . . ." Later, in their article, Heuss et al. quoted comments of Beard at a 1970 conference on Environmental Toxicology. Beard indicated that subsequent attempts to duplicate the original study of Beard and Wertheim (1967) were not successful when a double-blind methodology was used (Stewart et al., 1970). In addition to Beard and Wertheim, Stewart et al. (1970) were not able to verify the original findings of Beard and Wertheim in double-blind studies. Thus Stewart et al. concluded that the capability of their participants to determine time intervals was not diminished by carbon monoxide exposures that resulted in COHb levels several times greater than those which were thought to be present in the Beard and Wertheim study.

In an attempted rebuttal of the paper by Heuss et al. (1971), Barth et al. (1971) mentioned that other studies (such as Horvath et al., 1971; McFarland et al., 1944; Schulte, 1963) showed carbon monoxide-induced impairments of vigilance, sensor, perceptual, and cognitive functions at low concentrations of carbon monoxide. However, it should be noted that these three studies did not indicate any impairment below 5% COHb.

Table 10-7 summarizes the effects of carbon monoxide exposure on humans with respect to time perception, psychomotor function, visual threshold, and miscellaneous cognitive functions. The results of the various researchers are highly controversial. For instance, whereas different reseachers (McFarland et al., 1944; Halperin et al., 1959; Schulte, 1963; Bender et al., 1971; Groll-Knapp et al., 1972; Fodor and Winneke, 1972; Horvath et al., 1971) noted impairments of the

* In the Beard and Wertheim study, only the investigators were aware of the carbon monoxide levels; in a double-blind study, neither the investigators nor the subjects are aware of the treatment for any specific study group. The use of the double-blind study design eliminates the potential of observer bias.

Table 10-7. Effects of Experimental Carbon Monoxide Exposure on Humans

CO Concentration (ppm)	Duration	Effects	Comments	Investigators
		Time Perception and Psychomotor		
0,50,100,175,250	4 hr	At estimated COHb 2.5% sound duration discrimination decrement noted. No subjective effects.	Single blind. 18 subjects. Each isolated in audiometric booth during test. First effect noted after 90 min exposure to 50 ppm; proportionately shorter times required for higher concentrations. COHb not measured.	Beard and Wertheim, 1967
0,50,100,175,250	4 hr	At estimated COHb 2.5% significant overestimation of 30-sec time interval and sound duration discrimination decrement.	Single blind. 7 subjects. Same conditions and conclusions as above. COHb not measured.	Beard and Wertheim, 1969
0,50,125,200,250	3 hr	No decrement in estimating 10-sec interval or in psychomotor battery.	Double blind for concentrations < 200 ppm. 10 subjects. Each isolated in Thomas Dome during test. COHb measured.	Mikulka et al., 1970

Table 10-7. *(Continued)*

CO Concentration (ppm)	Duration	Effects	Comments	Investigators
<1,25,50,100,200, 500,1000	30 min to 24 hr	No decrement in estimating 1-, 3-, or 5-sec sound or light stimuli. Normal psychomotor battery.	Double blind for concentrations < 500 ppm. 18 subjects. Tested in group setting in 20′ × 20′ × 8′ environmental chamber. COHb measured.	Stewart et al., 1970
0,50,125	3 hr	No decrement in estimating 10-sec interval. Normal psychomotor battery.	Double blind. 9 subjects. Each isolated in Thomas Dome during test. COHb measured.	O'Donnell et al., 1971
0.75,150	9 hr	No decrement in estimating 10- or 30-sec time interval or performing Beard-Wertheim time discrimination test. Normal psychomotor battery.	Double blind. 4 subjects. Each isolated in Thomas Dome during test. COHb measured.	O'Donnell et al., 1971
<2,50,100,200,500	2.5–5 hr	No decrement in estimating 10- or 30-sec time interval or performing Beard-Wertheim time discrimination test.	Double blind. 27 subjects. Tested in audiometric booth, isolated in large environmental chamber and in group setting. COHb measured.	Stewart et al., 1973

		Visual Threshold		
Not stated	Not stated	Visual threshold decrement at 5% COHb.	COHb saturations of 5–20%.	McFarland et al., 1944
1×10^6	10–15 min	Visual disturbance at 4.5% COHb (statistical validity in doubt).	COHb saturations up to 20%.	Halperin et al., 1959
		Miscellaneous Cognitive Tests		
0,100	Varied	Decrement in arithmetic performance and in multiple cognitive tasks at 5% COHb.	Single blind. 49 subjects. Reported COHb saturations up to 20%, hence analytical method suspect.	Schulte, 1963
0,100	8 hr	Decrement in performance of an extensive battery of psychological tests while COHb at 7%.	42 subjects of both sexes.	Bender et al., 1971
0,50,100,150	110 min	Decrement in auditory vigilance test at 50 ppm.	20 subjects of both sexes.	Groll-Knapp et al., 1972
0,50	5 hr	Decrement in auditory vigilance test after 89 min of exposure. Normal ability to perform test regained after 35 min of testing. Normal psychomotor test performance immediately following exposure.	12 subjects of both sexes.	Fodor and Winneke, 1972

Table 10-7. *(Continued)*

CO Concentration (ppm)	Duration	Effects	Comments	Investigators
0,26,111	135–140 min	Decrement in vigilance test at 6.6% COHb. Normal vigilance performance at 2.3% COHb.	Single blind. 10 subjects.	Horvath, 1971
0,100	4 hr	Normal performance on critical tracking task and visual pursuit task.	Studied untrained subjects.	Hanks, 1970
< 100	8 hr occupational	Increased headache and general debility observed in exposed group. EEGs showed flat, low voltage with scanty alpha rhythm.	Mean COHb in exposed group was 7%. Mean COHb in control group was 3%.	Grudzinska, 1963
<1,25,50,100,200, 500,1000	30 min to 24 hr	Changes in visual evoked response at COHb > 20%. No gross alteration in spontaneous EEG at COHb saturations up to 33%.	Part of Stewart et al. study.	Hosko, 1970

		Driving		
960–1,060	58–76 min	Normal driving performance with COHb 25.1–30.4%.		Forbes, 1937
1,100–1,400	45–65 min	Marked decrement in performance at COHb > 30%.		
Not stated	Not stated	Decrement in driving performance with COHb > 10%.	Effect of COHb saturations of 0, 10, and 20% evaluated.	Ray and Rockwell 1970
1×10^6 (80 ml)	Not stated	Decrement in ability to drive at 3.4% COHb.	50 subjects.	Wright et al., 1973
0,700	Up to 1 hr to achieve desired COHb	Decrement in ability to drive at COHb saturations of 11 and 17%. Normal driving performance at 6% COHb.	Well-controlled experiment.	McFarland, 1973

Source: Stewart, R. D. (1975) *Ann Rev Pharmacol* 15:414.

functions, other researchers were not able to substantiate the observations (Stewart et al., 1970; Milkulka et al., 1970; O'Donnell et al., 1971, 1971a; Hosko, 1970; Forbes et al., 1937; Ray and Rockwell, 1970; McFarland, 1973). Based on the research of McFarland et al. (1944) and Halperin et al. (1959), Stewart (1976) concluded that a rapid increase in COHb to 5% affects the visual light threshold. In contrast, Stewart (1970), after a comprehensive review of literature, noted that the capability to perform complex behavior needing both judgment and motor coordination is not impaired by COHb levels below 10%.

Stewart (1976) tried to place these controversial data into an overall perspective. He reasoned that since the human brain is exceptionally efficient in compensating for decreased oxygen-carrying capacity of the blood, it does not follow that COHb levels of less than 5% can impair psychomotor and cognitive functions based on hypoxia. He finally concluded that the hypothesis that COHb saturations of less than 5% can result in impairment of cognitive task performance remains to be verified.

Since the original criteria document for carbon monoxide was published, it seems that considerable doubt has arisen with respect to the Beard and Wertheim studies and the perspective that psychomotor impairment begins to occur at about 2% COHb. At least in this respect, one of the bases on which the standard was established has been undermined to a certain extent. However, during the years since 1971, there has been mounting evidence that the most sensitive adverse effect relates to the lack of adaptive response by heart patients at COHb levels below 5%.

Anderson et al. (1973) and Aronow and Isbell (1973) conducted double-blind studies that investigated the influence of carbon monoxide exposures on the exercise tolerance of patients with angina. These studies indicated that even slight increases of COHb (to 1.6% saturation) adversely affect exercise performance in these individuals. A further study by Aronow et al. (1974) indicated that extremely low levels of COHb may adversely affect those persons with peripheral vascular disease. Specifically, patients with sclerotic arteries supplying their calf or thigh muscles experienced a diminished capability of performing a specific type of exercise activity prior to the sensation of pain when COHb saturation was only 1.7%. Radford (1976) tended to support this recent trend by noting that adverse health effects can be detected in high-risk individuals at levels as low as approximately 2.8% COHb.

If adverse health effects are actually recorded at levels as low as 2.8% COHb, does this imply that hypersusceptible individuals may be adversely affected by levels of COHb even lower than those presently recorded? According to Stewart (1976, discussion), each molecule of carbon monoxide that enters the body displaces a molecule of oxygen, thereby reducing the oxygen-carrying capacity of the blood. When this situation manifests itself, the normal person adapts by speeding up cardiac output or modifying blood flow to specific organs. Stewart indicated that this type of adaptation can be measured when COHb levels are in the range of 2%. He suggested that if the analytical monitoring were more sophisticated, the stress of even a single molecule could be measured. But the question remains, would this "stress" be an adverse health effect? It should be mentioned that humans naturally produce carbon monoxide in an endogenous fashion via heme catabolism such that COHb levels in nonsmokers are usually between 0.4 and 0.7% saturation. COHb levels may even be naturally higher in individuals who have various blood disorders such as anemia. Furthermore, hypermetabolism and drugs such as phenobarbital and dyphenylhydantoin are known to affect increases in the endogenous COHb levels (Stewart, 1976).

This presents an interesting problem. Could it be that our natural metabolism is such that it may produce adverse health effects in susceptible individuals? This is certainly possible. In fact, a quick reflection on the action of the detoxifying liver microsomal enzymes indicates that these enzymes can convert noncarcinogens (termed pre- or procarcinogens) to active carcinogenic substances. However, it must be realized that these actions have been selected for during the course of evolution. Hence their function is designed to enhance the survival of the species. Thus, even though it may be counterproductive in limited situations, the overall pattern of activity must be highly life supporting.

Thus one is faced with a situation in which humans naturally produce 0.4–0.7% COHb; at the same time, research indicates that the start of an adverse health effect in a susceptible individual occurs at levels as low as 2.8% COHb. There is also the theoretical concern that any amount of carbon monoxide may increase stress in cardiac patients. The present carbon monoxide standards are designed to prevent COHb saturation levels from exceeding 2%. Although these were originally geared to prevent the impairment of psychomotor functions, they are ideally suited to prevent an onset of carbon monoxide-induced heart stresses in high-risk patients. If it is true that adverse health effects may be experienced below 2% COHb, then the standard would not be totally protective of some of those persons at high risk. The questions then are, how many are at risk? What is the extent of the risk? With our present technology, can we sufficiently reduce the emissions of carbon monoxide? How much will it cost our society?

These are societal questions—the answers to which lie well beyond the scope of this book. However, in retrospect, nearly a decade after the promulgation of the carbon monoxide standard, it appears that the EPA made a reasonably good approximation of a generally safe level. The EPA may have been right for the wrong reasons. Yet it must be emphasized that it did the best it could with the data available. Time and further efforts have, of course, brought about new data and a new rationale for the standard.

Brief Note on New Lead Standard On December 14, 1977 the EPA proposed a national ambient air quality standard for airborne lead of 1.5 μg/cu m, monthly average, in response to a court order (Federal Register, 1977). The methodological protocol utilized by the EPA in developing the standard involved the following:

1. Identifying the most sensitive population or high-risk group (i.e., children aged 1–5 years).
2. Identifying the lowest detectable adverse health effect [i.e., increased levels of erythrocyte protoporphyrin (EP)].
3. Establishing the average population blood lead level that would offer sufficient protection to the most sensitive high-risk group considered (i.e., 15 μg Pb/dl).
4. Deriving a quantifiable relationship between air lead levels and blood lead levels. The EPA assumed that for each μg Pb/cu m there would be a concomitant increase of 2 μg Pb/dl of blood.
5. Deriving an acceptable blood lead increase from air exposure. Since the EPA assumed that children received 12 μg Pb/dl from nonair sources, they simply determined that 3 μg Pb/dl should be permitted from air (i.e., 15 μg Pb/dl − 12 μg Pb/dl = 3 μg Pb/dl). Since children could be permitted 3.0 μg Pb/dl from inhalation, it follows that 1.5 μg Pb/cu m would be an acceptable standard assuming a 1:2 air lead to blood lead ratio, as explained previously in number 4.

Several aspects of this methodology are worthy of comment:

1. The EPA based their proposed standard on a specific human high-risk group (i.e., young children). Although other high-risk groups were noted (e.g., those with nutritional and various enzyme deficiencies), they were not specifically considered in deriving the standard. Even if these groups are found to be even more sensitive than normal young children, there is not much additional protection any newer and stricter air quality standard could realistically offer,

especially when one considers that 80% of lead exposure would be presumably coming from nonair sources.

2. The selection of elevated levels by the EPA at 15 μg Pb/dl as an adverse health effect is a highly debatable point. The EPA is now using such changes in red cell biochemistry not only as bioindicators of exposure, but also as an actual adverse effect, since there is evidence that the EP elevation suggests an abnormal impairment of various cellular functions involving cells of the hematopoietic system and neural and hepatic tissues. This is a departure from previous attempts to prevent lead toxicity of a neurological and renal nature. If future research indicates that the body adapts to such low-level stress from lead, one would expect that the EPA's position on what constitutes an adverse health effect may be more difficult to defend. Even as it now stands, the EPA may have a very difficult time convincingly demonstrating that "substrate accumulation" and mitochondrial impairment are true adverse health effects with definable clinical implications. (See Chapters 5 and 11 concerning approaches to defining "what is an adverse health effect".)

REFERENCES

Amdur, M. O. and Corn, M. (1963) The irritant potency of zinc ammonium sulfate of different particle sizes. *Am Ind Hyg J* 24:326–333.

Anderson, E. W., Andelman, R. J., Strauch, J. M., Fortuin, N. J., and Knelson, J. H. (1973) Effect of low level carbon monoxide exposure on onset and duration of angina pectoris. *Ann Intern Med* 79:46–50.

Aronow, W. S. and Isbell, M. W. (1973) Carbon monoxide effect on exercise-induced angina pectoris. *Ann Intern Med* 79:392–395.

Aronow, W. S., Stemmer, E. A., and Isbell, M. W. (1974) Effect of carbon monoxide exposure on intermittent claudication. *Circulation* 49:415–417.

Ayres, S. M., Mueller, H. S., Gregory, J. J., Giannelli, S. Jr., and Penny, J. L. (1969) Systemic and myocardial hemodynamic responses to relatively small concentrations of carboxyhemoglobin (COHb). *Arch Environ Health* 18:699–709.

Barth, D. S., Romanovsky, J. C., Knelson, J. H., Altshuller, A. P., and Horton, R. J. M. (1971) Discussion of the Heuss et al. (1971) study. *J Air Pollut Control Assoc* 21(9):544–548.

Battigelli, M. C. and Gamble, J. F. (1976) From sulfur to sulfate: ancient and recent considerations. *J Occup Med* 18(5):334–337.

Beard, R. R. and Grandstaff, N. (1970) CO exposure and cerebral function. Paper presented at the New York Academy of Sciences Conference on Biological Effects of Carbon Monoxide, New York City.

Beard, R. and Wertheim, G. A. (1967) Behavioral impairment associated with small doses of carbon monoxide. *Am J Public Health* 57(11)2012–2022.

Bender, W., Gothert, M., Malorny, G., and Sebbesse, P. (1971) Effects of low carbon monoxide concentrations on man. *Arch Toxicol* 27:142–158.

Brasser, L. J., Joosting, P. E., and von Zuilen, D. (1967) Sulfur dioxide—to what level is it acceptable? Research Institute for Public Health Engineering, Delft, Netherlands, Report G-300.

Buckley, R. D., Hackney, J. D., Clark, K., and Posin, C. (1975) Ozone and human blood. *Arch Environ Health* 30:40.

Calabrese, E. J., Kojola, W., and Carnow, B. W. (1977) Ozone: a possible cause of hemolytic anemia in glucose-6-phosphate dehydrogenase deficient individuals? *J Toxicol Environ Health* 2:709–712.

Carnow, B. W., Lepper, M. H., Shekelle, R. B., and Stamler, J. (1968) The Chicago air pollution study: acute illness and SO_2 levels in patients with chronic bronchopulmonary disease. Presented at National Air Pollution Control Association Meeting, Minneapolis, Minn.

Chanlet, E. T. (1973) *Environmental Protection* p. 236. New York: McGraw-Hill.

Charles, J. M. and Menzel, D. B. (1975) Ammonium and sulfate ion release of histamine from lung fragments. *Arch Environ Health* 30:314–316.

Cooper, W. C. and Tabershaw, I. R. (1966) Biologic effects of nitrogen dioxide in relation to air quality standards. *Arch Environ Health* 12:522–530.

Criteria Document for Carbon Monoxide. (1971) USHEW, USPHS, USEHS. National Air Pollution Admin. Public Document No. AP-62.

Criteria Document for Sulfur Oxides (1971) USHEW, USPHS. Consumer Protection and Environmental Health Service, Washington, D.C.

EHRC (1975) Health effects and recommended standard for ambient manganese. Environmental Health Resource Center, Illinois Institute for Environmental Quality, Chicago, Ill.

Enterline, P. E. (1975) Not uniformly true for each cause of death. *J Occup Med* 17(2)127–128.

EPA (1971) Air quality criteria for nitrogen oxides. Summary and conclusions. EPA Air Pollution Control Office, Washington, D.C.

EPA (1974) Health consequences of sulfur oxides: a report from CHESS, 1970–1971. US EPA. Office of Research and Development, National Environmental Research Center, Research Triangle Park, N.C.

Federal Register (1971) vol. 36, p. 8186.

Federal Register (1977) Environmental Protection Agency Proposed National Ambient Air Quality Standard for Lead, Dec. 14, 1977.

Finklea, J. F., Shy, C. M., Moran, J. B., Nelson, W. C., Larson, R. I., and Akland, G. G. (1977) The role of environmental health assessment in the control of air pollution. In J. N. Pitts Jr. and R. L. Metcalf, eds., *Advances in Environmental Sciences and Technology*, vol. 7, pp. 315–389. New York: Wiley.

Fodor, G. G. and Winneke, G. (1972) Effect of low CO concentrations on resistance to monotomy and on psychomotor capacity. *VDI Berichte*, 180:98–106 (cited in Stewart, 1976).

Forbes, W. H., Dill, D. B., DeSilva, H., and Van Deventer, F. M. (1937) The influence of moderate carbon monoxide poisoning on the ability to drive automobiles. *J Indust Hyg Toxicol* 19:598–603.

Gaffey, W. R. (1976) A critique of the standardized mortality ratio. *J Occup Med* 18(3):157–160.

Goldsmith, J. R. (1975) What do we expect from an occupational cohort? *J Occup Med* 17(2):126–127.

Grennard, A. and Ross, F. F. (1974) Progress report on sulfur dioxide. *Combustion* 45:4.

Groll-Knapp, E., Wagner, H., Hauck, H., and Haider, M. (1972) Effect of low carbon monoxide concentrations on vigilance and computer-analyzed brain potentials. *VDI Berichte*, 180:116–120 (cited in Stewart, 1976).

Hackney, J. D., Linn, W. S., Karuza, S. K., Buckley, R. D., Law, D. C., Bates, D. V., Hazucha, M., Pengelly, L. D., and Silverman, F. (1977) Effects of ozone exposure in Canadians and southern Californians: evidence for adaptation? *Arch Environ Health* 32(3):110–115.

Halperin, M. H., McFarland, R. A., Niven, J. I., and Roughton, F. J. W. (1959) The time course of the effects of carbon monoxide on visual thresholds. *J Physiol* 146:583–593.

Heuss, J. M., Nebel, G. J., and Colucci, J. M. (1971) National air quality standards for automotive pollutants—a critical review. *J Air Pollut Control Assoc* 21:535–544.

Horvath, S. M., Dahms, T. E., and O'Hanlon, J. F. (1971) Carbon monoxide and human vigilance. *Arch Environ Health* 23:343–347.

Hosko, M. J. (1970) The effect of carbon monoxide on the visual evoked response in man. *Arch Environ Health* 21:174–180.

Izmerov, N. F. (1973) Principles underlying the establishment of air quality standards in the U.S.S.R. In *Control of Air Pollution in the U.S.S.R.*, pp. 42–60. Geneva: WHO.

JAPCA (1974) EPA proposes "significant harm" level for photochemical oxidant. *J Air Pollut Control Assoc* 25(5):493.

Lawther, P. J. (1958) Climate, air pollution and chronic bronchitis. *Proc R Soc Med* 51:262–264.

Lawther, P. J. (1963) Compliance with the Clean Air Act. Medical aspects. *J Inst Fuels (London)* 36:341–344.

Lunn, J. E., Knowelden, J., and Handyside, A. J. (1967) Patterns of respiratory illness in Sheffield infant school children. *Br J Prev Soc Med* 21:7–16.

McCarroll, J. and Bradley, W. (1966) Excess mortality as an indicator of health effects of air pollution. *Am J Public Health* 56:1933–1942.

McFarland, R. A. (1973) Low level exposure to carbon monoxide and driving performance. *Arch Environ Health* 27:355–359.

McFarland, R. A., Roughton, F. J. W., Halperin, M. H., and Niven, J. I. (1944) The effects of carbon monoxide and altitude on visual thresholds. *J Aviation Med* 15:381–394.

McIlvaine, P. M., Nelson, W. C., and Bartlett, D., Jr. (1969) Temporal variation of carboxyhemoglobin concentrations. *Arch Environ Health* 19(1):83–91.

Megonnell, W. H. (1975) Atmospheric sulfur dioxide in the United States: Can the standards be justified or afforded? *J Air Pollut Control Assoc* 25(1):9–15.

Menzel, D. B. (1976) Oxidants and human health. *J Occup Med* 18(5):342–345.

Merz, T., Bender, M. A. Kerv, H. D., and Kuller, T. J. (1976) Observations of aberrations in chromosomes of lymphocytes from human subjects exposed to ozone at a concentration of 0.5 ppm for 6 to 10 hours. *Mutat Res* 31:299.

Mikulka, P., O'Donnell, R., Heinig, P., and Theodore, J. (1970) The effect of carbon monoxide on human performance. *Ann NY Acad Sci* 174:409–420.

Morrow, P. E. (1975) An evaluation of recent NO_x toxicity data and an attempt to derive an ambient air standard for NO_x by established toxicological procedures. *Environ Res* 10:92–112.

NAS (1969) Effects of chronic exposure to low levels of carbon monoxide on human health, behavior and performance. National Academy of Sciences and National Academy of Engineering, Washington, D.C. (cited in Heuss et al., 1971).

National Academy of Sciences (NAS), National Research Council, and National Academy of Engineering (1973) News Report XXIII. No. 10, December, p. 5. Washington, D.C.

NIOSH (1972) Criteria for a recommended standard . . . occupational exposure to carbon monoxide. USHEW, Health Services and Mental Health Admin., NIOSH, HSM 73-11000.

O'Donnell, R. D., Chikos, P., and Theodore, J. (1971) Effect of carbon monoxide exposure on human sleep and psychomotor performance. *Toxicol Appl Pharmacol* 18:513–518.

O'Donnell, R. D., Mikulka, P., Heinig, P., and Theodore, J. (1971a) Low level carbon monoxide exposure and human psychomotor performance. *Toxicol Appl Pharmacol* 18:593–602.

Pearlman, M. E., Finklea, J. F., Creason, J. P., Shy, C. M., Young, M. M., and Horton, R. J. M. (1971) Nitrogen dioxide and lower respiratory illness. *Pediatrics* 47:391–398.

Petrilli, R. L., Agnesse, G., and Kanitz, S. (1966) Epidemiology studies of air pollution effects in Genoa, Italy. *Arch Environ Health* 12:733–740.

Radford, E. P. (1976) Carbon monoxide and human health. *J Occup Med* 18(5):310–315.

Ray, A. M. and Rockwell, T. H. (1970) An exploratory study of automobile driving performance under the influence of low levels of carboxyhemoglobin. *Ann NY Acad Sci* 174:396–408.

Redmond, C. K. and Breslin, P. P. (1975) Comparison of methods for assessing occupational hazards. *J Occup Med* 17(5):313–317.

Ross, F. F. (1971) What sulfur dioxide problem? *Combustion* 42:6.

Schimmel, H. (1976) Discussion. *J Occup Med* 18(5):338.

Schimmel, H. and Murawski, T. J. (1976) The relation of air pollution to mortality. *J Occup Med* 18(5).

Schulte, J. H. (1963) Effects of mild carbon monoxide intoxication. *Arch Environ Health* 7:524–530.

Shy, C. M., Creason, J. P., Pearlman, M. E., McClain, K. E., Benson, F. B., and Young, M. M. (1970) The Chattanooga school children study: Effects of community exposure to nitrogen dioxide. I. Methods, description of pollutant exposure, and results of ventilatory function testing. *J. Air Pollut Control Assoc* 20:539–545.

Shy, C. M., Creason, J. P., Pearlman, M. E., McClain, K. E., Benson, F. B., and Young, M. M. (1970a) The Chattanooga school children study: Effects of community exposure to nitrogen dioxide. II. Incidence of acute respiratory illness. *J Air Pollut Control Assoc* 20:582–588.

Stern, A. C., Wohlers, H. C., Boubel, R. W., and Lowry, W. P. (1973) *Fundamentals of Air Pollution* pp. 53–66. New York: Academic.

Stewart, R. D. (1975) The effects of carbon monoxide on humans. *Ann. Rev. Pharmacol.* 15:409–423.

Stewart, R. D. (1976) The effect of carbon monoxide on humans. *J Occup Med* 18(5):304–309.

Stewart, R. D., Peterson, J. E., Baretta, E. D., Bachand, R. T., Hasko, M. J., and Herrmann, A. A. (1970) Experimental human exposure to carbon monoxide. *Arch Environ Health* 21:154–164.

Stokinger, H. E. (1960) Toxicologic interactions of mixtures of air pollutants: review of recent developments. *Int J Air Pollut* 2:313–326.

Stokinger, H. E. (1965) Ozone toxicology: a review of research and industrial experience: 1954–1964. *Arch Environ Health* 10:719–731.

Stokinger, H. E. (1971) Sanity in research and evaluation of environmental health. *Science* 174:662–665.

Stokinger, H. E. (1972) Toxicity of airborne chemicals: air quality standards—a national and international view. *Ann Rev Pharmacol* 12:407–422.

Warner, P. O. and Stevens, L. (1973) Re-evaluation of the "Chattanooga school children study" in the light of other contemporary governmental studies: The possible impact of these findings on the present NO_2 air quality standard. *J Air Pollut Control Assoc* 23:769–772.

Wicken, A. J. and Buck, S. F. (1964) Report on a study of environmental factors associated with lung cancer and bronchitis mortality in areas of northeast England. Tobacco Research Council, London, Research Paper 8.

Wilkins, E. T. (1954) Air pollution aspects of the London fog of December 1952. *R Meteorol Soc J* 80:267–271.

Zelac, R. E., Cromray, H. L., Bolch, W. E., Dunavant, B. G., and Bevis, H. A. (1971) Inhaled ozone as a mutagen. I. Chromosome aberrations induced in Chinese hamster lymphocytes. Environ. Res. 4:262.

11 Comparison Of U.S. and Foreign Standards with Emphasis on Soviet Approaches

THIS BOOK PROVIDES THE framework by which one may understand the historical basis on which occupational and environmental health standards have been derived in the United States. However, to present a more complete perspective on the problems inherent within standard derivation procedures, it is necessary to consider the attitudes and approaches foreign countries and international organizations utilize in the control of environmental and occupational health problems, including standard setting. Chapter 6 compared the European, Soviet, and WHO-International approaches to drinking water standards. This chapter compares and contrasts American approaches to environmental and occupational air quality standard setting with those of foreign countries, with particular emphasis on Soviet methodology.

INFLUENCE OF U.S. AND U.S.S.R. ON STANDARDS IN INDUSTRIAL COUNTRIES

When one compares the different industrial countries of the world with respect to the process of developing occupational health standards, it is immediately obvious that each industrial nation follows the lead of either the United States or the Soviet Union. The United States and the Soviet Union assume the strong leadership positions in the environmental domain as a result of a combination of historical and political factors, in addition to economic considerations.

The vast majority of research on which occupational standards may be derived is conducted in the United States and the Soviet Union primarily because of their vast economic resources. Dinman (1976a), quoting from the Czechoslovak Commission for MAC, illustrates this point: "... we are fully aware that it is beyond the power of a small country to work out alone proposals for a large number of industrial chemicals. They are mostly dependent on information from the literature which they compare with their own experience." As one may expect, Socialist countries, on the whole, tend to follow the approach adopted by the Soviet Union, whereas Western industrial countries look to the United States for guidance. However, economic factors often preclude the implementation of many of the standards put forth by the Soviet Union. Tables 11-1 and 11-2 summarize the extent and scope of industrial standards in various Eastern Block and Western countries (not including the Soviet Union and the United States). The Bulgarian approach is nearly identical to that of the Soviet Union, whereas other Eastern countries show a certain degree of individuality in their approaches to the setting of occupational health standards. For example, Hungary has adopted the use of time-weighted averages (TWA), in marked contrast to the maximum permissible concentrations (MPCs) of the Soviet Union, and in agreement with the TWA approach of the United States. For a more detailed consideration, the reader should refer to Dinman (1976a,b,c), on which this discussion is principally based.

AIR QUALITY STANDARDS IN THE U.S.S.R.

Air quality standards in the Soviet Union and other Socialist countries are termed "maximum permissible concentrations" (MPCs). The Soviet Union issued their first MPCs in 1951; these were the first national community air standards developed anywhere in the world.* Subsequently, other countries including Bulgaria, Czechoslovakia, the German Democratic Republic, Poland, Romania, the Federal Republic

* The state of California adopted ambient air standards as early as 1947 (Stern et al., 1973).

Table 11-1. Comments on Occupational Health Standards In Eastern Block Countries (Dinman, 1976a)

Bulgaria	Czechoslovakia	Hungary	Poland	Romania
Lists MACs for 120 substances; for those unlisted substances, Soviet standards are considered legally binding.	In 1966, listed 68 MAC values—in all but 3, short-term single excursion was allowed.	In 1965, listed 104 MAC values for gases and vapors and 6 MPCs for dusts.	Listed 210 gases, vapors, and dusts and 14 compounds.	In 1966, listed 472 MAC values for gases and vapors plus 10 MAC values for dusts.
Governmental standards are considered legally binding.	No indication if these values are legally binding.	Used TWA instead of maximum limits.	No excursions above the MAC value are allowed.	Standards are legally binding.
		Standards are legally binding.	MACs for the 14 compounds are legally binding.	

Table 11-2. Comments on Occupational Health Standards In Western Countries (Dinman, 1976a)

France	Italy	Japan	Sweden	United Kingdom	West Germany
Use of MACs is not encouraged because they are not thought to be adequately predictive of potential health effects; margins of safety are questioned. Private voluntary organization (Association Interprofessionalle des Centres Medicaux et Sociaux of the Paris region) publishes an extensive MAC list that is largely adopted from the U.S. ACGIH (as guides to government).	There is very low activity with regard to setting industrial health standards; Dinman (1976a) reported only 7 industrial standards.	MACs are published by a nongovernmental group called the Japanese Association of Occupational Health. The MACs serve as a guide to government; they have no legal status.	MACs are published by the Swedish Institute of Occupational Medicine; these MACs serve as guides to the ministry of health; 76 MACs have been derived.	No legal status is granted to MAC values; however, values derived from ACGIH TLV lists have been used for control and surveillance functions by the H.M. Factory Inspectorate of the Ministry of Employment.	A nongovernmental group (Commission for the Evaluation of Toxic Materials in the Workplace) has published 144 MAK (as of 1972); these serve as a guide to governmental agencies.

of Germany, the United States, France, Italy, Japan, and others followed with their own ambient air standards.

According to Rjazanov (1965) and Izmerov (1973), air quality standards within the Soviet Union may be founded on hygienic, sanitary, or technological grounds. Hygienic standards are designed to reflect the ideal to which effective environmental controls should be directed. Such standards are designed to prevent the occurrence of (1) either direct or indirect harmful or unpleasant effects on humans, (2) impairment of working capacity, and (3) detrimental effects on physical and psychological well-being. Rjazanov (1965) indicated that since this ideal cannot be attained in all cases, it has been necessary to supplement hygiene standards with sanitary standards. Sanitary standards are of a temporary nature and consider the present economic and technological realities. Finally, the third type of MPC is called the technological standard. It is designed to regulate the release of various toxic substances into the ambient air as well as to control the release of valuable materials. Technological standards are not concerned with health considerations, but are focused on technological and economic factors. Hygienic standards are subdivided into maximum average concentrations over a 24-hour period and maximum permissible single concentrations. The intention of the daily maximum average standard is to prevent the occurrence of chronic effects caused by the action of the variously regulated toxic substances. The maximum permissible single concentration standard is designed to prevent an exposure that may cause reflex reactions through irritation of the receptors in the respiratory tract, especially in the olfactory region of the nasal cavity. Table 11-3 lists the maximum permissible concentrations of 114 potentially harmful substances in the air of population centers. It is interesting to note the striking contrast between the number of ambient air standards (six) in the United States as compared to greater than 100 in the Soviet Union.* Table 11-4 gives a comparison of air standards for five pollutants adopted by the U.S.S.R. It represents the experimental (animal) and epidemiologic data that affected the adoption of the specific standard. It is interesting to note that the standard for arsenic trioxide was not based on neurological changes as compared to standards for other substances. Furthermore, the thresholds for resorption effects in animals proved to be very comparable to similar effects in humans. According to Izmerov (1973), these results of experimental research concerning the resorptive effects in animals may be legitimately extrapolated to humans and are reliable for use in the derivation of MPCs (mean 24-hour concentration) for air pollutants.

* With the recent addition of lead, there are now seven such standards in the United States.

Table 11-3. Maximum Permissible Concentrations of Harmful Substances in the Air of Population Centers

Pollutant	Maximum Permissible Concentration (mg/cu m)		Pollutant	Maximum Permissible Concentration (mg/cu m)	
	Maximum on One Occasion	Average Over 24 Hours		Maximum on One Occasion	Average Over 24 Hours
1. Nitrogen dioxide	0.085	0.085	16. Butane	200.0	—
2. Nitric acid			17. Butyl acetate	0.10	0.10
calculated as HNO	0.4	0.4	18. Butylene	3.0	3.0
calculated as hydrogen	0.006	0.006	19. Butanol	0.1	—
3. Acrolein	0.03	0.03	20. Tributyl triphosphate (butiphos)	0.01	0.01
4. α-Methylstyrene	0.04	0.04	21. Valeric acid	0.03	0.01
5. α-Naphthoquinone	0.005	0.005	22. Vanadium pentoxide	—	0.002
6. Amyl acetate	0.10	0.10	23. Vinyl acetate	0.15	0.15
7. Amylene	1.5	1.5	24. Hexamethylenediamine	0.001	0.001
8. Ammonia	0.2	0.2	25. Hexachlorocyclohexane	0.03	0.03
9. Aniline	0.05	0.03	26. Divinyl	3.0	1.0
10. Acetaldehyde	0.01	0.01	27. Diketone	0.007	—
11. Acetone	0.35	0.35	28. Dimethylaniline	0.0055	0.0055
12. Acetophenone	0.003	0.003	29. Dimethyl sulfide	0.08	—
13. Benzene	1.5	0.8	30. Dimethylamine	0.005	0.005
14. Benzine, from petroleum (low sulfur fraction, calculated as C)	5.0	1.5	31. Dimethyl disulfide	0.7	—
			32. Dimethylformamide	0.03	0.03
15. Benzine, from shale (calculated as C)	0.05	0.05	33. Dinyl (24% of diphenyl + 76% of diphenyl oxide)	0.01	0.01

Table 11-3. *(Continued)*

Pollutant	Maximum Permissible Concentration (mg/cu m): Maximum on One Occasion	Maximum Permissible Concentration (mg/cu m): Average Over 24 Hours
34. Dichloroethane	3.0	1.0
35. 2,3 Dichloro-1,4 naphthoquinone	0.05	0.05
36. Diethylamine	0.05	0.05
37. Isopropyl benzene	0.014	0.014
38. Iso-octanol	0.15	—
39. Isopropyl benzene hydroperoxide	0.007	0.007
40. Propane-2-01 (isopropyl alcohol)	0.6	0.6
41. Caprolactam (fumes, aerosol)	0.06	0.06
42. Caproic acid	0.01	0.005
43. Carbophos	0.015	—
44. Xylene	0.2	0.2
45. Intrathion (M-81)	0.001	0.001
46. Maleic anhydride (fumes, aerosol)	0.2	0.05
47. Manganese and its compounds (calculated as MnO_2)	—	0.01
70. Mercury, metallic	—	0.0003
71. Soot	0.15	0.05
72. Lead and its compounds (except tetraethyl lead) calculated as Pb	—	0.0007
73. Lead sulfide	—	0.0017
74. Sulfuric acid		
calculated as H_2SO_4	0.3	0.1
calculated as hydrogen	0.006	0.002
75. Sulfur dioxide	0.5	0.05
76. Hydrogen sulfide	0.008	0.008
77. Carbon disulfide	0.03	0.005
78. Hydrogen cyanide	—	0.01
79. Hydrochloric acid		
calculated as HCl	0.2	0.2
calculated as hydrogen	0.006	0.006
80. Styrene	0.003	0.003
81. Tetrahydrofuran	0.2	0.2
82. Thiophene	0.6	—
83. Tolyl di-isocyanate	0.05	0.02
84. Toluene	0.6	0.6

48. Butyric acid	0.015	0.01
49. 2,4,6-Trimethylaniline (mesidine)	0.003	0.003
50. Methanol	1.0	0.5
51. Parathion-methyl (metaphos)	0.008	—
52. Metachlorophenyl isocyanate	0.005	0.005
53. Methyl acrylate	0.01	0.01
54. Methyl acetate	0.07	0.07
55. Methyl mercaptan	9×10^{-6}	—
56. Methyl methacrylate	0.1	0.1
57. Monomethylaniline	0.04	0.04
58. Monoethylamine	0.01	0.01
59. Arsenic (inorganic compounds, except hydrogen arsenide calculated as As)	—	0.003
60. Naphthalene	0.003	0.003
61. Nitrobenzene	0.008	0.008
62. Nitrochlorobenzene (o- and p-)	—	0.004
63. Parachloroaniline	0.04	0.01
64. Parachlorophenyl isocyanate	0.0015	0.0015
65. Pentane	100.0	25.0
66. Pyridine	0.08	0.08
67. Propylene	3.0	3.0
68. Propanol	0.3	0.3
69. Dust, nontoxic	0.5	0.15

85. Triethylamine	0.14	0.14
86. Trichloroethylene	4.0	1.0
87. Carbon monoxide	3.0	1.0
88. Carbon tetrachloride	4.0	2.0
89. Acetic acid	0.2	0.06
90. Acetic anhydride	0.1	0.03
91. Phenol	0.01	0.01
92. Formaldehyde	0.035	0.012
93. Phosphorus pentoxide	0.15	0.05
94. Phthalic anhydride (fumes, aerosol)	0.1	0.1
95. Fluorine compounds calculated as F gaseous compounds (HF, SiF_4)	0.02	0.005
readily soluble inorganic fluorides (NaF, $NaSiF_6$)	0.03	0.01
sparingly soluble inorganic fluorides (AlF_3, $NaAlF_6$, CaF_2)	0.2	0.03
admixed with gaseous fluorine and fluorides	0.03	0.01
96. Furfural	0.05	0.05
97. Chlorine	0.10	0.03
98. Chlorobenzene	0.10	0.10
99. m-Chloroaniline	—	0.01
100. Chloroprene	0.10	0.10
101. Trichlorfon (chlorophos)	0.04	0.02
102. Chlorotetracycline (for mixing with animal feed)	0.05	0.05

Table 11-3. *(Continued)*

Pollutant	Maximum Permissible Concentration (mg/cu m): Maximum on One Occasion	Average Over 24 Hours	Pollutant	Maximum Permissible Concentration (mg/cu m): Maximum on One Occasion	Average Over 24 Hours
103. Hexavalent chromium (calculated as CrO_3)	0.0015	0.0015	109. Ethanol	5.0	5.0
104. Cyclohexane	1.4	1.4	110. Ethyl acetate	0.1	0.1
105. Cyclohexanol	0.06	0.06	111. Ethyl benzene	0.02	0.02
106. Cyclohexanone	0.04	—	112. Ethylene	3.0	3.0
107. Cyclohexanone oxine	0.1	—	113. Ethylene oxide	0.3	0.03
108. Epichlorohydrin	0.2	0.2	114. Ethylenimine	0.001	0.001

When the following substances are simultaneously present in the atmosphere, their combined effect is equal to the sum of their individual effects.

1. acetone and phenol
2. valeric acid, caproic acid, and butyric acid
3. carbon dioxide and phenol
4. carbon dioxide and nitrogen dioxide
5. carbon dioxide and hydrogen fluoride
6. carbon dioxide and sulfuric acid aerosol
7. hydrogen sulfide and dinyl
8. isopropylbenzene and isopropylbenzene hydroperoxide
9. furfural, methanol, and ethanol
10. cyclohexane and benzene
11. strong mineral acids (sulfuric, hydrochloric, and nitric in terms of hydrogen ion concentrations)

12. ethylene, propylene, butylene, and amylene
13. 2,3 dichloro-1,4-naphthaquinone and 1,4-naphthaquinone
14. acetic acid and acetic anhydride
15. acetone and acetophenone
16. benzene and acetophenone
17. phenol and acetophenone
18. sulfur dioxide, sulfur trioxide, ammonia, and nitrogen oxides

The sum of the ratios of the recorded concentrations to the maximum permissible concentrations should not exceed unity. This may be expressed mathematically by the following formula

$$\frac{a}{m_1} + \frac{b}{m_2} + \frac{c}{m_3} \cdots \leq 1$$

where $a,b,c\ldots$ are the recorded concentrations of the individual pollutants.
$m_1,m_2,m_3\ldots$ are the corresponding maximum permissible concentrations for exposure to the single substances.

When the following substances are present simultaneously in the atmosphere the maximum permissible concentration for each substance separately continues to apply.

1. hydrogen sulfide and carbon disulfide
2. carbon monoxide and carbon dioxide
3. phthalic anhydride, maleic anhydride, and a-naphthaquinone

When p-chlorophenyl isocyanate and m-chlorophenyl isocyanate are present simultaneously in the atmosphere, the criterion to be applied is that for the more toxic substance, i.e., p-chlorophenyl isocyanate. This is a temporary measure, to be used until a method is found of determining the substances separately.

Source: Izmerov, N. F. (1973) *Control of Air Pollution in the U.S.S.R.*, pp. 129–132. Geneva: WHO.

Table 11-4. Comparative Assessment of Experimental and Full-Scale Studies on the Biologic Effects of Atmospheric Pollutants[a] (all concentrations expressed in mg/cu m)

		Experimental Studies						Full-Scale Studies[b]				
		Odor		Reflex Effect		Resorptive Effect		Physiologic and Biochemical Changes		Change in Morbidity, Physical Development, or Blood		
Substance	Approved MPC	Threshold	Subthreshold	Threshold	Subthreshold	Threshold	Subthreshold	Present	Absent	Present	Absent	Reference
Monoethylamine	0.01/ 0.01	0.05	0.04	0.004 (EEG)		0.05 Fall in sulfhydryl compounds and cholinesterase activity in the blood; changes in the coproporphyrin content of the urine	0.01	0.028/ 0.011 Changes in the threshold of smell, cholinesterase activity, and coproporphyrin in urine	0.0006/–	0.028/ 0.011 Increase in child morbidity from acute infections	0.006/–	Tkacev, 1969

Arsenic trioxide (calculated as As)	—/ 0.003	—	—	—	—	0.005 Fall in sulfhydryl compounds and cholinesterase in the blood; accumulation of arsenic in organs	0.0013	0.0027/	0.005–0.0009/–0.0041/– Arsenic found in children's hair	—	—	Rozenstein, 1970
Lead oxide (calculated as Pb)	—/ 0.0007	—	—	—	—	—/0.1. Modification of higher nervous activity; change in the coproporphyrin content of urine; lead in bones —/0.0039 Modification of higher nervous activity; lead in the bones	—/0.0011	—/0.001 0.001 and 0.0025 Increase of coproporphyrin and lead in children's urine	—/ 0.0027/-	—	—	Gusev, 1961

Table 11-4. *(Continued)*

		Experimental Studies						Full-Scale Studies[b]				
		Odor		Reflex Effect		Resorptive Effect		Physiologic and Biochemical Changes		Change in Morbidity, Physical Development, or Blood		
Substance	Approved MPC	Threshold	Subthreshold	Threshold	Subthreshold	Threshold	Subthreshold	Present	Absent	Present	Absent	Reference
						0.22	0.088	—/0.26	—/0.13	—	—	
Chloroprene	0.1/0.1	0.3	0.25	0.2	0.1	Fall in sulfhydryl compounds, cholinesterase, and 17-ketosteroids		Change in coproporphyrin and 17-ketosteroid content of urine				Mnacakajan, 1964 Mnacakajan and Musegjan, 1964

[a] In the table, the figure preceding the oblique stroke refers to the one-time concentration and that after the oblique stroke to the mean 24-hour concentration.
[b] Carried out on groups of children selected on the basis of socioeconomic background and housing and living conditions.
[c] See Table 9-3.

Source: Izmerov, N. F. (1973) Principles underlying the establishment of air quality standards in the U.S.S.R. In *Control of Air Pollution in the U.S.S.R.*, p. 56. Geneva: WHO.

PREDICTION OF MPCs

In contrast to methodologic procedures in the United States, Soviet scientists devote considerable effort to trying to predict toxicity for organic compounds on the basis of physical and chemical properties (Lyublina and Rabotnikova, 1971; and Sanotskiy, 1969). These researchers have presented their attempts to derive regression equations of physical properties (molecular weight, specific gravity, vapor pressure, boiling and melting points, refractional index, and surface tension) for a variety of related substances whose MPCs are known. Based on such calculations, it is hoped that the potential toxicity of a substance may be predicted. This procedure may offer a tool to administrators in developing priority lists of compounds intended to be given toxicity tests. In light of the stringency of procedures in the United States with respect to standard setting, it is difficult to see a very strong role for this technique in the standard derivation process. It should be emphasized, however, that this technique is not widely adopted in the United States and that its potential usefulness must be explored more fully. Yet one must agree with Dinman (1976a) in concluding that this approach offers only approximations and should not replace animal studies in the derivation of MPCs. This statement does not suggest that Soviets do substitute this procedure for animal testing. Such studies are usually conducted concurrently with toxicologic investigations.

TOXICOLOGIC APPROACHES IN THE U.S.S.R

An extremely heavy reliance is placed on the information gathered from animal and human experimentation in controlled settings. Rjazanov (1965) and Izmerov (1973) outlined the experimental protocol used in the course of animal experimentation (Table 11-5). In the process of standard derivation, profound importance is placed on long-term modifications in higher nervous system activity of the experimental models. The researchers first establish what is called a "dynamic stereotype" of baseline level of neurophysiologic activity in the animals. Once the baseline is established, the animals are exposed to the toxicants. Every day or so during exposure the animals are examined to discern whether any changes in the functions of the cerebral cortex have occurred. According to Izmerov (1973), the changes in neurophysiologic activity are not specific with regard to the toxic substance; the only difference between the substances lies in the threshold response.

Table 11-5. Toxicologic Procedures Employed To Support Standard Setting In the Soviet Union

Step 1.	Identification of the chemical and physical characteristics of the toxicant substances (e.g., description of vapor pressure, solubility, flammability, etc.)
Step 2.	Acute Toxicity A. Animal studies 1. Mice studies —Static exposures —Two hours/day for 3 weeks 2. Rat studies —Static exposures —Four hours/day for 3 weeks B. Purpose —Define lethal concentrations —Determine the clinical sequence —Characterize changes in the conditioned and direct reflexes
Step 3.	Subacute exposure—four hours/day for 6 days/week for one to two months
Step 4.	Chronic Exposure —Use of multiple species —Exposure for 4 hours/day, 6 days/week for 5-6 months —Chronic studies include (1) microscopic changes in various organ systems including brain, heart, liver, spleen, kidney, or lung; (2) biochemical indicator tests
Step 5.	Accumulation Studies

In addition to monitoring changes in nervous system activity, consideration is also given to other sensitive indicators of pollutant exposure. For example, urinary excretion levels of coproporphyrin are used to monitor exposures to a variety of substances such as carbon dioxide, dinile, styrene, and dimethyl formanide, which decrease coproporphyrin levels, whereas lead and toluene diisocyanate increase them. Other studies of biologic monitoring include examining the potential effects of changes in the degree of dispersion of the serum proteins such as the ratios of globulins and albumins. The results of these long-term animal exposure experiments are used to provide the fundamental basis on which the MPCs are derived.

EPIDEMIOLOGIC STUDIES

Information derived from morbidity studies is not considered to be of significant help in the derivation of MPCs, since they relate information of harmful levels and not how low a level must be to be "safe." Furthermore, Rjazanov (1965) noted that there often is a very wide range of concentrations between the toxic and "safe" levels. This transitional zone, although not invoking pathological responses, does cause the occurrence of adaptational and protective mechanisms. Rjazanov (1965) emphatically stated that such adaptational changes cannot be permitted. He concluded that the mere existence of such protective changes is clear evidence of the presence of an unacceptable environment. MPCs must be set at levels that are sufficiently low so as not to evoke such adaptive or protective responses. Thus, even though morbidity studies can help to substantiate that pollutant levels may be too high, they cannot tell what the MPCs should be.

PHILOSOPHICAL PERSPECTIVES

According to Magnuson et al. (1964), Magnuson (1965), and Glass (1975), the derivation of air quality standards in the U.S.S.R. is based on four general principles:

1. Exposure at the permissible dose shall not cause any variation from the normal condition of the individual nor cause any disease condition.
2. The setting of standards (i.e., hygienic) should be based on health effects without regard to economic and technical feasibility.
3. The values should be maximum concentration instead of time-weighted average values.
4. The ideal standard to be ultimately achieved is a zero level of exposure.

Table 11-6 compares approaches to standard setting and implementation in the Soviet Union and the United States.

As discussed in the chapter on threshold responses, a major point of departure between Soviet and American approaches to standard setting involves the answer each provides to the question, what is an adverse health effect? Historically speaking, Americans have always assumed the presence of a healthy environment. It was only adverse health effects such as death or disease that caused action to be taken. Our whole environmental movement, including occupational health, is clearly one of reaction to adverse health effects. Elkins (1961) noted that many of our occupational health standards have been based on primitive criteria ranging from death to changes in body or organ weight. Hatch (1972) concurred with this perspective and stated that in the United States the

Table 11-6. Soviet Union and United States: A Contrast in Approaches to Standard Setting (Based on Magnuson et al., 1964)

U.S.S.R.	U.S.A.
A. Approaches to Setting Industrial Standards	
1. Maximum allowable concentration will not permit the development of any disease or deviation from normal.	1. Minor physiologic adaptive changes are permitted to be developed.
2. Standards should be based entirely on health and not technological and economic feasibility.	2. Economic and technologic feasibility are important considerations in the development of U.S. standards. (Note: OSHA is required to perform "inflationary impact statements" for proposed standards.)
3. Concentrations are maximum values.	3. Values are time-weighted averages.
4. Emphasis on nervous system testing.	4. Research emphasis on pathology.
5. Goal to be achieved is a zero level of exposure.	5. With the exception of carcinogens in the workplace, goals of zero exposure have not been seriously discussed.
B. Manner of Obtaining Experimental and Clinical Data for Setting Limits	
1. Highly structured—the only data available are those for which the government has planned.	1. Highly varied sources—government, industry, universities.
2. More efficient exchange of scientific data.	2. Inefficient exchange of scientific information.
C. Formal Mechanism for the Actual Adoption of a Standard	
1. Government selects scientific review committee; they recommend a standard to the Ministry of Health; if Ministry agrees, they institute a standard.	1. As a result of the OSHAct of 1970; —NIOSH is charged to develop a review of the literature with respect to a pollutant. —NIOSH submits this to OSHA with a recommended standard.

Table 11-6. (*Continued*)

U.S.S.R.	U.S.A.
	—OSHA then creates an independent committee (outside of OSHA and NIOSH) to critically review the evidence supporting the proposed standard. Following the review process, OSHA decides whether to promulgate a standard.
D. Enforcement of Standards	
1. Function of: —inspectors of M.H. (medical public health training) —inspectors from the All Union Councils (engineers) —inspectors from local trade union committee	1. The Occupational Health and Safety Administration is responsible for the enforcement of U.S. industrial health standards.

overwhelming thrust in standard setting has been an adoption of lower and lower values as a result of the development of increasingly more sensitive indicators of preclinical, physiologic, biochemical, and other indices of functional disturbances. The case of the carcinogen vinyl chloride demonstrates how industrial health standards employed in the United States have markedly changed as a result of biomedical studies indicating more enhanced risk than previously thought (Table 11-7). In contrast, Hatch (1972) noted that the Soviets initiate their studies at the opposite end of the dose-response relationship spectrum by proceeding from zero dose and the initial characterization of baseline physiologic-biochemical parameters in the experimental individuals. These standards are then theoretically established such that the MPC is below the lowest level that causes a statistically significant deviation in extremely sensitive indicators of behavioral or biochemical responses.

The passage in 1976 of the Toxic Substances Control Act, which demands pretesting of all new chemicals planned to be used in interstate commerce, seems to be a major shift within the United States to adopt the "overall orientation" of the Soviets with respect to the development of accurate "toxicologic" dose-response relationships. The adoption of pretesting procedures offers what has been widely referred to as a front

Table 11-7. Vinyl Chloride Regulations Since 1962

1962 American Conference of Governmental Industrial Hygienists Adopts 500-ppm TLV

↓

1971 Occupational Safety and Health Administration (OSHA) Adopts 500-ppm TLV

↓

April 1974 OSHA Issues Temporary Emergency Standard 50-ppm Ceiling

↓

May 1974 OSHA Proposes Standard No Detectable Level

↓

October 1974 OSHA Issues Final Standard Reduction of Exposures To 1-ppm/8-hr Time-Weighted Average With 5-ppm Max and 0.5-ppm Action Level

end control process. Such a perspective, of necessity, leans heavily on the development of a broad variety of toxicologic testing procedures, employed prior to the granting of manufacturing approval of the substance in question so that adequate safety testing can be completed before any significant human exposure occurs. Consequently, the basis of many future industrial standards in the United States may be based entirely on toxicologic studies. Certainly, later epidemiologic studies will serve to validate the previous toxicologic test results and interpretations. This procedure, which seems likely to be the orientation of the future, looks very similar to that of the U.S.S.R. Epidemiologic studies have played a tremendous role in the derivation of such standards as lead, arsenic, asbestos, vinyl chloride, mercury, and others and undoubtedly will continue to be a vital force in the development of future standards. However, the long-term role of epidemiologic studies in standard derivation seems to be that of applying the "fine tuning" to the conclusions derived from previous toxicologic studies. Furthermore, in some cases, the epidemiologic studies will undoubtedly discern diseases that slipped through the predictive testing procedures of the toxicologic methods. It appears that testing in the United States will probably be more like that of the U.S.S.R. than the other way around. Even though the United States is adopting front end control measures for future industrial pollutants, it seems unlikely that there will be much agreement between American and Soviet toxicologists with respect to defining an adverse health effect. Future pretesting of industrial chemicals in the United States will involve the traditional toxicologic testing procedures as well as tests for mutagenicity, teratogenicity, and carcinogenicity. The results of these studies will yield data that may serve as the initial basis on which standards may be derived. However, similar testing procedures by Soviet scientists may result in lower standards because the Soviets base their hygiene MPCs on changes in neurophysiologic measures that are usually exceptionally sensitive indicators of exposure.

Dinman (1972) was very critical of the adoption of standards based on the occurrence of changes in sensory input. He contended that to keep concentrations of environmental agents below one's level of perception may create "a creature [that] would have a Faustian life span but [one that] could receive no new knowledge, an existence barely removed from that of the amoeba in its ability to experience that world around it. The danger that a world oblated of sensory inputs imposes has been clearly demonstrated by sensory deprivation experiments. It seems that man is destined to fail and challenge, to perceive and appreciate his world, but at the cost of mortality. Otherwise, what an immortality, what a life!"

Perhaps the most significant question is not, what is an adverse health effect, but why are the Soviet scientists so insistent on placing extraordinary emphasis on nervous system effects? Why are changes in nervous activity viewed in such a negative way? Since neurophysiologic changes may be much more sensitive than alterations of other cellular functions, it automatically sets up a situation in which the use of a neurologic bioindicator in standard setting results in a drastically lower standard. This difference in approach between American and Soviet scientists provides for much of the divergence in actual standard levels between the two countries.

In a series of articles assessing the development of workplace standards in foreign countries, Dinman (1976a,b,c) addressed the issue of the role of alterations of nervous function in the development of Soviet health standards. To understand why such stress is placed on neurological indicators of pollutant exposure, Dinman (1976a) contended that it is necessary to first understand the inseparable relationship of political considerations with scientific thought in socialist countries. Their approach to the development of criteria for standard derivation follows from their philosophical views on the nature of man, which were derived in large part from the writings and teachings of Marx and Lenin. Perhaps the critical factor in this process is the synthesis of the writing of Lenin with the work of Pavlov on higher nervous sytem function.

The unity of thought with respect to Marx, Lenin, and Pavlov stems from the fact that they tried to explain man totally in mechanistic terms. Thus the adoption of Darwin's theory of evolution by Marx offered Lenin the geo-biologic-historic foundations for considering man in terms of biologic adaptation. Thus the scientific foundations for Lenin's rejection of the dualism of body and mind as developed by Descartes were founded in the acceptance of a mechanistic concept of evolutionary theory. At the experimental level, Pavlov also rejected the Cartesian dualism of body and mind, and, according to Dinman, this suited Lenin's needs.

According to Pavlov, the brain is in a constant state of balance, regulating its functioning to acquire a balance between external and internal stimuli by effecting inhibitory and stimulatory responses. The cerebral cortex is considered the principal regulator of such activity; it is also in dynamic equilibrium with subcortical, stem, and visceral regulatory centers that control the body's homeostatic processes. Thus Pavlov considered nearly all of man's activity as fundamentally under the control of the cortex (Dinman, 1976a).

Various attempts have been made to translate Lenin's thought into physiologic perspectives analogous to Pavlov's terminology. For example, Dinman (1976a), quoting Kupalov et al., indicated that "according to Lenin . . . the dialectical way to consciousness of objective reality procedes from active contemplation (similar to Pavlov's First Signal System) to abstract thought (Pavlov's Second Signal of Activity) and thence to practical application, i.e., the experimental testing of abstract thought." Consequently, the relationship between the dialectic-materialistic theory of consciousness of Marxism-Leninism and Pavlov is intimate and reinforcing. According to Dinman, it is the unique synthesis of Lenin's and Pavlov's thought that ultimately influenced the importance of reflex behavior in the standard derivation process in the U.S.S.R. He further stated that experimental procedures that question the validity of the Pavlovian perspective can be logically thought to be a threat to the foundations of Soviet biologic science. Thus maximum allowable concentrations not primarily derived from Pavlovian principles are not seriously considered.

These comments lead back to the initial question of why Soviet scientists are so concerned with changes in neurological activity. The reason why such responses should be considered as "toxic or adverse" effects is that they are related to, and affect, "instability of nervous regulation, arising against the background of a changed functional state of the cortex and subcortical region. These are believed in turn to lead to altered reactivity of the vascular and endocrine systems and instability of neuro-circulatory processes. The manifestations of such changes in the affected individual are the general, non-specific symptoms of toxicity arising from such autonomic dysequilibrium"

(Dinman, 1976a). Therein lies the principal philosophical and scientic argument used by Soviet toxicologists to support the need for using neurophysiologic indicators of pollutant toxicity.

From a scientific perspective, it would appear that the Soviet emphasis exploring the effects of altered CNS function (via pollutant activity) on health is clearly an area in which Western scientists could learn from their Soviet counterparts. A cursory knowledge of psychosomatic illness, including the effects of stress on health, suggests that environmental health scientists should closely examine this area. Further Western developments in this area may in fact result in greater areas of agreement between American and Soviet approaches to standard setting. However, unless pollutant-induced CNS instability can be directly tied to the induction of some disease process, it is likely that such data will not play a significant role in standard derivation in the United States.

INDUSTRIAL STANDARDS

Elkins (1961) compared American and Soviet occupational health standards in the early 1960s when it was generally recognized that the standards of these two countries were often at variance with each other; that is, the Soviet standards are, in general, considerably lower than those of the United States. Elkins noted that standards (i.e., based on 1960 standards) for irritant gases and vapors in the United States and the U.S.S.R. were in generally close agreement, with several notable exceptions such as acetaldehyde and ethylene oxide, for which the ACGIH-recommended TLVs were 67 (200 vs. 3 ppm) and 80 (50 vs. 0.6 ppm) times higher than the Soviet standards, respectively. In contrast to the irritants, the Soviet standards for both hydrocarbon solvents and especially the chlorinated hydrocarbons are often considerably lower than those standards proposed by the ACGIH. Furthermore, there is also reasonably good agreement between the Soviet and ACGIH values for both the organic dusts and fumes as well as the inorganic dust and fumes. Once again, there are exceptions such as chlordane, dimeton, and lindane for organics and lead, manganese, metallic-mercury and tellurium for inorganics (Elkins, 1961). After 1960 there were a considerable number of changes in the ACGIH TLVs. These changes, as previously noted, almost invariably resulted in lowered levels (i.e., stricter standards). Soviet standards have also reflected several reductions (e.g., ammonia) in MAC levels. However, in several instances (e.g. hydrogen fluoride and SO_2), there have been increases. In a general

Table 11-8. Comparison of Soviet and ACGIH Industrial Standards (1976) (in ppm) For Irritant Vapors[a]

	ACGIH	U.S.S.R.	MAC Ratio ACGIH/USSR
Acetaldehyde	100(200)	3.0	33
Acetic acid	10	2.0	5
Acrolein	0.1 (0.5)	0.3	0.33
Ammonia	25(100)	15(30)	1.6
Butylamine	5	3	1.6
Chlorine	1	0.3	3.3
Diethylamine	25	10	2.5
Dimethyl formamide	10	3	3.3
Ethylene oxide	50	0.5 (0.6)	100
Formaldehyde	2(5)	0.8	2.5
Furfural	5	2.5	2
Hydrochloric acid	5	4(7)	1.3
Hydrogen fluoride	3	0.7 (0.6)	4.3
Iodine	0.1	0.1	1
Ozone	0.1	0.05	2
Nitrogen dioxide	5	2	2.5
Phosgene	0.1 (1.0)	0.1	1
Sulfur dioxide	5	4(3)	1.3
Toluene diisocyanate	0.02 (0.1)	0.07	0.3

[a] Where there are two values listed for a substance in the U.S. or U.S.S.R. the value in 1960 was changed. The number in parenthesis indicates the 1960 value; the adjacent number is the 1976 value. For those substances which have only one value per country, this value represents the 1976 standard which is identical with its 1960 value (i.e., it has not been changed since 1960).

sense, the extent of the differences between these standards does not appear to have markedly diminished, although certain individual MAC ratios have changed considerably. Table 11-8 is an updated comparison of ACGIH and Soviet standards for selected irritants from an earlier (1961) table by Elkins and the 1976 RCGIH Documentation of TLVs.

SAFETY FACTORS

In the U.S.S.R. the term "reserve co-efficient" is used to represent the difference between acceptable concentrations (MPCs) and threshold levels for physiologic responses. In American terms, the "reserve co-efficient" may be viewed as the safety factor that is typically built into each standard. Several approaches have been proposed in the U.S.S.R. for the derivation of safety factors. According to Izmerov (1973), MPCs should be provided with a safety factor of 30% in all cases. This means that if a presumed threshold is found, the next level tested is 30% lower. If there is no pollutant-induced activity at that level, the experimentation can stop, with that final concentration becoming the MPC. This may seem like a very small safety factor as compared to the traditional factors of 10 or 100 in the United States, but it must be recalled that the Soviets are trying to prevent the occurrence of the most sensitive indicator of exposure and not necessarily the truly pathologic response.

Medved and Kagan (1966), commenting on Talokantsev (1964), noted other types of safety factors. For example, they suggested that to obtain a proper safety factor, one should subtract the value of three standard deviations from the value of the threshold concentration. These are entirely empirical approaches to safety factor derivation. Hence they lack a sound biologic basis for their "reason to be." This was noted by Medved and Kagan (1966), who suggested a more flexible and more comprehensive approach including special considerations of the inherent toxicity of the substance, and the extent of the toxic effects, bioaccumulation, and so on.

Medved and Kagan (1966) also mentioned the use of a mathematical derivation based on set formulae that take into account the quantitative correlations between toxic effects, chemical structure, and physiochemical properties of the toxicant considered. Magnuson et al. (1964) presented the following formula that represents the theoretical basis on which the safety factor in the Soviet Union is often based and the relationship of the safety factor to the MAC.

$$\mathrm{MAC} = \mathrm{L1M_{CH}}/\mathrm{K}$$

$$K = \text{a safety factor}$$

Derivation of K

$$K = a\,\frac{\dfrac{\mathrm{L1M_{AC}}}{\mathrm{L1M_{CH}}} \times \dfrac{\mathrm{C_{20^\circ}}}{\mathrm{LC}-50}}{\dfrac{\mathrm{LC}-50}{\mathrm{L1M_{AC}}}}$$

$$= a\,\frac{(\mathrm{L1M_{AC}})^2 \times \mathrm{C_{20^\circ}}}{(\mathrm{LC}-50)^2 \times \mathrm{L1M_{CH}}}$$

$\mathrm{L1M_{AC}}$ = lowest single dose giving an effect
$\mathrm{L1M_{CH}}$ = lowest repeated dose giving an effect

$C_{20°}$ = vapor concentration at 20°
(LC − 50) = 2 hr LC − 50
a = 1 for vapor? for nonvolatile materials
$L1M_{AC}/L1M_{CH}$ = zone of chronic action
$C_{20°}/(LC - 50)$ = coefficient of possible inhalation exposure
$(LC - 50)/L1M_{AC}$ = zone of acute action

Magnuson et al. (1964) reported that the use of this formula with respect to safety factor derivation may provide considerable insight in explaining some of the striking differences between certain American and Soviet MAC values, especially with regard to aromatic hydrocarbons and chlorinated hydrocarbons, which tend to have high volatility. Magnuson et al. (1964) thought that the use of vapor pressure in the derivation of MACs is an erroneous approach for the following reason. Since the MAC is considered to be a level of an atmospheric concentration of a substance, it does not make any difference how that concentration was reached. The question of volatility is an important one with respect to engineering controls and in selecting appropriate materials, but it is not a factor per se in the development of toxicity.

STANDARDS FOR CARCINOGENS

Whereas considerable controversy has been engendered in the United States concerning the development of an acceptable approach to the derivation of health standards for carcinogens, the U.S.S.R. has acted very strongly. It is the first country in the world to establish the MPC for 3,4-benzo[a] pyrene (Sanotskiy, 1974; Izmerov, 1974). The principle on which the Soviets based their approach to carcinogens is identical to that proposed by Druckrey (1967) and Jones and Grendon (1975), which indicated that the latent period varies inversely with the dosage. Thus the Soviets have chosen to select a MPC that, in theory, will cause an effect (i.e., cancer) after the expected normal life time of the individual (Sidorenko et al., 1974).

The original research supporting the present Soviet standard for 3,4-benzo[a] pyrene was conducted with groups of rats exposed to a variety of doses (0.005, 0.02, 0.1, 0.5, and 2.5 mg) by using a tenfold intratracheal administration (Yanysheva and Antomonov, 1976). The results indicated that the number of rats with tumors decreased as the benzo[a]pyrene dose was lowered. Furthermore, there was an inverse relationship of latent period and dose. The data of this experiment revealed that a total dose of 0.1 mg is weakly carcinogenic, whereas at 0.02 and 0.005 mg, an increased cancer incidence was not evident.

Yanysheva and Antomonov (1976) mathematically estimated the appearance of the initial lung tumor after administration of the different total doses of benzo[a]pyrene (Table 5-3). At 0.05 mg the first lung tumor is projected to occur toward the very end of the animal's life span. Lower doses (0.02–0.002 mg) are projected to initiate lung tumor development progressively beyond the animal's natural life span. On the assumption that humans may respond in similar fashion to these rats with respect to dose-time-effect relationships, Yanysheva and Antomonov (1976) recommended a dose of 0.02 mg as the permissible dose used in the dose-concentration formula.

It should be noted that the research presented here on which the 3,4-benzo[a]pyrene standard is said to have been based used only a relatively small number of animals per experimental group (i.e., the number ranged from 16 to 40 per group). The authors discussed the statistical constraints such a sample size places on subsequent conclusions.

The data on the benzo[a]pyrene-induced carcinogenesis in rats were impressive with respect to developing a dose-time-effect relationship. Whether data derived from rat experiments should be directly extrapolated to humans (despite generous safety margins) is highly debated. The arguments against such attempted derivations (such as cocarcinogenesis, the presence of high-risk groups, multiple carcinogens, synergism, species variations, etc.) are addressed elsewhere (Chapter 5). A number of Western environmental/occupational toxicologists and epidemiologists (Jones and Grendon, 1975; Thomas and Busick, 1976; Druckrey, 1967; Enterline, 1976) seem to support the approach adopted by the Soviets.

ROLE OF HIGH-RISK GROUPS

In the process of standard setting, most Soviet authors reviewed here have not provided noticeable emphasis on the role of identifying and quantifying individuals who may be at high risk, perhaps because such a strong emphasis is placed on animal experimentation as compared to epidemiologic studies. However, Izmerov (1973), commenting on Ciziko (1970), reported that animal models of various high-risk conditions such as pulmonary insufficiency and hormonal insufficiency are used to predict toxicity on these subgroups of the population. Krasovskii (1976) also noted the Soviet use of animal models with epilepsy, nephritis, and toxic myocarditis in studies designed to predict the enhanced toxicity of individuals biologically predisposed to the adverse effects of pollutants. He concluded by stating that "the increased vulnerability of population samples suffering from chronic disorders must be considered in studying the problem of the reliability of experimentally established health standards." In the development of the standard for the carcinogen 3,4-benzo[a]pyrene, Yanysheva and Antomonov (1976) decided to conduct their studies on random-bred animals, because they are considered genetically heterogeneous. The authors argued that since this animal species was highly variable it most closely simulated the highly diverse human population. Although this may be true, one may conclude that the standard for 3,4-benzo[a]pyrene was based on an animal model that did not give any special consideration to potential high-risk groups. It is interesting to note that Yanysheva and Antomonov (1976) did mention that it would be advisable to employ animals with high cancer susceptibility in testing the effect of a small carcinogen dose; however, they rejected this reasoning in their actual experimentation.

RECOMMENDED SAFE CONCENTRATION ZONES

The Report of the Sixth Session of the Joint ILO/WHO Committee on Occupational Health (1968) stated that the TLVs of the ACGIH and the MACs of the U.S.S.R. are very similar for 24 industrial and/or agricultural chemicals. In these cases, the values were within a factor of 2. Furthermore, most of the countries adopting industrial health standards have followed the recommendations of either the ACGIH and/or the U.S.S.R. Thus the WHO committee developed a list of "safe concentration zones" for the 24 industrial chemicals. However, when one considers that there are over 500 toxic substances that have TLVs and MACs, agreement on 24 represents only approximately 5% of the total. The committee noted that rather large differences often exist between the ACGIH and U.S.S.R. values, with a difference of a factor of 90 in several instances. They suggested that attempts should be made to understand and eliminate such differences. The implication is that it is nearly impossible to adopt an approved WHO standard without agreement between the American and Soviet perspectives. However, based on the fundamental differences between the Soviet and American approaches to setting standards for noncarcinogens, the outlook for future widespread agreement is not optimistic.

REFERENCES

ACGIH (1976) Documentation of the Threshold Limit Values. Cincinnati, Ohio.

Dinman, B. D. (1972) "Non-Concept" of "no-threshold": chemicals in the environment. *Science* 175:495–497.

Dinman, B. D. (1976a) Development of workplace environment standards in foreign countries. Pt. 1—Historical perspectives; Criteria of response in the U.S.S.R. *J Occup Med* 18(6):409–417.

Dinman, B. D. (1976b) Development of workplace environment standards in foreign countries. Pt. 2—Concepts of higher nervous function in the U.S.S.R. *J Occup Med* 18(7):477–484.

Dinman, B. D. (1976c) Development of workplace environment standards in foreign countries. Pt. 3—Procedures for the development of MAC values in the U.S.S.R. *J Occup Med* 18(8):550–560.

Druckrey, H. (1967) Quantitative aspects in chemical carcinogenesis. In R. Truhaut, ed., *Potential Carcinogenic Hazards from Drugs,* Vol. 7, p. 60. *Evaluation of Risks.* VICC Monograph Series, Berlin: Springer.

Elkins, H. B. (1961) Maximum acceptable concentrations. *Arch Environ Health* 2:45.

Enterline, P. E. (1976) Pitfalls in epidemiological research. *J Occup Med* 18(3):150–156.

Glass, R. I. (1975) A perspective on environmental health in the U.S.S.R. *Arch Environ Health* 30:391–395.

Hatch, T. F. (1972) Permissible levels of exposure to hazardous agents in industry. *J Occup Med* 14:134–137.

Izmerov, N. F. (1973) Principles underlying the establishment of air quality standards in the U.S.S.R. In *Control of Air Pollution in the U.S.S.R.,* pp. 42–60, 129–132. Geneva: WHO.

Izmerov, N. F. (1974) Estimating the maximum permissible intensity of the complex effect of chemical factors in the production, municipal and domestic environment on man. Proceedings of 1st U.S./U.S.S.R. Symposium on a Comprehensive Analysis of the Environment. March 25–29, 1974, pp. 45–48.

Jones, H. B. and Grendon, A. (1975) Environmental factors in the origin of cancer and estimation of the possible hazard to man. *Food Cosmet Toxicol* 13:251.

Krasovskii, G. N. (1976) Extrapolation of experimental data from animals to man. *Environ Health Perspect* 13:51–58.

Lyublina, E. I. and Rabotnikova, L. V. (1971) The possibility of prognosticating the toxicity of volatile organic compounds judging by physical constants. *Gig i Sanit* 36:33.

Magnuson, H. J. (1965) Soviet and American standards for industrial health. *Arch Environ Health* 10:542–545.

Magnuson, H. J., Fassett, D. W., Gerarde, H. W., Rowe, V. K., Smyth, H. F., Jr., and Stokinger, H. E. (1964) Industrial toxicology in the Soviet Union—theoretical and applied. *Indust Hyg Assoc J* 25:185–197.

Medved, L. I. and Kagan, J. S. (1966) Toxicology. *Ann Rev Pharmacol* 6:293–304.

Rjazanov, V. A. (1965) Criteria and methods for establishing maximum permissible concentrations of air pollution. *Bull World Health Org* 32:389–398.

Sanotskiy, I. V. (1969) Ways and means of devising methods for setting MPC of noxious substances in the air of production areas. *Gig i Sanit Prof Zabol* 13:4 (cited in Dinman, 1967c).

Sanotskiy, I. V. (1974) The concept of the threshold nature of the reaction of living systems to external actions and its consequences in the problems of protecting the biosphere against chemicals. *Proceedings of the 1st U.S./U.S.S.R. Symposium on a Comprehensive Analysis of the Environment.* March 25–29, 1974, pp. 49–54.

Sidorenko, C. I., Korenevskaya, Ye. I., Pinigin, M. A., and Krasovskiy, G. N. (1974) Hygienic bases for protecting the environment. *Proceedings of the 1st U.S./U.S.S.R. Symposium on a Comprehensive Analysis of the Environment.* March 25–29, 1974, pp. 63–66.

Stern, A. C., Wohlers, H. C., Boubel, R. W., and Lowry, W. P. (1973) *Fundamentals of Air Pollution* New York: Academic.

Thomas, R. H. and Busick, D. D. (1976) Reducing patient exposure to ionizing radiation—is it really necessary? *Am Ind Hyg Assoc J* 37:657–664.

Yanysheva, N. Y. and Antomonov, Y. G. (1976) Predicting the risk of tumor occurrence under the effect of small doses of carcinogens. Environ Health Perspect 13:95–99.

12

Asbestos: A Case Study

THE PREVIOUS CHAPTERS HAVE shown how standards are derived for both environmental and occupational pollutants. At this point, it is appropriate to synthesize these principles and concepts by focusing on how this entire process may be applied to one pollutant by way of a case study. This chapter presents in a detailed fashion the way in which the occupational health standard for asbestos was derived and implemented within society, including (1) the historical basis of how asbestos was used and became a "pillar" of modern society, (2) the physical and chemical properties of asbestos, (3) biologic effects including asbestosis, lung cancer, and mesothelioma, and (4) how industry, government (OSHA and NIOSH), and labor unions interacted from their own perspectives in dealing with the health and economic issues inherent in the problem. Finally, considerable emphasis is placed on how the government actually derived its standards as well as what this means for the health of the worker and the economic impact on the industries involved.

HISTORICAL PERSPECTIVE

Man's fascination with asbestos has been known for nearly 5000 years. In fact, there is reliable evidence from archaeological finds from Finland that asbestos was employed 4500 years ago to bind pottery; about 2500 years ago, the ancient Greeks used it to aid in the collection of ashes from cremated corpses (Villecco, 1970). Anecdotal stories about the life of Charlemagne relate that he had his tablecloth woven from asbestos for use in "witchcraft" to frighten his rivals and enemies. Thus, after a meal, the asbestos tablecloth would be tossed into the fire, and several moments later it would be withdrawn unburnt and even whiter and cleaner than before. As a result of this "mysterious" quality, the substance was called "asbesta" which meant "unquenchable or the inextinguishable" (Safety Standards, May–June 1972).

In our modern civilizations, asbestos has become a vital part of many industrial processes because of its ability to resist heat stresses of up to 500°C without decomposition. Its fibers are used to produce fireproof draperies, textiles, brake linings, boiler blankets, valve packings, building materials, and appliance cords. Asbestos is sprayed on steel foundations, girders, and ships for both thermal insulation and fireproofing. In addition, it does not shrink or crack, and it dries and hardens in a minimum of time (Safety Standards, May–June 1972).

The unique benefits with which asbestos has provided society must be considered now in light of the adverse health effects exposure to asbestos may cause. Since the 1950s there has been a progressive and (now) rapidly developing fund of knowledge concerning the carcinogenic nature of the human response to asbestos exposure. These diseases, including respiratory cancer and mesothelioma, which have resulted from contact with airborne (or perhaps waterborne) asbestos fibers, are for the most part irreversible and often fatal. It has been suggested that only through strict regulation and enforcement of the mining, processing, and use of this mineral or perhaps by its functional extinction will the exposure of asbestos fibers to future populations be kept to a minimum. Thus the question this chapter addresses is, how does society deal with a known carcinogen, especially a substance on which society itself has become highly dependent?

PHYSICAL/CHEMICAL PROPERTIES—INDUSTRIAL APPLICATIONS

Asbestos is a general term applied to a group of naturally occurring fibrous mineral silicates that differ in chemical composition. The asbestos silicates can be further divided into two subgroups: (1) the pyroxenes, which include chrysotile asbestos, the most widely used type of asbestos in United States industry and (2) the amphiboles, which include amosite, crocidolite, tremolite, anthophyllite, and actinolite asbestos.

These varieties of asbestos have different origins, properties, and different functions in industry. Three principal types of asbestos used in the United States are chrysotile, crocidolite, and amosite. Chrysotile, often called white asbestos, comes from the mineral serpentine and is the most widely used form of asbestos. Accounting for over 90% of the world's consumption of asbestos, chrysotile is mined worldwide; the largest deposits are found in Canada, Russia, and Rhodesia. Chrysotile is easily crushed during milling and produces fine, silky fibers that can be processed into many heat-resistant products (Hendry, 1965; Gaze, 1965).

A second type of asbestos not mined or milled in the quantities of chrysotile, but still of great importance to industry, is crocidolite, which accounts for 5% of the world's use of asbestos. Mined in South Africa, Australia, and Bolivia, this form of asbestos is derived from the mineral niecidolite, which has fine, resilient fibers. Niecidolite is a more specialized form of asbestos and is used in the manufacture of asbestos cement sheets and pressure pipes (Federal Register, 1975; Hendry, 1965; Gaze, 1965).

Amosite is the third major type of asbestos used in the United States and accounts for nearly 5% of the total use of asbestos in the world. Derived from the mineral grunerite, it is mined solely in South Africa. Amosite is used especially as a component in producing asbestos cement and heat-insulating products (Federal Register, 1975; Hendry, 1965; Gaze, 1965).

The remaining types of asbestos, anthophyllite, tremolite, and actinolite, are not used in American industries, but are used in other countries. Anthophyllite, mined in Kenya and Finland, is relatively rare and is used primarily as an inexpensive insulatory filler. Tremolite, which ironically is mined mainly in three states, New York, Vermont, and Montana, is found variably as a component of both industrial and commercial talcs. Actinolite is very rare and is seldom used in industry because of its low fiber strength (Federal Register, 1975).

The mining and milling of asbestos in the United States is a relatively small industry compared to those of other countries; less than 1000 workers are employed. However, in the United States, it is in the production of asbestos products that the fibers are released, and in that way, most of the occupational exposure to asbestos occurs.

The quantity of asbestos used in 1972 in the construction industry amounted to 186,000 short tons or 77% of the total American asbestos consumption (Bureau of Mines, 1973). Furthermore, in the construction industry, 92% of asbestos used in the manufacturing processes is locked into the product, in materials like floor tiles, cements, roofing felts, and shingles. The other 8%, used in powdered form, is present in such materials as insulation, cement products, and acoustical products. It is the loose powdered form of asbestos that presents the greatest danger to health in the construction environment, because the manufacture and use of these products result in more airborne fibers than the firmly bonded products containing asbestos. However, even in the manufacture of "locked in" asbestos products, workers in the industry are exposed directly to fibers as they mix, dump, and clean asbestos in their occupational chores (Federal Register, 1975).

In nonconstruction industries, asbestos is used to produce textiles, paper, paints, plastics, roof coatings, floor tiles, and friction materials like brake linings and clutch facings. Here, too, asbestos fibers are often released like a cloud into the air as paints are sprayed or as brake linings encounter resistance. Those products such as textiles in which asbestos is "locked in" present little hazard, but production itself, unless safely controlled, could expose workers to dangerous levels of asbestos (Federal Register, 1975).

RECOGNITION OF ASBESTOS AS A HEALTH HAZARD

The hazards of breathing asbestos dust have been recognized for many years. In 1906 a physician in London noted the death of a 33-year-old man, employed for 14 years as a textile worker, whose main responsibility was stirring asbestos. The death was due to severe lung scarring (asbestosis), and the man was the last survivor of a group of ten men working in the same factory (Departmental Committee, 1907). In 1924 a British physician reported on the occurrence of "curious bodies" throughout the lungs of a female asbestos worker. In this case, the cause of death was related to lung scarring (Cooke, 1924). In addition to development of asbestosis, evidence emerged during the 1930s that asbestos may also be a human carcinogen. The first evidence for such an association came in 1935 when Lynch and Smith reported a death from lung cancer in association with asbestos.

The first attempt to derive a standard in the United States for controlling worker exposure to asbestos occurred as a result of a study of 541 employees from several asbestos textile plants in 1938. The results indicated that there were few, if any, actual cases of pneumoconiosis where dust concentrations were less than 5 million particles per cubic foot (mppcf), whereas numerous cases were found when levels exceeded 5 mppcf. The study had its limitations in that many of those people screened for various disorders (333) had not been working with asbestos for more than 5 years, and only 66 had worked there more than 10 years. Therefore, the authors of this standard stated that. . . . "5 mppcf may be tentatively regarded as the threshold limit value (TLV) for asbestos dust exposure until better data are available" (PHB, 1938). This figure was later adopted by the ACGIH as their recommended TLV (Schall, 1965).

A major uncertainty with this original TLV was that it lumped all particles together via the use of the impinger technique. The chief culprit in asbestosis is the asbestos fiber itself in contrast to the numerous other dust particles that are common in any workplace. To quantify more precisely the number of asbestos fibers in an air sample, several analytic techniques, including the membrane filter method and phase contrast microscopic analysis, are now used.

Following these developments in analytic techniques, it became possible to base standards on the number of respirable fibers in the air, rather than the total dust counts. A "respirable fiber" is defined as any fiber of a length of 5μ or more with a length-to-diameter ratio of 3 to 1 and a maximum diameter of 5μ (Federal Register, 1975). Applying this definition of a fiber to the old TLV of 5 mppcf, a value of approximately 30 asbestos fibers per cc is obtained. The first standard using fiber counts lowered the standard to 12 fibers per cc in the late 1960s. This limit remained until December 1971, when OSHA issued an emergency standard of 5 fibers per cc (Vare, 1972).

Although the earliest considerations of the adverse health effects of asbestos emphasized the development of asbestosis, information of a cancer association continued to grow. In 1947, 12 years after the initial study of an asbestos-associated lung cancer, it was reported in England that greater than 50% of patients diagnosed as having asbestosis ultimately died of lung cancer (Waldbott, 1973). A later British study in 1955 by Richard Doll noted 18 deaths from lung cancer from 105 postmortems of workers from a single asbestos works. In addition to lung cancer, there were increased instances of cancers of the stomach, colon, and rectum. These results prompted more extensive research efforts in different countries. Consequently, from the late 1950s through the mid 1960s, there was a proliferation of epidemiologic studies concerning the role of asbestos as an industrial carcinogen. The next section reviews the historical development of the derivation of occupational health standards for asbestos exposure in Britain and the United States. The role of the previous health studies as well as the epidemiologic and toxicologic research of the 1950s to the present is reviewed with respect to these standards.

A STANDARD IS DEVELOPED

The first official asbestos standard affecting all American industries covered by the OSHAct was an emergency standard of 5 fibers per cc put forth by OSHA in 1971. This emergency standard was promulgated because it was well-established that workers in the United States who were regularly exposed to asbestos dust had an enhanced risk of contracting asbestosis, which could result in permanent disability and/or death. However, to understand the development of the OSHA standard for asbestos, it is necessary to consider first the British experience with asbestos and its relationship to the development of subsequent asbestos standards in the United States.

In Britain during the 1960s a committee was organized to study morbidity and mortality due to asbestos exposure as well as technological approaches to the prevention of asbestos-related disease. The committee, operating under the British Occupational Hygiene Society (BOHS), published a report in 1968 entitled "Hygiene Standards for Chrysotile Asbestos for the Prevention of Asbestosis." The report suggested that the incidence of asbestosis could be eradicated if occupational exposure to asbestos was identified and monitored in both factories and end-product use.

The committee study in Britain was initiated because several commercial forms of asbestos were demonstrated to have a positive association with the incidence of a variety of diseases in numbers greater than expected. The diseases (asbestosis, bronchogenic carcinoma, mesothelioma, and cancers of the stomach, colon, and rectum) did not become apparent until at least 20 years after the first exposure, at which time the asbestos-related diseases became evident by way of clinical signs and/or x-ray films.

The British study was based on data derived from a single large asbestos mill, which indicated that there was little clinical or x-ray evidence of asbestos disease at that mill in workers who had been employed there at least 10 years. It was thought that this mill would produce an ideal model from which to determine the level of asbestos exposure accounting for the low prevalence of asbestos disease in the worker population. Further, the BOHS hoped that the level of fiber exposure in that mill could be used to determine a safe standard for asbestos exposure. Thus, using modern fiber counting methods, the

Society used the mill as a model for a standard in determining a "safe" exposure level to asbestos dust. According to the survey conducted in 1966 that correlated modern fiber counting methods (i.e., membrane filter method) to those used previously (i.e., impinger) to assess dust levels, the British mill workers' exposure to asbestos dust from 1933 to 1966 was estimated to have been 4–15 fibers per cc. These data, along with the low prevalence of asbestos-related disease at the mill, carried considerable influence in the development of the British standard, since information on both exposure and effects of asbestos was available (BOHS, 1968).

The BOHS committee came to the conclusion that since there was infrequent disease at asbestos fiber levels of 4–15 fibers per cc, the lowering of the level to 2 fibers per cc as a time-weighted average (TWA) would afford workers occupationally exposed to asbestos little or no risk of developing asbestosis. It was calculated that workers exposed to 2 fibers per cc for 50 years or 4 fibers per cc for 25 years would have a 1% chance of developing early signs of asbestosis (BOHS, 1968).

In setting this standard, the British realized that asbestosis was not the only disease that could result from asbestos exposure; some information had been gathered that related an increased risk of cancer development to asbestos exposure. However, the BOHS committee, although it recognized this problem, developed its standard based on information related to the risk of developing asbestosis, noting that it was "not possible at this time to specify an air concentration of asbestos which is known to be free of cancer risk" (BOHS, 1968).

In developing a standard for asbestos, the United States relied heavily on the evaluation by the BOHS committee, because there were few comparable data collected in the United States. Even though there were a variety of epidemiologic and clinical studies (Dunn et al., 1960; Mancuso and Coulter, 1963; Selikoff et al., 1964; Enterline, 1965; Cooper and Balzer, 1968) in the United States concerning asbestos exposure, few dust counts had ever been taken over the years; those which had been taken did not use the more modern and valid membrane filter counting method. Thus, when regulatory agencies in the United States realized that there was a need for a permanent standard based on clinical and epidemiologic studies, NIOSH affirmed the British recommendation of 2 fibers per cc for use in the United States.

As previously indicated, an emergency standard for asbestos was promulgated by OSHA in 1971. Although occupational studies from Great Britain and the previously mentioned American studies had suggested a standard level of 2 fibers per cc for the prevention of asbestosis, OSHA decided not to immediately lower the standard as recommended by NIOSH. Instead, it appeared to OSHA that the asbestos hazard could be controlled by limiting exposures to 5 fibers per cc TWA for 4 more years, until 1976, when it would be permanently lowered to 2 fibers per cc. In putting forth this standard, OSHA hoped that the reduction of asbestos fiber levels to prevent asbestosis would also lead to the control of asbestos-related cancers, an assumption that was openly debated (Brodeur, 1974). The promulgation of this standard, which was not as strict a standard as NIOSH had recommended, was an avenue of compromise for OSHA because of political, economic, and social pressures surrounding the asbestos issue (Brodeur, 1974; Safety Standards, May–June, 1972; Little, 1972).

Prior to the promulgation of the emergency standard, industry had been subject to a voluntary standard of 12 fibers per cc adopted by the ACGIH (a volunteer organization that recommends safety standards for hazardous substances in industry; see Chapter 7). Despite governmental encouragement for industry to achieve the voluntary standard of the ACGIH, violations of this limit were common. In fact, even though the government set forth an emergency standard in 1971 of 5 fibers per cc, violations continued to occur (Brodeur, 1974). A case in point involved the asbestos product manufacturing plant in Tyler, Texas.

THE TYLER, TEXAS INCIDENT

Union Asbestos and Rubber Company (UNARCO) opened an asbestos plant at Tyler, Texas in 1954 and later sold it to the Pittsburg Corning Corporation (PCC) in 1962. The following year, the new owners asked the Industrial Hygiene Foundation of America to evaluate Tyler's asbestos dust hazard; the Foundation complied by sending engineers in July and August of that year. The report from this investigation indicated that the plant was well below the TLV of 5 mppcf recommended by the ACGIH in 1946. However, there was a significant error in this report. The foundation apparently interpreted 5 million particles to mean 5 million fibers. Normally, fibers constitute approximately 10% of total airborne dust, but at Tyler, they were later discovered to average about 56% (Brodeur, 1974). In 1966 PCC conducted its own survey of the asbestos levels. The results revealed that of the 16 areas of the plant that were analyzed, seven were above the TLV and three had reached an incredible 20 mppcf (Brodeur, 1974).

These results were so alarming that in 1967 Dr. Lee Grant, medical director for PCC, asked the USPHS Division of Occupational Health to conduct another survey. The results were even more disturbing than those of the previous survey. Using modern methods of fiber counting, the TLV was found to have been exceeded in 44 of 55 samples taken. Despite this evidence, there was no mention of a health hazard at Tyler, and no recommendations were made to improve its inadequate ventilation system (Brodeur, 1974).

On February 13, 1969, the U.S. Department of Labor, Wage and Labor Standards Administration held a safety and health inspection at Tyler, and they found the TLV confusing because the 5 million particles included many nontoxic dusts. This was the last inspection of any kind under the old standard. Throughout these inspections and surveys, it is not known if any of the employees at the factory received the benefit of medical interpretation, but it seems unlikely. Two years after the 1967 survey, one of the workers died of mesothelioma (Brodeur, 1974).

After the standard was lowered to 12 fibers per cc of air, Tyler came under even closer scrutiny. In October of 1971, Dr. William M. Johnson of the newly created NIOSH conducted a survey and found the asbestos levels much higher than the 12 fiber standard. For PCC, this information was not surprising, but this time the results of the survey were made known to union representatives. As a result of this survey, considerable pressure was put on Secretary of Labor James Hodgson to declare an emergency standard of 2 fibers per cc; Hodgson set the emergency standard at 5 fibers per cc in December, 1971 (Brodeur, 1974).

Why was such a blatant violation tolerated for so long? In this case, it was due to lack of medical interest by the company doctor as well as the potential economic concerns of the company and ignorance of the potential hazard by the workers (Brodeur, 1974).

In addition to the violations of the asbestos standard at the Tyler factory, epidemiologic investigations of workers from other asbestos industries revealed the existence of a very grave and widespread health problem. From 1963 to 1966 a considerable amount of data concerning the health effects of asbestos had been amassed. For example, insulators who spent half their time working with a product composed of 15% asbestos had radiologic evidence of asbestosis. Out of a total of 1117 workers belonging to the Union of Heat and Frost Insulators and Asbestos Workers, half had asbestosis. In addition, of 392 men with more than 20 years experience with asbestos, 339 had asbestosis. In more than 50% of the cases, the disease was of a moderate or extensive nature (Selikoff et al., 1965). In 1966 the New York Academy of Sciences, after its Conference on the Biological Effects of Asbestos, sent copies of its report on the health hazards of asbestos to health agencies, doctors, and medical libraries throughout the country in an effort to inform persons of the risks of asbestos (Brodeur, 1974).

The standard of 5 fibers per cc adopted by OSHA was not decided on without many hearings and stormy debates. The hearings, involving testimony from industry and its medical advisors as well as representatives from NIOSH and the Unions and their medical researchers, were sponsored by OSHA. Of primary consideration was the fact that the past standards and those used in other countries were designed to prevent asbestosis, but they did not give equal consideration to asbestos-related cancers. At the hearings, Dr. Nicholson, an associate of Dr. Selikoff, stated that no knowledge presently existed on a "safe" working level exposure to asbestos and that "there is evidence that even a standard of 2 fibers/cm^3 of air will be inadequate for the prevention of asbestos disease" (Brodeur, 1974). Many believed that OSHA had to ultimately reduce the level of fiber exposure to zero to comply with regulations cited in the OSHAct of 1970, which stated "no employee will suffer diminished health, functional capacity or life expectancy as a result of his work experience."

ECONOMIC FACTORS AFFECTING THE ASBESTOS STANDARD

Why didn't OSHA accept the NIOSH-recommended standard? Possible reasons include both political and economic interests. OSHA may have been influenced by the pressures exerted by industrial leaders who did not wish the standard to be lowered for economic reasons. In the hearings, certain industries made it clear to OSHA that they believed the lowered standard would price American-made asbestos products out of the domestic and international markets, because of the expense of buying the technology necessary to achieve the proposed standard, a technology that had not been perfected. Instead, industry advocated the reinstatement of the threshold value of 12 fibers per cc for asbestos since there was an unresolved controversy concerning what constituted a "safe" level of exposure. In further testimony, the Johns-Manville Company, a large asbestos products producer, indicated that to achieve a standard of 2 fibers per cc, it would require a capital expenditure of $12 million per year. The company stated that such an outlay would not be possible and would result in the closure of five plants and cause unemployment for 1600 workers (Brodeur, 1974). Industrial leaders stated that the economy of America would suffer if the standard were lowered and argued that it would be complete social irresponsibility to adopt a standard of 2 fibers per cc for occupational exposure to asbestos.

Arthur D. Little Co. Economic Impact Statement

Adverse Health Effects Estimate

For a better indication of why OSHA did not accept the NIOSH recommendation, a closer consideration of the economic impact of such a proposed standard should be made. OSHA contracted with the internationally recognized consulting firm of Arthur D. Little, Inc. to undertake a study to identify the impact of possible standards (2, 5, 12, and 30 fibers per cc) for asbestos. The conclusion of the Little Report clearly spelled economic trouble for the standard of 2 fibers per cc proposed by NIOSH. In fact, their first conclusion was that with a standard of 5 fibers per cc, a marked reduction in the asbestos-related diseases (including asbestosis, lung cancer, and mesothelioma) would occur that would result in less than a 1% incidence of such diseases following a 40-year working lifetime. The cost of implementing a standard of 5 fibers per cc was viewed as highly variable, depending on the industry, with only the textile industry and on-board ship repair seriously effected (i.e., these two industries faced possible elimination). However, if a 2-fiber standard were implemented and enforced, the cost to industry would be approximately double that of complying with the 5-fiber standard. It was also suggested that the increased cost to consumers of American asbestos-related goods would markedly increase foreign competition. Finally, the "possible" elimination of the textile and ship repair industries would become a reality. The states in which asbestos manufacturing industries were concentrated would be the hardest hit. These included New Jersey, Pennsylvania, North Carolina, South Carolina, California, Illinois, Ohio, Connecticut, and Massachusetts. Thus they concluded that the 5-fiber standard and not the 2-fiber standard would be the most cost effective. Furthermore, in addition to offering the most economic benefits, the 5-fiber standard was also intended to provide greater than 99% of the health benefits obtainable from total dust control.

With this information in hand, it can easily be seen why OSHA proposed the standard of 5 fibers per cc. However, since the acceptance of either the 2- or 5-fiber standard hinges, for the most part, on this report, it is necessary to consider it in more detail to evaluate the soundness of its methodology.

According to the Little Report (1972), there is considerable evidence implicating the toxicologic and carcinogenic effects of asbestos as a serious hazard for man. However, despite the overwhelming evidence of asbestos-related adverse effects, there is considerable disagreement among respected experts in the field on the relationship of dose to effect. Probably the most significant factor in this disagreement is that there is uncertainty as to the asbestos exposure level of workers in the 1930–1940s who have subsequently exhibited asbestosis and various types of cancer. Consequently, it was reasoned that as long as there was an insufficient data base from which to derive reliable and generally consistent conclusions, it would be necessary to rely on the professional judgment of experts.

To arrive at a best guesstimate of the dose-response relationship, the Little Company sought a consensus conclusion from 12–15 qualified experts according to a methodologic application of the Delphi process. In this process, the confidential individual judgments are gathered; next, the combined data values are made known to the participants (individual judgments are not identified with any person). The participating members, however, are listed along with the combined data. A second phase involves sending the combined results to the experts for their reevaluation. A third review, identical to the second phase, is usually undertaken. Table 12-1 gives the data obtained from eight members of the panel of experts. The data indicate that marked improvements in health benefits occur when the standard is significantly reduced from 30 to 12 fibers per cc and from 12 to 5 fibers per cc, but the effects are minimal when the standard is changed from 5 to 2 fibers per cc. However, there is a considerable range in the projected response—clearly an indication of the lack of agreement, even among experts, concerning the dose-response relationship.

In a situation in which experts vary to the extent that one says that there will be no excess disease in 40 per 100 workers as compared to 100 per 100 workers, what confidence can one have in any decision? Yet it is from this group of experts, with this range of opinions, that OSHA had to make its decision. It is obvious that the composition of the committee is absolutely critical. How were the panelists selected? Was there a selection bias favoring either the management or labor perspective? Since most experts also have socioeconomic-political orientations that may influence their specific judgments, how was this factor dealt with in the selection process? Questions such as these are of critical significance, since the selection of any numerical standard depends on the collective opinions of individual panel members. The Little Report (1972) noted that the panel members were chosen for their expertise on the subject matter. There is little question that the individuals have

Table 12-1. Asbestos-Related Disease Status of 100 Workers (Excess Morbidity and Mortality) at the End of 40 Years of Exposure to Asbestos (Predominantly Chrysotile)

	Judgments Concerning Number of Instances of Excess Disease[a]			
If Average Exposure Is f/cc	No Excess Disease	Asbestosis	Bronchogenic Cancer	Mesothelioma
2	99.9[b]	0	0	0
	(100–80)[c]	(20–0)	(1–0)	(0.1–0)
5	99	1	0.3	0
	(100–40)	(55–0)	(5–0)	(0.2–0)
12	89	9	1.5	0.1
	(93–20)	(70–6)	(10–.05)	(1–0)
30	78	19	3.4	0.1
	(90–9)	(75–9)	(15–1)	(2–0)

[a] Columns 2 + 3 + 4 + 5 = 100 for estimates of panel members. Use of median values results in some rows that do not sum to 100. Numbers in parentheses indicate full range of data.
[b] The median value is the middle number when all values are ordered from the largest to the smallest. For example, of the eight estimates, 100, 100, 100, 99.9, 99.9, 98.9, 98.9, 94.0, 80.0, the midpoint falls at 99.9.
[c] The range is the spread between the largest and smallest estimates, in this case, 100 and 80.

Source: Little, A. D. and Company (1972). Economic Assessment for an Occupational Asbestos Standard.

national reputations in this area. However, the Little Report does not address the very real problem of political bias on the part of panelists. As distasteful as this matter may be, it must be addressed. This is not necessarily intended to be a direct criticism of the panel in question, but as a critique of the methodologic process with respect to the selection of an expert group.

Another point that the conclusions of the Little Report tended to overlook is that when comparing the 5-fiber and 2-fiber standards, a no-excess disease rate of between 99 and 99.9% was considered insignificant. However, this is a matter of perspective. Considering a total American asbestos worker population of 200,000 to 500,000, this could result in several thousand additional diseases. Based on one's perspective, this number may not be considered important, but the opposite opinion is certainly held by many. In any case, the projected absolute number as well as relative percentages of excess diseases should be mentioned.

Economic Costs

The estimation of the actual cost to industry if certain standards were to be implemented was determined in a manner similar to that for the estimated health effects. The panel members consisted of experts from the asbestos industry. The panel was asked to estimate the feasibility of meeting alternative standards and the costs for compliance and allocations of resources for achieving the different standards. During the course of study, asbestos users were separated into four general groupings:

1. Manufacturers (including asbestos cement products, high and low pressure pipe, corrugated and flat board products, vinyl asbestos floor tile, friction materials, packing and gaskets, insulation, piping materials, textiles, certain plastics, paints, roof coating, and other miscellaneous materials)
2. Applicators and installers of asbestos insulation
3. Private shipyards
4. U.S. Navy shipyards

Manufacturers The Little Report made use of the comments of the industrial panel, which included 10 individuals representing the different manufacturing areas; to serve as a check on this methodologic approach, the Little Company conducted their study of expenses at a few industrial locations (see Appendix B of the Little Report). The report compared three different plants based on the number of employees with respect to the actual itemization of equipment purchases and installation costs. Table 12-2 is taken from the Little Report and represents the finding of the panel. A comparison between the two assessments with respect to a 5-fiber standard was fairly similar, with a $750–4100 per worker cost range ($1,116.34 median) for the panel, whereas the Little Company obtained a range of $740–1770 per worker. Although there was a seemingly large difference between the two estimates, they are actually in the same general range. The concept of utilizing several different techniques is an excellent approach to solving complex problems, especially when the end result is based on judgment. However, it should be noted that the use of only one

Table 12-2. Manufacturing: Panel Estimates of Dollar Cost for Compliance with Standards (cost increment per worker/per industry segment)

Standard Fiber Level f/cc	Textiles		Other Manufacturing		Applicators		Total		Percentage Gross 1971	
	Capital in 2 Years	Annual Operating	Capital in 2 Years	Annual Operating	Capital in 2 Years	Annual Operating	Capital in 2 Years	Annual Operating	Capital in 2 Years	Annual Operating
2	70,000 70MM	1,500 1.5MM	2,860 86MM	1,573 47MM	1,000 36MM	2,000 72MM	192MM	121MM	12	8
5	4,100 4.1MM	750 .8MM	1,634 49MM	970 30MM	750 27MM	1,000 36MM	80MM	37MM	5	2
12	1,770 1.8MM	450 .5MM	1,125 34MM	500 1.5MM	500 18MM	500 18MM	54MM	18MM	3	1

	Workers	Gross 1971 Sales	Cost as Percentage of Gross: 2 Fibers Capital 2 Years	2 Fibers Operating	5 Fibers Capital 2 Years	5 Fibers Operating	12 Fibers Capital 2 Years	12 Fibers Operating
Textiles	1,000	$ 38MM	184	4	11	2	5	1
Other Mfg.	30,000	954MM	9	5	5	3	4	2
Applicators	36,000	570MM	6	13	5	6	3	3
	67,000	$1,562MM						

Compare these estimates of the panel with a separate analysis by ADL at specific plant locations (See Appendix B, pages 74–76).

Source: Little, A. D., Inc. (1972). Economic Assessment for an Occupational Asbestos Standard.

representative plant of the three general size categories is an extremely limited sample size.

Shipbuilding and Repair. The data on shipbuilding and ship repair in the Little Report were obtained via the Navy and the Shipbuilding Council, with the latter utilizing a representative panel that employed the Delphi methodology through two phases. Table 12-3 gives the results from the private shipbuilding/repair panel. As indicated earlier, the main compliance report did not mention how the "representative" group was selected for the estimates. It could be legitimately argued by some that this panel had a vested interest in projecting higher than real costs to encourage OSHA not to adopt a stricter standard. It would seem that OSHA should allow provisions for outside consultants to make their own estimations of costs for compliance. Of course, it may be objected that the most knowledgeable people are those within the industry. Consequently, as in the case of health effects, economics by consensus also revolves around choosing the panel members.

It is also important to note that small firms (i.e., those with less than 20 workers) were not considered in the Little Report. Yet 49% of all asbestos-related manufacturing companies have fewer than 20 workers; also, in this regard, the asbestos applicator and installer industrial segment is composed of a large number of companies with less than 20 workers. For these companies, compliance with the 5- or 2-fiber standard will be even more difficult. It is an accepted fact of life that small companies have a comparatively more difficult time obtaining capital and resources. Ashford (1976) indicated that "attempts to diminish occupational hazards are often complicated by other kinds of economic problems, not necessarily identifiable as marketable imperfections. Safe conditions are often more expensive to achieve for smaller firms than for larger ones since the cheapest control technologies often require a minimum size that is larger than the operations of many production units. Also, these smaller firms may often find it difficult to locate and pay for the expert advice or information they may require." Thus much of the bias that may have entered into the calculations on the basis of potential "stacking" of the panel in favor of a managerial perspective was certainly diminished by a lack of specific consideration of the economic constraints of the small companies.

BASIS FOR THE NIOSH RECOMMENDATION OF DECEMBER 1976

Because of additional studies concerning the adverse health effects of asbestos and the extensive legal controversies surrounding the initial

Table 12-3. Private Shipbuilding: Panel Estimates of Dollar Cost for Compliance with Standards (amount/production worker for 102,000 workers in 1971)

Standard Fiber Level f/cc	Shipbuilding (65,000 production workers) Capital to Meet in 2 Years Per Worker	Total (MM)	Annual Operating Per Worker	Total (MM)	Repair and Conversion (36,700 production workers) Capital to Meet in 2 Years Per Worker	Total (MM)	Annual Operating Per Worker	Total (MM)	Industry Total (102,000 production workers) Capital to Meet in 2 Years (MM)	Annual Operating (MM)
2	45	2.9	42	2.7	100	3.7	70	2.6	6.6	5.2
5	40	2.6	40	2.6	50	1.8	54	2.0	4.4	4.6
12	25	1.6	22	1.4	28	1.0	50	1.8	2.6	3.2

	Workers	Gross 1971 Sales	Cost as Percentage of Gross: 2 Fibers Capital 2 Years	Operating	5 Fibers Capital 2 Years	Operating	12 Fibers Capital 2 Years	Operating
Shipbuilding	65,000	$1950MM	0.15	0.14	0.13	0.13	0.08	0.07
Repair and conversion	36,700	870MM	0.42	0.30	0.21	0.23	0.11	0.21
Total	102,000	$2820MM	0.23	0.18	0.16	0.16	0.09	0.11

1971 Annual Report, Shipbuilders Council of America.

Comments: from p. 30. (1) one specialty shipyard believes that the 2-fiber level will result in foreign competition that could mean the loss of 1200 jobs; (2) another panel member felt that a 2-fiber per cc standard would shut down 20% of insulation work.

Source: Little, A. D., Inc. (1972). Economic Assessment for an Occupational Asbestos Standard.

promulgation of the asbestos standard of June 7, 1972, OSHA sought to lower the existing standard (NIOSH, 1976). Thus, in October 1975, OSHA proposed* to lower the 1972 standard to an 8-hour TWA concentration of asbestos fibers of 0.5/cc with a ceiling concentration of 5 fibers per cc. Although this proposal was not followed to completion, OSHA requested NIOSH in December of that year to reevaluate all relevant information with respect to developing an occupational exposure standard for asbestos.

Although the 1972 standard advocated by the Criteria Document was believed to be able to "prevent" asbestosis and perhaps "prevent" asbestos-induced neoplasms, new information was becoming available that cast doubt on the "safe" standard level recommended in 1972. Thus, with the publication of new toxicologic and epidemiologic studies implicating asbestos as a serious health hazard and a human carcinogen, NIOSH proposed a new exposure limit for asbestos in December 1976.

In discussing the evidence to support a new standard, NIOSH concerned itself with three main areas of asbestos research: (1) biologic effects of asbestos on animal models, (2) epidemiologic studies, and (3) the reevaluation of former studies in light of recent data. The next section first considers the animal data, after which the epidemiologic data is discussed according to adverse effect (i.e., asbestosis, bronchogenic carcinoma, mesothelioma, pleural calcifications, and other cancers). Next, the role of fiber characteristics (i.e., length, width, shape, and type) in the etiology of disease for both animal models and humans is evaluated. Finally, the development of a dose-carcinogenic response is considered.

Animal Studies

Analyses of animal data with respect to asbestos exposure involved experimental studies concerned with intratracheal, intraperitoneal, and intrapleural injections of asbestos, ingestion, inhalation, fiber analysis within tissues, and mutagenicity (NIOSH, 1976). Studies conducted on hamsters and rats receiving intratracheal injections of chrysotile with benzo[a]pyrene demonstrated that the effect of chrysotile was addictive to that of benzo[a]pyrene in causing tumors of the respiratory tract

* "On February 4, 1976, 65 representatives from companies and trade associations representing manufacturers and processers of asbestos products in the U.S. participated in a meeting held in Washington, D.C. and overwhelmingly endorsed that the 2 fiber level has been or can be attained by application of engineering technology but the proposed 0.5 fiber level is unnecessary, impracticable and lacks medical justification" (Rajhans, 1977).

(Miller et al., 1965; Shabad et al., 1974). However, no lung tumors or mesotheliomas occurred in different experiments in which rats received chrysotile or benzo[a]pyrene alone (Shabad et al., 1974). Intraperitoneal studies have shown that mesotheliomas of the peritoneum occurred with exposure to amosite, crocidolite, and chrysotile asbestos (Reeves et al., 1971, Maltoni and Annoscia, 1974). However, these studies were conducted without adequate controls; therefore care was taken by NIOSH in drawing conclusions about asbestos carcinogenicity (NIOSH, 1976). When asbestos was administered intrapleurally to rats, all commercial types of asbestos were found to produce mesotheliomas (NIOSH, 1976). In these intrapleural studies, a long latent period was discovered in which response from initial exposure to the appearance of tumors was dose related. In one study by Shabad et al. (1974) in which Russian chrysotile was injected intrapleurally into 67 rats, 31 developed mesotheliomas within 2 years.

From experiments in which rats ingested varying amounts of asbestos, a variety of results were reported. Gross et al. (1974) reported that the feeding of chrysotile and crocidolite to rats produced no significant difference in tumor incidence as compared to controls. This study had an extremely small sample size, and much information was lacking including survival rates and pathological details. In contrast, Wagner et al. (1977) reported a reduced life span and increased tumor incidence among rats fed chrysotile in malted milk, as compared to the groups fed only malted milk.

In inhalation studies with rats performed by Wagner et al. (1974), all types of asbestos fibers were found to produce asbestosis. In comparing a control group of rats to those exposed to chrysotile and the amphiboles, it was discovered that an increasing incidence of cancerous tumors was observed with increasing exposures to each form of asbestos. In addition to the aforementioned carcinogenic effects of asbestos, it has been shown to have mutagenic properties. For example, Sincock and Seabright (1975) found that both chrysotile and crocidolite asbestos induced chromosomal aberrations in Chinese hamster cells (*in vitro*).

Epidemiologic Studies

Asbestosis

Asbestosis or asbestotic pneumoconiosis was the first clearly demonstrated morbid effect of asbestos exposure in man. At present, there is general agreement that all forms of asbestos in commercial use can produce fibrotic effects in the lung (Karachova et al., 1969). With asbestosis, the disease is usually classified and diagnosed according to x-rays, with clinical signs being evidenced as lung crepitations, finger clubbing, and dyspnea. Physiologic changes occur that resemble the effects of any restrictive lung disorder such as decreased forced vital capacity, decreased diffusion capacity, and decreased pulmonary compliance.

Smoking is one factor that in conjunction with asbestos exposure, dramatically increases the risk of acquiring asbestosis. In a study of 100 asbestos textile workers, it was found that the prevalence of pulmonary fibrosis was 40% in smokers and 24% in nonsmokers (Selikoff et al., 1964). The prevalence of pulmonary fibrosis rose with increasing amount and duration of cigarette smoking and with increasing exposure to asbestos.

These facts were known before the 1972 United States standard took effect. However, they were viewed in a different light when workers in the British "model" factory, on whose morbidity and mortality experience both the United States and British standards of asbestos were based, experienced greatly increased amounts of asbestotic disease. X-rays of workers in the factory showed abnormalities in the lung or in the coverings of the lung pleura (Howard et al., 1976). Thus, in 1972, there was an increased difference between the prevalence of abnormal x-ray findings among workers than there was in 1966.

In addition, clinical data became available in which asbestos lung scarring and mesothelioma were shown to occur in individuals who were exposed to much lower levels of asbestos than those who worked with asbestos in their occupations. One study in particular examined 210 family contacts of former asbestos workers, and 38% of the *contacts* were reported to have had x-ray changes characteristic of asbestos exposure (Anderson et al., 1976).

Bronchongenic Carcinoma

A second biologic effect whose risk is enhanced by occupational exposure to asbestos is bronchogenic carcinoma. During the 1950s and 1960s, an increased association between lung cancer and asbestos exposure was noted in a number of studies (Braun and Traun, 1958, Selikoff et al., 1964). However, a series of studies in the early 1970s considerably strengthened the relationship between past occupational exposure to asbestos and higher than expected incidence of bronchogenic carcinoma. Table 12-4 summarizes studies from 1958 to 1972 that indicate the enhanced relative risk of asbestos workers to the development of respiratory cancer.

Further research indicates that there may be differences in the degree of risk of lung cancer among different occupationally exposed groups working with asbestos. A study conducted by Enterline et al. (1972) noted differences in mortality between maintenance workers in the same asbestos factory. In this study, the maintenance workers were found to have a higher incidence of mortality by lung cancer than the production workers, a discrepancy that was in need of explanation. Two tentative explanations have been put forth. One was that maintenance workers were intermittently exposed to very high levels of asbestos dust (for example, when a mixing machine was malfunctioning), and that this concentrated high-dose exposure resulted in greater numbers of lung cancers. Another explanation concerned the types of asbestos to which the two groups were exposed. Sixty-four percent of the maintenance workers were exposed to high concentrations of crocidolite asbestos, whereas only 8% of production workers had been exposed to crocidolite, possibly indicating that crocidolite was more carcinogenic than other asbestos dusts.

It has been observed in studies of asbestos workers that 20% of all deaths are caused by lung neoplasms (for both factory workers and direct users of asbestos products (Selikoff et al., 1972). This percentage, however, varies with exposure, age, duration of exposure, and especially duration from onset of asbestos work. Besides these variables, another critical variable affecting the incidence of lung cancer (as well as asbestosis) among asbestos workers is smoking. In 1968, Selikoff et al. found that the incidence of lung cancer was not significantly increased among asbestos workers with no history of cigarette smoking, but when there was a history of cigarette smoking, the incidence of lung cancer increased markedly. In fact, an asbestos worker who smoked had 92 times the risk of dying from lung cancer compared to individuals who did not smoke and work with asbestos, and eight times the risk of developing lung cancer compared to other smokers.

Mesothelioma

The third biologic effect is mesothelioma, which is usually regarded as an exceedingly rare cause of morbidity and mortality. However, among asbestos workers, its frequency is abnormally high. It is known that mesothelial tumors affect both the pleura and peritoneum, but the exact risk of death to workers has not been well defined. Available data

Table 12-4. Prospective Epidemiologic Studies of Asbestos Exposure and Respiratory Cancer in the United States and Canada

Primary Author	Date Published or First Presented	Population Studied	Deaths from Respiratory Cancer			Time Since First Exposure on Entry to Study	Years Cohort Followed	Size of Cohort	Source of Cohort	Basis for Expected Deaths
			Obs	Exp	RR					
Braun	June 1958	Asbestos miners	9	6	1.5	5 yr +	6	5958	Company records	Quebec experience
Dunn	Oct 1959	Asbestos insulators, California	10	2.8	3.6	Not stated	7	529	Union records	California experience
Mancuso	Feb 1963	Mfg. brake lining and textiles	19	5.6	3.4	Unknown	20½	1462	BOASI	Ohio experience
Selikoff (1)	June 1963	Asbestos insulators, NY, NJ	92	10	9.2	20 yr +	29	623	Union records	U.S. experience
Enterline (1)	Oct 1964	Building mtls.,	46	35.4	1.3	Unknown	13½–16½	12402	BOASI	U.S. experience
		friction mtls.,	32	26.6	1.2			7510		
		textiles	14	6.0	2.3			1843		

Cooper	April 1968	Asbestos insulators, San Francisco	14	1.8	7.8	20 yr +	10	250	Union records	California experience
McDonald	June 1971	Asbestos miners	134	93	1.4	1 month or more	Not stated	11107	Company records	Quebec mining region experience
Enterline (2)	Feb 1972	Retired asbestos prod. workers	58	21.7	2.7	25 yr avg.	2–29	1348	Company records	U.S. experience
Wagoner	May 1972	Mfg. brake lining and textiles	14 31	9 11	1.6 2.8	< 20 yr	28	3367	Company records	U.S. experience
Selikoff (2)	Sept 1972	Mfg. insulation	27 49	5 6	5.4 8.2	< 20 yr 20 yr +	25½–30	928	Union records	U.S. experience
Selikoff (3)	Oct 1972	Asbestos insulators, U.S.	24 215	7 37	3.4 5.8	< 20 yr 20 yr +	5	17800	Union records	U.S. experience

Source: Enterline, P. E. (1976) Pitfalls in epidemiological research. *J Occup Med* 18(3):150–156.

suggest that 5–11% of deaths of all asbestos workers may be due to mesotheliomas (Newhouse and Berry, 1975).

Research has established a relationship between the occurrence of mesotheliomas in asbestos workers and the type of asbestos to which a worker is exposed (Wagner, 1965). Presumably, the difference lies in the shape of the particular fiber, that is, the difference in shape between amphibole (amosite, crocidolite, hemolite, etc.) and chrysotile fibers. It has been hypothesized that the amphibole fibers, which are straight, can more easily reach the deepest areas of the lung as compared to the chrysotile fibers which are coiled and more easily intercepted by the bronchi. Although amphibole fibers seem to have a predilection for deeper lung tissue and seem to be associated with a greater number of mesotheliomas than the chrysotile fibers, not all amphibole fibers show this increased prevalence to mesothelioma development.

In addition to fiber shape, fiber diameter may also affect the development of mesotheliomas (Wagner, 1972). Differences in fiber diameter may explain the variations in frequency of mesothelioma between two groups of workers exposed to the same general type of asbestos. For example, a team of researchers noticed that there was a difference in the frequency of mesothelioma between two regions of Africa where crocidolite asbestos was mined. On investigation, they found that the mean fiber diameters in the Transvaal mining region (where the frequency of mesothelioma is low) were three times the diameter of the fibers of the Northwest Cape area (where frequency was much higher). It was suggested that because of their greater size, a higher proportion of Transvaal fibers do not penetrate as deeply into the lungs as compared to the shorter, smaller fibers of the Northwest Cape. Thus perhaps the greater penetration potential of the smaller fibers could account for the differences in the incidence of mesothelioma.

Not all mesotheliomas are related solely to dust exposure; other factors in addition to type of fiber and fiber size must be examined. Factors such as metallic cations, which may be present in asbestos dust, still remain to be evaluated in terms of mesotheliomal development.

Like asbestosis and lung cancer, mesothelioma often has a latency period greater than 30 years. However, unlike the other two diseases, cigarette smoking does not appear to enhance the development of mesotheliomas (Federal Register, 1975).

Pleural Calcifications

Another major health effect of asbestos is the development of pleural calcifications in the lungs. These plaques have been reported to occur among workers exposed to any type of asbestos or asbestos mixtures. Pleural calcifications, like other asbestos-related diseases, have a latency period and usually do not appear until 20 years after the first exposure. An interesting feature of pleural calcification, not so evident in the other asbestos-related diseases, is that it is reported as common in large segments of populations not directly exposed to asbestos (Kiviluoto, 1960). People not occupationally involved with asbestos who show pleural calcifications have lived in or near regions in which asbestos fibers were either produced or used.

Other

Asbestos has also been associated with an increase in gastrointestinal tract cancers (stomach, colon, and rectum) and cancers of the larynx, pharynx, and esophagus. Although the increase in these cancers is less than the increase for lung cancer or mesotheliomas in asbestos workers, the increased risk is still significant (Selikoff et al., 1964, 1968; Elmes and Simpson, 1971). Selikoff (1976) reported that amosite workers with 3 months employment or less in the asbestos industry had excess cancer risks of 3.87 for lung cancer, 1.68 for colon and rectal cancer, and 1.6 for cancer for all sites when compared to expected values in the general population.

Reevaluation of British Data

The British data, on which both the British and OSHA standards for asbestos were originally based, have been made clearer and support the development of a new and more stringent standard (Howard et al., 1976; BOHS, 1968). In the British data of 1968 used to determine the standard, only eight of 290 individuals were found to have x-ray evidence of asbestotic disease. Later investigations revealed, however, that the study consisted primarily of workers employed at the factory for less than 20 years and that the means of measurement of disease were not sufficient. Recent investigations have shown that nearly one-half of the workers now show radiologic evidence of asbestos disease, fibrotic or carcinogenic. In addition, the need for a new standard is supported by the fact that both the British and American standards for asbestos were primarily based on data relating to asbestosis, because a valid dose-response relationship between asbestos exposure and cancer was not known.

Fiber Characteristics and Disease Etiology

Fiber Type

There has been a great deal of speculation on the ability of different types of asbestos to cause lesser or greater amounts of fibrosis. The results of studies to date on the relative fibrogenicity of different asbestos dusts on animals are not consistent, and human data are still inconclusive (Selikoff, 1970). Two studies were conducted in this regard, one involving the particular effects of chrysotile and amosite dust on pipe workers, another involving tremolite in talc. No conclusions applicable to workers in general could be reached, however, since the studies only took into account peak concentrations (greatest amount of fibers per cubic centimeter in 15 minutes during an 8-hour day) and not the total exposure to fibers accumulated by the worker over an 8-hour day.

A good deal of epidemiologic research has been done since 1972 in relation to the types of fibers present in asbestos-related diseases. It is almost impossible to implicate one fiber type as most "pathogenic," since in most industrial processes different types of fibers are mixed; thus exposure to a single type of fiber is infrequent. Most mortality studies of workers in asbestos manufacturing plants, shipyards, and insulation plants have been based on exposure to mixed fiber types, and these studies provided the strongest evidence for the implication of asbestos as a causative agent in bronchial cancer, pleural and peritoneal mesotheliomas, and asbestosis (Newhouse, 1969; Selikoff, 1970). All fiber types have been shown to produce asbestos-related diseases in man (Enterline and Henderson, 1973; Wagoner et al., 1973; Seidman et al., 1976).

Research concerned with fibers in tissues and migration of fibers in humans has shown some interesting findings. A study of asbestos fibers in the lungs of 12 occupationally exposed men with fibrosis and/or carcinoma of the lung revealed the presence of numerous amphibole fibers in the upper lobes of the lungs. Among those workers with carcinoma, but not fibrosis of the lung, fibers were found mostly in the lower lobes of the lung; these fibers were predominantly of the chrysotile type (Pooley, 1972; Fondimare et al., 1974).

Although differences in both the carcinogenic and fibrotic tendencies of asbestos fibers have been postulated, *all* types have been shown to be both fibrotic and carcinogenic (Karachova et al., 1969; Shin and Firminger, 1973). In contrast to these studies, evidence indicates that different types of asbestos have different toxic effects. For example, McDonald et al. (1971) reported that workers exposed only to chrysotile asbestos had no incidence of mesothelioma. This study has therefore been offered in support of maintaining the present asbestos standard, since 90% of the asbestos used in the United States is chrysotile.

Fiber Migration

Asbestos in humans has also been shown to have some migratory potential. Studies on patients with pleural plaques and mesothelioma have shown that fibers associated with these diseases were also found in the lymph nodes, spleen, abdomen, and intestinal mucosa (Goodwin and Jagatic, 1970; Gross et al., 1973; Taskinen et al., 1973). These studies reveal the potential of asbestos to penetrate tissue and to be transported to sites other than the initial contact site. The occurrence of fiber migration clearly supports the need for asbestos standards that prevent the access of fibers into the body by any route.

The controversy is still unresolved concerning the question of whether asbestos fibers can migrate from their original point of deposition to other sites and cause pathogenic disease. Gross et al. (1974) concluded that there was no satisfactory evidence that transmigration of fibers occurs. However, a number of other studies (Volkheimer, 1973; Shreiber, 1974; Westlake et al., 1965; and Cunningham and Pontefract, 1973) have shown that transmigration of asbestos fibers can occur. Cunningham and Pontefract (1973) found asbestos fibers in the blood and a variety of tissues after feeding a group of rats a diet containing 6% asbestos. In attempting to repeat an experiment to confirm or deny transmigration of asbestos fibers in rats, Morgan et al. (1971) found that asbestos injected intraperitoneally in rats was capable of moving to various local tissues if the fibers were less than 20 μ long. Other investigators have found that asbestos fibers injected subcutaneously are also capable of transmigration; they were found in the spleen, liver, kidney, and brain of mice. In studies of placental transmigration of asbestos fibers, it was found that chrysotile asbestos intravenously injected into pregnant rats was able to cross the placenta and appeared in the liver and lungs of the fetuses (Cunningham and Pontefract, 1973, 1974).

Dose-Response Relationship

It has been shown that the cancer risk afforded by asbestos is dose related. For instance, in a study of women factory workers exposed to a mixture of chrysotile, amosite, and crocidolite dusts, those with heavier exposures showed a sixfold excess of cancer with a 15-year latency period, whereas women with moderate or low exposures showed a 25-year latency period (Newhouse et al., 1972). It is interesting to note that the rate of development of mesothelioma, like lung cancer, increased with both the quantity and length of exposure, but even those persons with as few as 2 years exposure to asbestos exhibited an increased risk of mesothelioma development.

Enterline (1976) indicated that the marked variability in the latent periods is most likely related to the degree or intensity of exposure; that is, the latent period is also a function of dose. As mentioned in several earlier chapters, Jones and Grendon (1975) estimated that the latent periods of several types of carcinogens (they did not specifically study asbestos) is a function of the cube root of the dose. Enterline illustrated this dose-latent period relationship (Figure 5-5). The figure shows that those workers with low exposures to asbestos (2 fibers per cc) exhibit very little response prior to 20–25 years, whereas those with more elevated levels (20 fibers per cc) show a marked response considerably earlier. Although precise dose-response relationships with respect to the development of cancer must be developed, excessive cancer risks are still found at all measurable fiber concentrations (NIOSH, 1976). In one study in which miners were exposed to amphibole fibers (amosite) in concentrations less than 2 fibers per cc (the average concentration was 0.25 fiber per cc), a threefold increase in the risks of mortality from both malignant and nonmalignant respiratory diseases was shown (Gillan et al., 1976).

There is additional evidence of increased frequency of asbestos-related diseases, especially cancer, occurring in individuals who have been exposed to low levels of asbestos (in the environment) or to intermittent exposures of asbestos at higher levels. One example of this concerns asbestos disease among shipbuilding and ship repair workers who had little or no contact with workmates who were using asbestos products (Harries, 1968). According to data from NIOSH (1976), there seems to be no threshold level at which asbestos exposure can be deemed safe based on human data studies. No available data specify a level of asbestos exposure at which there is no increased risk of developing cancer.

CRITIQUE OF EPIDEMIOLOGIC METHODOLOGIES

Despite the impressive general agreement of the vast majority of epidemiologic studies that asbestos is a human lung carcinogen, it should be noted that there is considerable variation with respect to relative risk to respiratory cancer according to different epidemiologic studies. Based on a comparison of 11 epidemiologic studies, Enterline (1976) reported that the relative risks for respiratory cancer ranged from 3.4 to 9.2 for five studies derived from union records, 1.3–2.8 for four studies from company files, and 1.2–3.4 based on social security records. Table 12-4 presents a comparative summary of these 11 studies. According to Enterline, the differences in the results of these reports have been caused in part from the differences in research goals. For example, some researchers tried to determine the maximum effect of asbestos by permitting only those persons with greater than 20 years exposure into the study cohort (see Table 12-4; Selikoff, June 1963, September, 1977, and October 1972 as examples). Other studies were structured to determine when the effects on high-risk groups could first be determined. Thus the first type of study usually results in a high relative risk, whereas the second dilutes the relative risk to a considerable degree.

Enterline (1976) also pointed out that several of these studies incorporated methodologic errors that either decrease or increase the reported relative risks. Several of these "pitfalls" include "correcting death certificates and then making comparisons with published vital statistics based on uncorrected death certificates; deriving expected deaths from populations that do not reflect the mortality experience of the area in which the study population lives; overlapping exposure and follow-up periods in dose-response studies; and failure to adjust for competing causes of death." Examples of two of these methodologic "pitfalls" (i.e., the correcting of death certificates on only the study population and using the national population instead of the local population as the standard) are shown in Table 12-5. Although most of these studies tend to overestimate the relative risk to respiratory cancer as affected by asbestos, it is quite clear that asbestos exposure is strongly implicated as a human carcinogen.

NIOSH'S RECOMMENDED STANDARD

Thus, taking all accumulated data into account, in 1976 NIOSH developed a recommended standard for asbestos, which was set at the lowest fiber level concentration able to be detected by available instrumentation (HEW, 1976). Since analysis by phase-contrast microscope is the only generally available tool, NIOSH developed its asbestos recommendation with the limitations of this tool in mind. Therefore, the level

Table 12-5. Studies of Asbestos Exposure and Respiratory Cancer 20 Years Since First Exposure on Entry to Chart

Primary Author	Relative Risk[a]	Adjusted for Death Certificate Correction	Plus Adjusted for Local Death Rate
Selikoff (1)	9.2	7.6	6.2
Enterline (2)	3.0	3.0	2.9
Cooper	7.8	7.8	6.7
Wagoner	2.8	2.8	3.6
Selikoff (2)	8.2	6.8	5.5
Selikoff (3)	5.8	4.7	4.7

[a] Based on U.S. experience.

Source: Enterline, P. E. (1976) Pitfalls in epidemiological research. *J Occup Med* 18(3):150–156.

of exposure for the new standard has been defined as 100,000 fibers greater than 5 μ in length per cubic meter or 0.1 fibers per cc on an 8-hour TWA basis, with peak concentrations not to exceed 0.5 fibers per cc in any 15 minutes.

According to NIOSH, this new recommended standard is intended to

1. Protect against the noncarcinogenic effects of asbestos.
2. Materially reduce the risk of asbestos-induced cancers (although only a total ban on asbestos can protect against cancer caused by asbestos).
3. Be measured by techniques that are valid, reproducible, and available (HEW, 1976).

The process of implementing occupational health standards in a democratic society such as the United States does not follow any simple, defined course with a present time table. There are competing interests within our society that view the world from different vantage points. The role of OSHA in establishing health standards must involve a serious consideration of all reasonable interest groups. The task is not an easy one. Not only must health be protected, but jobs and the economy must also be protected. For if it is true that one of the contributing factors in crime is unemployment, one does not need much imagination to write a string of scenarios concerning how irresponsible a regulatory agency may be if it eliminates thousands of jobs under the banner of protecting the workers' health. Without jobs, many of the unemployed face greater dangers to health than the increased risk of dying 5 years sooner than expected from an occupationally related cancer. In fact, a study by Brenner (1976) for the Joint Economic Committee of the U.S. Congress revealed a stunning series of positive associations of unemployment from 1940 to 1973 with seven stress indicators (total mortality, homicide, suicide, cardiovascular-renal disease mortality, cirrhosis of the liver mortality, total state imprisonment and state mental hospital admissions). Table 12-6 indicates the association of unemployment rates with these seven stress indicators. It is quite clear from Brenner's data that unemployment has a markedly adverse effect on the health of the individual person as well as society. Such results, if independently verified, have enormous political, social, and economic implications for our society. With regard to environmental and occupational health policies and practices, it suggests that our regulatory agencies (e.g., OSHA) must be aware of the widespread implications of their actual and proposed regulations and standards. Developing comprehensive health-effects criteria and economic impact statements, although necessary in any standard-setting process, are not enough. Society needs accurate predications of all significant effects of far-reaching proposals in advance of any final decisions. Society requires a truly

Table 12-6. Impact of a Sustained 1% Change in Unemployment

Social Stress Indicator	Data Period	Change in the Stress Indicator (%)
Suicide	1940–1973	4.1
State mental hospital admissions	1940–1971	3.4
Males		4.3
Females		2.3
State prison admissions	1935–1973	4.0
Homicide	1940–1973	5.7
Cirrhosis of the liver mortality	1940–1973	1.9
Cardiovascular-renal disease mortality	1940–1973	1.9
Total mortality	1940–1974	1.9

Source: Brenner, H. (1976). Estimating the social costs of national economic policy: Implications for mental and physical health, and criminal aggression. A study prepared for the use of the Joint Economic Committee Congress of the United States. U.S. Government Printing Office, p. V.

comprehensive and expanded concept of the environmental impact statement such that it is conceptually viewed as a societal impact statement.

This type of scenario, however realistic, must not be used to extort OSHA to establish lax standards. In fact, many of the traditional arguments used by various industries in standard-setting procedures have often exaggerated the economic and technological difficulties. OSHA must deal effectively with these issues because society is not truly served if health or jobs are lost.

REFERENCES

Anderson, H. A., Lilis, R., Daum, S. M., Fischbein, A. S., and Selikoff, I. J. (1976) Household-contact asbestos neoplastic risk. *Ann NY Acad. Sci* 127:311–323.

Ashford, N. A. (1976) *Crisis in the Workplace: Occupational Disease and Injury.* Cambridge: MIT Press.

Braun, D. C. and Traun, T. D. (1958) An epidemiological study of lung cancer in asbestos miners. *Arch Ind Health* 17:634–653.

Brenner, H. (1976) Estimating the social costs of national economic policy: Implications for mental and physical health, and criminal aggression. U.S. Government Printing Office, Washington, D.C.

British Occupational Hygiene Society (1968) Sub-committee on asbestos hygiene standards for chrysotile asbestos dust. *Ann Occup Hyg* 11:47–49.

Brodeur, Paul (1974) *Expendable Americans,* pp. 28–162. New York: Viking Press.

Bureau of Mines Mineral Yearbook (1973).

Cooke, W. E. (1924) Fibrosis of the lungs due to the inhalation of asbestos dust. *Br J Med II:* 147.

Cooper, W. C. and Balzer, J. L. (1968) Evaluation and control of asbestos exposures in the insulating trade. *From Proceedings of a Working Conference on the Biological Effects of Asbestos,* Dresden, Germany.

Cunningham, H. M. and Pontefract, R. D. (1973) Asbestos fibers in beverages, drinking water and tissues: their passage through the intestinal wall and movement through the body. *J Assoc Analyt Chem* 56:976–981.

Cunningham, H. M. and Pontefract, R. D. (1974) Placental transfer of asbestos. *Nature* 249:177–178.

Department of Employment and Productivity Her Majesty's Factory Inspectorate (1969) Standard for Asbestos Dust Concentration for Use with the Asbestos Regulations. Technical Note, 13, 1970.

Departmental Committee on Compensation for Industrial Disease (1907) H. M. Stationary Office, London.

Doll, R. (1955) Mortality from lung cancer in asbestos workers. *Br J Ind Med* 12:81–86.

Dunn, J. E., Linden, G., and Breslow, L. (1960) Lung cancer mortality of men in certain occupations in California. *Am J Public Health* 50:1475–1487.

Elmes, P. C. and Simpson, J. J. C. (1971) Insulation workers in Belfast. 3. Mortality 1946–66. *Br J Indust Med* 28:226–236.

Enterline, P. E. (1965) Mortality among asbestos workers in the U.S. *Ann NY Acad Sci* 132:156–165.

Enterline, P. E. (1976) Pitfalls in epidemiological research. *J Occup Med* 18(3):150–156.

Enterline, P. E., DeCoufle, P., and Henderson, V. (1972) Mortality in relation to occupational exposure in the asbestos industry. *J Occup Med* 14:897–903.

Enterline, P. E. and Henderson, V. (1973) Type of asbestos and respiratory cancers in the asbestos industry. *Arch Environ Health* 27:312–317.

Federal Register, Dept. of Labor, OSHA. Occupational exposure to asbestos. October 9, 1975. Vol. 40, No. 197, pp. 47652–47658.

Fondimare, A. et al. (1974) Etude semi quantitative de l'empoussierage par l'amiante. *Arch Anat Pathol (Paris)* 22:55.

Friedrichs, K. H., Hilscher, W., and Sethi, S. (1971) Fiber and tissue studies on rats after intraperitoneal injection of asbestos. *Arch Arbeitsmed* 28:341–354.

Gaze, R. (1965) The physical and molecular structure of asbestos. *Ann NY Acad Sci* 132:23–30.

Gillam, J. D., Lemen, R. A., Archer, V. E., Wagner, J. K., and Dement, J. (1976) Mortality patterns among hard rock gold miners exposed to an asbestiform mineral. *Ann NY Acad Sci* 271:336–344.

Godwin, M. C. and Jagatic, J. (1970) Asbestos and mesotheliomas. *Environ Res* 3:391–416.

Greenberg, M. and Davies, L. (1974) Mesothelioma Register 1967–1968. *Br J Indust Med* 31:91–104.

Gross, P. (1974) Is short fibered asbestos dust a biological hazard? *Arch Environ Health* 29:115–117.

Gross, P., Davies, J. M. G., Harley, R. A. Jr., and de Treville, R. T. P. (1973) Lymphatic transport of fibrous dust from the lungs. *J Occup Med* 15:186–189.

Gross, P., Harley, R. A., Swinburne, L. M., Davis, J. M., and Green, W. B. (1974) Ingested mineral fibers. Do they penetrate tissue or cause cancer? *Arch Environ Health* 29:341–347.

Harries, R. G. (1968) Asbestos hazards in naval dock yards. *Ann Occup Hyg* 11:135–145.

Hendry, N. W. (1965) The geology, occurrences and major uses of asbestos. *Ann NY Acad Sci* 132:12–22.

Howard, S., Kinlein, L. T., Lewinshohn, H. C., Peto, J., and Dole, R. (1976) *A Mortality Study Among Workers in an English Asbestos Factory.* Oxford University Press.

Jones, H. B. and Grendon, A. (1975) Environmental factors in the origin of cancer and estimation of the possible hazard to man. *Food Cosmet Toxicol* 18:251–268.

Karachova, V. N., Olshvang, R. A., and Kugan, F. M. (1969) Changes in certain organs after experimental intraperitoneal injection of asbestos-containing dust in experiment. *Byull Eksp Biol Med* 67:117–120.

Kiviluoto, R. (1960) Pleural calcification as a roentgenologic sign of non-occupational endemic anthophyllite-asbestosis. *Acta Radiol Suppl* 194:1–67.

Little, A. D. and Company (1972) Economic Assessment for an Occupational Asbestos Standard.

Lynch, K. M. and Smith, W. A. (1935) Pulmonary asbestos. III. Carcinoma of asbestos silicosis. *Am J Cancer* 24:56–64.

McDonald, J. C., MacDonald, A. B., Gibbs, G. W., Siemiatychi, J., and Rossiter, C. E.

(1971) Mortality in the chrysotile asbestos mines and mills of Quebec. *Arch Environ Health* 22:677–686.

Maltoni, C. and Annoscia, C. (1974) Mesotheliomas in rats following the intraperitoneal injection of crocidolite. In Davis, W. and Maltoni, C., eds., *Advances in Tumor Prevention, Detection and Characterization* vol. 1, pp. 115–116. *Characterization of Human Tumors,* Amsterdam: Excerpta Medica.

Mancuso, T. F. and Coulter, E. J. (1963) Methodology in industrial health studies. *Arch Environ Health* 6:210–226.

Miller, L., Smith, W. E., and Berliner, S. W. (1965) Tests for effect of asbestos on benzo(a)pyrene carcinogenesis in the respiratory tract. *Ann NY Acad Science* 132:489–500.

Morgan, A., Holmes, A., and Gold, C. (1971) Studies of the solubility of constituents of chrysotile asbestos *in vitro* using radioactive tracer techniques. *Environ Res* 4:558–570.

Newhouse, M. L. (1969) A study of the mortality of workers in an asbestos factory. *Br J Indust Med* 26:294–307.

Newhouse, M. L. and Berry, G. (1975) The risk of developing mesothelioma tumors among workers in an asbestos textile factory. XVIII International Congress on Occup. Health, Bristol, England.

Newhouse, M. L., Berry, G., Wagner, J. C., and Turok, M. E. (1972) A study of the mortality of female asbestos workers. *Br J Indust Med* 29:134–141.

NIOSH (1976) NIOSH, Evaluation of Data on Health Effects of Asbestos Exposure and Revised Recommended Numerical Environmental Limits. Department of HEW.

Pooley, F. D. (1972) Asbestos bodies, their formations, composition and character. *Environ Res* 5:363–379.

Public Health Bulletin (1938) A study of asbestos in the asbestos textile industry. No. 241.

Rajhans, G. S. (1977) Here's an update on asbestos. *Occup Health Safety* November–December 38–42.

Reeves, A. L., Puro, H. E., Smith, R. G., and Vorwald, A. J. (1971) Experimental asbestos carcinogenesis. *Environ Res* 4:496–511.

Safety Standards (May-June 1972) Asbestos: airborne danger. U.S. Dept. of Labor, Division of Occupational Safety and Health Administration.

Schall, E. L. (1965) Present threshold limit value in the U.S.A. for asbestos dust: a critique. *Ann NY Acad Sci* 132:316–321.

Seidman, H., Lilis, R., and Selikoff, I. J. (1976) Short-term asbestos exposure and delayed cancer risk. Third International Symposium of Detect. Prevent Cancer, New York, April 26–May 1, 1976.

Selikoff, I. J. (1970) Partnership for prevention—the insulation industry hygiene research program. *Indust Med* 39:4.

Selikoff, I. J. (1976) Asbestos disease in the United States, 1918–1975. Environmental Science Laboratory, Mt. Sinai School of Medicine of the City University of New York. *Rev Frac Mal Resp 4* (Suppl):7–24.

Selikoff, I. J., Churg, J., and Hammond, E. C. (1964) Asbestos exposure and neoplasia. *JAMA* 188:22–26.

Selikoff, I. J., Churg, J., and Hammond, E. C. (1965) The occurrence of asbestosis among insulation workers in the U.S. *Ann NY Acad Sci* 132:139–155.

Selikoff, I. J., Churg, J., and Hammond, E. C. (1968) Asbestos exposure, smoking, and neoplasia. *JAMA* 204:106–112.

Selikoff, I. J., Hammond, E. C., and Churg, J. (1972) Carcinogenicity of amosite asbestos. *Arch Environ Health* 25:183–186.

Shabad, L. M., Pylev, L. N., Krisosheeva, L. V., Kuligina, T. F., and Nemenko, B. A. (1974) Experimental studies on asbestos carcinogenicity. *J Natl Cancer Inst* 52:1175–1180.

Shin, M. L. and Firminger, H. I. (1973) Acute and chronic effects of intraperitoneal injection of two types of asbestos in rats with a study of the histopathogenesis and ultra structure of resulting mesothelioma. *Am J Pathol* 70:291–314.

Shreiber, G. (1974) Ingested dyed cellulose in the blood and urine of man. *Arch Environ Health* 29:39–42.

Sincock, A. and Seabright, M. (1975) Induction of chromosome changes in chinese hamster cells by exposure to asbestos fibers. *Nature* 257:56.

Taskinen, E., Ahlman, K., and Wiikeri, M. (1973) A current hypothesis of the lymphatic transport of inspired dust to the parietal pleura. *Chest* 64:193–196.

Vare, R. (July 8, 1972) Asbestos under fire. *New Republic.*

Villecco, M. (December 1970) Technology: danger of asbestos. *Architectural Forum.*

Volkheimer, G. (1973) Persorption. *Acta Hepato-Gastroenterology* 20:361–362.

Wagner, J. C. (1965) Epidemiology of diffuse mesothelial tumors: evidence of an association from studies in South Africa and the United Kingdom. *Ann NY Acad Sci* 132:575–578.

Wagner, J. C. (1972) Current opinions on the asbestos cancer problem. *Ann Occup Hyg* 15:61–65.

Wagner, J. C. Berry, G., Skidmore, J. W., and Timbrell, V. (1974) The effects of the inhalation of asbestos in rats. *Br J Cancer* 29:252–269.

Wagner, J. C., Berry, G., Cooke, T. J., Hill, R. J., Pooley, F. D., and Skidmore, J. W. (1977) Animal experiments with talc. In Walton, W. C., ed., *Inhaled Particles and Vapours,* vol. IV. New York: Pergamon. (Cited in IARC, Monographs on the Evaluation of Carcinogenic Risk of Chemicals to Man: Asbestos. Lyon, 1977.)

Wagner, J. C., Gilson, J. C., Berry, G., and Timbrell, V. (1971) Epidemiology of asbestos cancers. *Br Med Bull* 27:71–76.

Wagoner, J. K., Johnson, W. M., and Lemen, R. (1973) Malignant and non-malignant respiratory disease mortality patterns among asbestos production workers. U.S. Government Printing Office, Washington, D.C., Vol., 119, Part 6, pp. 7828–7830.

Waldbott, G. L. (1973) *Health Effects of Environmental Pollutants.* St. Louis: Mosby.

Westlake, G. E., Spjut, H. J., and Smith, M. N. (1965) Penetration of colonic mucosa by asbestos particles. *Lab Invest* 14:2029–2033.

Index